# Insect Behavior

# INSECT BEHAVIOR

**ROBERT W. MATTHEWS**
**JANICE R. MATTHEWS**

University of Georgia

A WILEY-INTERSCIENCE PUBLICATION

JOHN WILEY & SONS, New York • Chichester • Brisbane • Toronto

**Library of Congress Cataloging in Publication Data:**

Matthews, Robert W.        1942–
  Insect behavior.
  "A Wiley-Interscience publication."
  Includes bibliographies and index.
  1.  Insects—Behavior.   I.   Matthews, Janice R.,
1943–    joint author.   II.   Title.
QL496.M39        595.7'05        78-7869
ISBN 0-471-57685-9

Printed in the United States of America

10 9 8 7 6 5 4 3 2

# Preface

This book has been written for all students of the biological sciences. Besides being of interest to those who have had some basic introduction to entomology and/or animal behavior, we hope it will appeal to newcomers interested in insect behavior from the perspective of the science educator or comparative psychologist, for whom most books on insects are either oversimplified or too technical.

An outcome of a senior level course offered in the Department of Entomology at the University of Georgia since 1970, the book has been guided by the same objectives that have shaped the course. The first of these has been to help the student understand how a number of major behavioral systems function. Thus, this is not an encyclopedia; it does not document numerous strings of examples merely for "completeness of coverage," but offers a comparative evolutionary approach to processes and fundamental concepts. The second objective has been to help the student gain insight into the ways in which behavioral research can be conducted. Whenever possible, we have included discussions of important experiments and investigations rather than presenting a simple rhetoric of conclusions. In addition, selected principles are interwoven with case studies which explore them in relation to specific situations, presenting actual examples in a manner compatible with the dynamic, open-ended field and laboratory experiences in which they have arisen.

Each chapter concludes with a list of selected references (predominantly published in the past decade) for those wishing more detail on a particular subject. These materials are accessible in most college libraries; many of the suggested readings are reviews or articles from *Scientific American,* the majority of which are available as offprints.

Like any writer of a general textbook, we recognize a deep obligation to many others—to those of whose work we write, to other authors whose ideas we use, to our own teachers who have shaped our perspective and interests, and to our students and colleagues with their many stimulating and invaluable suggestions and criticisms. We would like to express special gratitude to M. C. and M. L. Birch, Murray S. Blum, Paul Decelles, Thomas Eisner, Howard E. Evans, Darryl T.

Gwynne, John F. MacDonald, Glenn K. Morris, Kevin O'Neill, and E. O. Wilson for major constructive criticism. Other readers who critically reviewed selected parts of the manuscript include Lincoln P. Brower, R. Hugh Dingle, W. G. Eberhard, George C. Eickwort, Bert Hölldobler, H. B. D. Kettlewell, Louis M. Roth, Lee Ryker, Robert L. Smith, and James W. Truman. For any errors or failure to communicate effectively with readers, we alone are responsible.

We also thank the many scientists and editors of scientific journals who have freely granted permission for the use of published material. Numerous colleagues have generously provided us with photographs; special thanks are due to Robert L. Jeanne, Carl W. Rettenmeyer, and Robert E. Silberglied. We are especially grateful to Joan W. Krispyn for the numerous original drawings. Figure captions carry the appropriate figure credits.

Finally, we sincerely thank the editorial and design staffs of John Wiley & Sons, Inc., for their efforts on behalf of the book.

ROBERT W. MATTHEWS
JANICE R. MATTHEWS

*Athens, Georgia*
*February 1978*

# Contents

# Insect Behavior

# Insect Behavior: an Introduction

An overview of the insect world reveals two paradoxical characteristics: great diversity and equally great constancy. On the one hand, there are over one million named insect species, with estimates ranging up to three million. How can such a great diversity be explained? Study of this basic question has become the domain of evolutionary biology. On the other hand, each kind of organism tends to reoccur in virtually the same form with the same basic features for generation after generation. Why do they tend to show such constancy, such resistance to change? The study of this question, in turn, is largely the domain of genetics. Together, these two great branches of biology—evolution and genetics— form a powerful tool for the investigation of nearly every aspect of life. This introductory chapter deals briefly with their application to the study of behavior and then turns to an historical overview of behavior as a field of study to provide a perspective for the chapters that follow.

## THE BIOLOGICAL BASIS OF INSECT BEHAVIOR

Insects belong to the phylum Arthropoda, a very large assemblage of animals with jointed legs and a hard outer skeleton. One major group in this phylum, the Chelicerata, have sickle-shaped chelicerate jaws and lack antennae; they include the Arachnida (spiders, mites, scorpions, etc.) and two smaller marine groups. The other major group, the Mandibulata, possess antennae and have mandibles that work against each other. They include six classes: the insects, the crustaceans (a predominantly aquatic group), the centipedes, the millipedes, and two smaller classes, Symphyla and Paurapoda. Of all the land arthropods, insects are by far the most abundant (Fig. 1-1), followed by spiders and mites.

Ordinal divisions within the class Insecta remain a matter of dispute but in general reflect present understanding of the evolutionary history

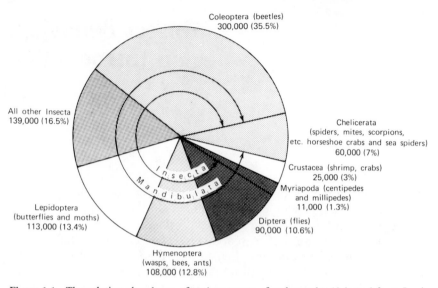

Coleoptera (beetles)
300,000 (35.5%)

Chelicerata
(spiders, mites, scorpions,
etc. horseshoe crabs and sea spiders)
60,000 (7%)

All other Insecta
139,000 (16.5%)

Crustacea (shrimp, crabs)
25,000 (3%)

Myriapoda (centipedes
and millipedes)
11,000 (1.3%)

Lepidoptera
(butterflies and moths)
113,000 (13.4%)

Diptera (flies)
90,000 (10.6%)

Hymenoptera
(wasps, bees, ants)
108,000 (12.8%)

**Figure 1-1**   The relative abundance of major groups of arthropods. (Adapted from Levi, Levi, and Zim, 1968.)

of the class. Broadly speaking, four important stages are distinguished. First was the appearance of primitively wingless insects (Apterygota), probably during the Devonian Period. Silverfish (Thysanura) and springtails (Collembola) are living representatives of these earliest insect forms. Second was the development of wings, thought to have occurred during the Lower Carboniferous. These early winged insects had a wing-hinging mechanism that did not permit them to fold, so wings had to be held out from the body. The Odonata (dragonflies and damselflies) and the Ephemeroptera (mayflies) are surviving remnants of these ancient groups.

The third evolutionary stage, the development of the wing flexion mechanism, occurred during the Upper Carboniferous. Now able to fold their wings down tightly over their abdomens, insects could more easily run and hide from predators and move into a wide variety of previously inaccessible niches. Of contemporary insects, roughly 97% have flexing wings, and this mechanism is in large part responsible for the dominance of insects today.

Fourth was the development of complete metamorphosis (holometaboly), which seems to also have arisen in the Upper Carboniferous. The earliest insects remained essentially similar in their wingless body form throughout their entire lives. More advanced groups developed the simple metamorphosis exhibited by insects such as grasshoppers today, where immature stages resemble miniature adults but wings are lacking (although external wing buds are plainly visible) until the last molt, when the insect becomes sexually mature. The most highly advanced groups, however, evolved the complete metamorphosis illustrated by the life cycle of a butterfly or honey bee. The immature stages, the larvae, bear no resemblance to adults, and wing buds are developed internally, becoming visible only when the larva transforms into the pupal stage, from which the winged adult emerges.

A summary of the evolutionary relationships of the insect orders is depicted in Fig. 1-2. A discussion of the geological history and evolution of insects is provided by Carpenter (1973).

## What Is Behavior?

Behavior can be simply defined as what animals do. More precisely, it is the ways in which an organism adjusts to and interacts with its environment. Admittedly, the term covers a very wide range of activities, and it can be helpful to recognize some subcategories. General locomotion, grooming, and feeding, for example, are essentially individual matters. These **maintenance activities** keep an insect in good shape but usually have little influence on others of his kind. On the other hand, a broad range of **communicatory activities** are "other oriented." They are concerned with conveying information to, and influencing the moods and activities of, others of the same species. Often such actions are conspicuous and stereotyped, and not surprisingly they have been a favorite study material for behaviorists.

Although movement is essential for most behavior, stillness is also behavior. For example, while many insects react to danger by fleeing, the survival of many others depends upon camouflage attained by freezing in some posture so completely as to seem to vanish (Fig. 1-3). As this example shows, behavior operates in circumstances that vary from one species to another and in ways that have **survival value.** In any situation insects that respond more appropriately have a better chance of living—and of leaving more progeny—than individuals responding less appropriately. Thus, a weeding-out process is continually in progress, a natural selection of behaviors that enhance the chances of survival and success for individuals of a species. The behavior patterns one observes today have had a long history of evolutionary development.

Natural selection dictates that everything an animal does should ultimately contribute to the optimization of its reproductive success. The rather limited amount of energy available to an individual must be

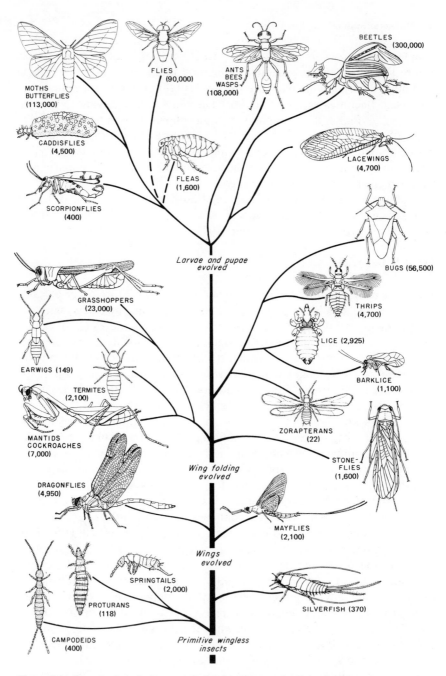

**Figure 1-2** Presumed evolutionary relationships among the living insect orders. Numbers of described species based on data taken from Borror, DeLong, and Triplehorn, 1976. (Modified from Ross, 1962; courtesy of Illinois Natural History Survey.)

4

**Figure 1-3** A geometrid moth as it rests motionless by day on a tree limb. The scalloped wing margins and mottled coloration blend into its background, making visual discovery difficult unless the moth moves. (Courtesy of R. E. Silberglied.)

divided between maintenance and reproductive activities, the latter being basically matters of communication. This division is broadly reflected in the book's organization. The first several chapters concern behavior of the individual insect—how it moves, orients, disperses, and feeds, including the role of the nervous and endocrine systems in integrating behavioral responses. The next three chapters on communication in a sense form the core of the book and lead logically to consideration of defensive, reproductive, and social behaviors, all of which are mediated by communicative codes.

## The Phylogeny of Behavior

Again and again, research has demonstrated a strong concurrence between behavior and phylogeny based on morphology and taxonomy. That is, behavioral differences and similarities within a group of species reflect phylogenetic relationships as determined morphologically. As Tinbergen (1968) has said, ". . . comparison of present-day species can give us a deep insight, with a probability closely approaching certainty, into the evolutionary history of animal species."

Perhaps one of the most elegant examples supporting this assertion is found among the cockroaches, an ancient but highly successful group that reached its highest development during the Upper Carboniferous. (Indeed, there has been relatively little morphological change in the intervening 250 million years; a fossil roach generally resembles a recently swatted roach except that early forms tended to have visible exerted ovipositors much like those of present-day katydids.) If you have ever tried to swat a cockroach, you know the creature's main defense is extreme proficiency at quick getaways. What you may not appreciate is that the common household varieties of roach represent only a miniscule proportion of some 5000 living blattid species that have invaded nearly every conceivable ecological niche from desert to aquatic habitats (see Roth and Willis, 1960). Despite a comparatively good fossil record the classification and phylogeny of these were subject to dispute by various authorities. As a result, three classifications based on different sets of morphological characteristics arose (Fig. 1-4).

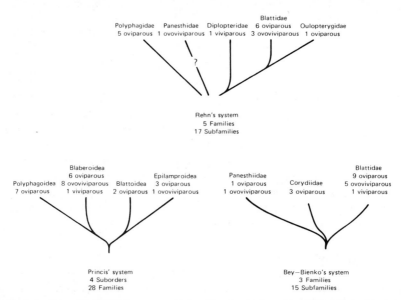

**Figure 1-4**  Comparison of three recent classifications of cockroaches (Blattaria), based solely on external features observed in museum specimens. Rehn's system, based primarily on wing venation, automatically excludes all wingless forms, of which there are a great many. Princis' system, based on diverse and apparently unrelated structural differences. Bey-Bienko's system, based similarly to that of Princis. (Redrawn and adapted from Roth, 1970 with permission, from the Annual Review of Entomology, volume 15. Copyright © 1970 by Annual Reviews Inc. All rights reserved.)

Behavioral characteristics and internal morphology were given little attention in any of these schemes, yet cockroach reproductive biology is exceedingly diverse. Although it is unusual to find more than one mode of oviposition behavior in a particular insect group, cockroaches display an almost complete spectrum. Many species are "ordinary"; like most insects and birds, they lay eggs in packets (**oothecae**). Such reproduction, where fertilized eggs develop outside the female's body, is termed **oviparity.** Other cockroach species exhibit **ovoviviparity**; the eggs are first extruded and then retracted into a brood sac until the embryos mature, at which time the ootheca is again extruded and the nymphs hatch and drop free from their mother. A third type of reproduction, **viviparity**, is found in other roach species. This is superficially similar to the previous type, but the eggs lack sufficient yolk when first formed to allow complete development. Both nutrients and water are absorbed during embryonic development in the brood pouch, whereas in ovoviviparous species only water is absorbed from the female.

When the distribution of species having these three types of reproduction is superimposed on the classifications in Fig. 1-4, they intermingle throughout the suborders, families, and subfamilies rather than sorting out into any consistent pattern. This puzzling fact prompted McKittrick (1964) to undertake still another detailed investigation of recent cockroaches. For comparative study, she chose four character systems: (1) female genitalia, (2) male genitalia, (3) the proventriculus (a portion of the intestinal tract specialized for grinding food, also referred to as the gizzard), and (4) oviposition behavior. Upon analyzing her results, McKittrick proposed a new classification (Fig. 1-5) with two phyletic lineages based primarily on reproductive behavior. In one lineage, the superfamily Blattoidea, all species remained oviparous, undoubtedly the ancestral form of reproduction. This group is exemplified by the notorious American roach, *Periplaneta americana*. The other lineage, the Blaberoidea, which includes the equally notorious German roach, *Blattella germanica*, evolved ovoviviparity and viviparity.

In the years since her work was published, McKittrick's conclusions have gained widespread acceptance and support from a variety of other studies, especially behavioral work (for an excellent summary, see Roth, 1970). However, she must be given credit for being the first to stress reproductive behavior as the unifying character of cockroach classification.

Comparative behavioral studies which uncover a seeming progression in one or more characters within a set of related species make it tempting to conclude that evolution has actually proceeded through such a series. However, this conclusion should be tempered with extreme

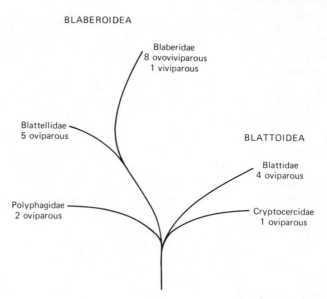

**Figure 1-5** The currently accepted higher classification and phylogeny of the Blattaria, based on McKittrick, 1964. The numerals represent the number of subfamilies. Compare with Figure 1-4. (Redrawn from Roth, 1970 with permission from the Annual Review of Entomology, volume 15. Copyright © 1970 by Annual Reviews Inc. All rights reserved.)

caution. The fact that intermediate forms are possible and do exist indicates only that evolution *might* have proceeded through a similar (but probably not identical) series. It does not, in the absence of a fossil record or some compelling logical argument, even indicate the direction of the evolution of the trait in question.

Prey transport, for example, varies considerably among predatory solitary wasps. Many species carry paralyzed prey back to the nest in their jaws; others always hold the prey with their middle legs (see Fig. 1-8) and still others use their hind legs. Some fly hunters even carry the prey impaled on their stinger; but perhaps the most remarkable prey-carrying adaptation is found in a sphecid wasp, *Clypeadon laticinctus,* which transports paralyzed ants on a specialized "ant clamp" formed by the modified apical abdominal segments (Fig. 1-6). While prey carriage seems to be constant among all members of a particular morphologically defined genus, there appears to be no correlation of prey carriage type with wasp phylogeny beyond this, nor with nest type or phylogenetic position of the prey.

What other factors might have molded the evolution of prey carriage? Evans (1962) has suggested some answers. In the soil-nesting species,

**Figure 1-6** A female of *Clypeadon laticinctus* digging into her nest while holding a worker harvester ant by the "ant clamp" on her abdomen. (Courtesy of A. Steiner and H. E. Evans.)

mandibular prey carriage tends both to obstruct the major sense organs and to impede the wasps' ability to dig into their nest entrance, Evans points out; increased efficiency would be afforded by shifting the prey ventrally and posteriorly. Species that use the mouthparts in prey transport either tend to leave the nest open between foraging trips or use such large prey that they must drag them back to the nest. Mandibular carriers that closed the nest between prey trips would be forced to lay the prey down temporarily while they reopened the nest, thereby leaving the prey vulnerable to attack by predators and parasites.

What evidence is there for such a shift? No wasps are known to actually carry prey in their front legs, but a few embrace the prey with one or more pairs of legs while holding it in their jaws. Most species utilize their middle or hind pair of legs, holding the prey lengthwise beneath their body. Only a very few carry the prey on their sting or with an abdominal clamp. Perhaps the advantage of having all three pairs of legs free is offset by the relatively greater risk of exposure of the prey to parasites, especially to attacks by satellite flies (Fig. 1-7). Also involved

**Figure 1-7**  A satellite fly, *Senotainia trilineata* (Sarcophagidae), that has landed on a nest-marker nail. Such flies typically follow just behind prey-laden wasps, awaiting an opportunity to deposit their live larvae on the wasp's prey. (Courtesy of H. E. Evans.)

in the evolution of prey carriage behavior is a trend toward smaller prey sizes. Smaller prey permit a greater foraging range with minimal energy while still allowing quick returns to the nest and seems to be prerequisite for pedal or abdominal prey transport mechanisms to evolve.

Not surprisingly, behavior does not fossilize well. Known examples tend to demonstrate the antiquity of the particular behavior pattern preserved. These include an entire colony of weaver ants, including larvae, from the Oligocene in Africa, fossil dung balls fashioned by scarab bettles in the Lower Oligocene, and fossil lepidopterous leaf mines from the early Eocene of Wyoming (Hickey and Hodges, 1975). The fossil ant nest demonstrates that distinct morphological castes in ants evolved very early and that social life in ants has apparently changed but little over the past 60 million years (Wilson and Taylor, 1964). The fossil dung balls similarly demonstrate scarab beetle nesting behavior (see Chapter 10) to have been at a fairly advanced level very early in the history of the group. There can be no doubt as to the correct identity of the dung balls because many clearly show scratches that look just like the tibial strokes made by emerging adults, and most also show the round exit opening corresponding to the location of the original egg chamber (Halffter and Matthews, 1966).

To deduce the evolutionary direction of apparent behavioral progressions, another procedure is to try plotting the behavioral evidence against the "family tree." One example of this practice involved the aquatic insect order Trichoptera. Larvae of certain caddisfly families construct a fixed retreat; those of others construct a portable case; and still others are free living. Through superposition upon a morphologically based phylogeny it became apparent that fixed retreat construction was the ancestral behavior and portable case building the derived state, with a series of steps leading from one to the other (Ross, 1964). Such an examination of behavioral characteristics depends upon the concept of **homology,** that is, the assertion that similarities in pattern are explainable on the hypothesis of common ancestry. In theory such a concept is relatively simple. However, since it depends upon past events, which can never be known with absolute certainty, in actual practice it can be quite controversial.

Related to this concept is a second fundamental one—that of **evolutionary convergence.** Often a similar trait appears in two or more clearly unrelated groups. For example, sociality appears in both ants and termites, which are distantly related groups. Tiger beetles, solitary wasps, and trap-door spiders all nest in underground burrows. Mud-daubing wasps and mason bees both construct nests of mud (see Chapter 10). In each case, although the selection pressures under which the trait evolved were presumably similar, the trait did not arise in a common ancestor.

Not only may behavioral traits differ between species or through time but with geography as well. When they show geographical variation within a given species, the term cline, or **ethocline,** may be employed.* Although such variations have not yet received due emphasis in behavioral studies, some examples are known. One involves the Australian sand wasp, *Bembix variabilis* (see Fig. 10-10), a widely distributed species studied in 20 localities in five states and two territories, including some 370 records of prey (Evans and Matthews, 1975). Throughout most of the continent, *B. variabilis* females consistently hunt Diptera (Fig. 1-8); prey belonging to at least 14 families of flies were recorded. In the Northern Territory near Darwin, however, a population of *Bembix* wasps was found that combine both flies and damselflies (Odonata) in their prey. Some 300 miles southwest of Darwin, along the Ord River, the population sampled used exclusively damselflies.

The *Bembix* wasps of Australia also illustrate a behavioral example of another well-known evolutionary phenomenon—**adaptive radiation.** While flies are the prey of nearly all of the 300-plus species of *Bembix*

---

* Ethocline is also applied to variation between related species as, for example, an ethocline in prey carriage in the solitary wasps mentioned earlier.

**Figure 1-8** Variation in prey preference by the common Australian sand wasp, *Bembix variabilis*: (*a*) A female opening her nest entrance with her front legs while holding typical prey, a small fly, by her middle legs. (*b*) A recently discovered population in northwest Australia has switched from flies to damselflies but still transports them in the typical manner. (From Evans and Matthews, 1973.)

throughout the world, eight out of the 22 Australian species for which biological data exist have switched to other prey (Fig. 1-9). This unique radiation in prey preferences in Australia is presumably related to the absence of related genera of sand wasps that occupy these food niches elsewhere in the world. Other factors may also be involved, such as

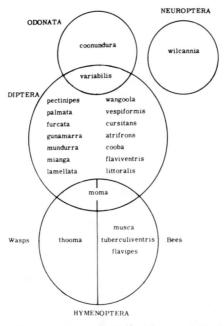

**Figure 1-9**   Adaptive radiation in prey preferences of 22 Australian *Bembix* species. Most species hunt only flies (Diptera), the prey used exclusively by studied species of this genus elsewhere in the world. Here, six species are specialists on other orders while two species—*variabilis* and *moma*—hunt both flies and other insects. (From Evans and Matthews, 1973.)

scarcity of usual prey in some habitats, especially arid ones. Species such as *B. variabilis* may be in the earliest stages of splitting into two distinct species. When such speciation first is fully complete may not always be discernible by the use of strictly morphological criteria.

Differences in mating behavior often constitute the strongest sort of species-isolating devices. For example, consider the larger fireflies of eastern North America, which belong mainly to the genus *Photuris*. This is a confusing group in which the males show almost no differences in structure or body coloring but much variation in flash pattern used to attract females during courtship (Barber, 1951). For many years this problem bothered Barber, and eventually he came to recognize 18 species mainly on the basis of male flash patterns (Fig. 1-10). Ten of these were named as new since they had not been previously recognized.

Behavioral traits may be used not only to confirm previously uncertain phylogenetic relationships but to discover new ones. Lloyd (1966), studying the common smaller fireflies of the United States, also found

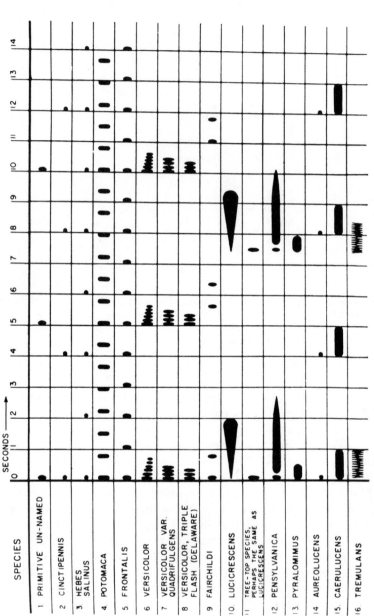

**Figure 1-10** Flash patterns of *Photuris* fireflies. Species 2 and 14 have similar flash patterns, but probably do not occur in the same geographical area and are quite different morphologically. Other species at least partially overlap in range in the eastern United States, and adult morphologies are quite similar. (Reprinted by permission of the Smithsonian Institution Press from "North American Fireflies of the Genus Photuris," Herbert S. Barber, page 12. Figure 1 in Smithsonian Miscellaneous Collections, Volume 117: Washington, D.C. © 1951.)

several such "hidden species" in *Photinus*. First recognized by consistent differences in flash signals, these fireflies later were found to differ in minor details of body color. In many places, two or more species of *Photinus* fly together but are prevented from interbreeding by their specific courtship patterns (see Chapter 6). The males fly at different heights and in different flight patterns; their flashes differ in length, number of pulses per flash, and sometimes in color or intensity of the light.

Although morphological criteria usually form the basis for taxonomic studies, there is validity in the question as to which usually comes first, the structure or the behavior for which the structure is used. As Mayr (1970, p. 363) views it, behavior is the pacemaker of evolution: "A shift into a new niche or adaptive zone is, almost without exception, initiated by a change in behavior. The other adaptations to the new niche, particularly the structural ones, are acquired secondarily."

There are certainly many good examples to support his contention. As illustration, consider the halictid bee, *Lasioglossum (Dialictus) zephyrum*, where subsets of morphologically similar individuals within a nest can be differentiated quite easily by comparing the relative frequencies of various behaviors (Brothers and Michener, 1974). Nudging of other bees in the burrow is more frequently done by queens than by others, and the queen is less often nudged herself than are others. The most characteristic bit of queen behavior often follows; if the nudged bee turns to face the nudger, the latter, usually the queen, rapidly backs down the burrow, followed by the former who moves head first. However, if the nudged bee is one whose main activity is nest guarding, it will rarely turn and follow the queen. In contrast, other workers, especially those most active in work on cells and in foraging, readily turn and are often led back deeper into the nest.

## Genetics and Behavior

In fruit flies (*Drosophila*), probably the single best genetically studied insects, a host of intriguing behavioral abnormalities have been traced to mutations of single genes. Intensive inbreeding of laboratory stocks has produced populations identical genetically in every respect except for a single mutant gene induced by radiation or certain chemicals. Such populations have been demonstrated to differ behaviorally, often in several traits (Benzer, 1973).

Ultimately, the behaviorist has, as one goal, the identification of a particular behavioral component with a particular gene, determining the actual site at which the gene wields its influence and how it does so.

However, the task is complicated by the fact that the behavioral impact of a mutant gene may not be its primary effect. A functional nervous or muscular system must be constructed in a precise way for behavior to be "typical." Since genes code for enzymes, even a single gene mutation may have broad ramifications. A single enzyme's presence or absence as a result of an error in coding by the genes may have large-scale impact on the construction of such systems.

Throughout the chapters that follow we repeatedly assume the genetic determination of behaviors. For this reason, it is worth noting two useful methods by which such assumptions might be verified: *crossing experiments* and *selection experiments*. However, the field of behavioral genetics is beyond the scope of this book; for a comprehensive look at this field, the interested reader is referred to Fuller and Thompson (1960), Parsons (1967), Hirsch (1967), McClearn and Defries (1973), and Ehrman and Parsons (1976). For an entry into current insect behavioral genetics, the papers by Benzer (1973) and Hotta and Benzer (1972) on *Drosophila* and by Bentley and colleagues (1972, 1975) and Hoy and Paul (1972) on crickets are particularly interesting. Manning (1967) and Ewing and Manning (1967) review earlier studies on the behavioral genetics of insects.

A **crossing experiment** depends upon mating individuals that differ in a particular kind of behavior and then examining the behavior of their offspring. (For such experiments to be meaningful, the environment must be constant or very well controlled.) The simplest type of crossing experiment would be to use individuals which exhibit certain behaviors that are known (by previous crossing experiments) to differ at only a single gene locus. These are only rarely detectable in normal animals, but when encountered they can give elegant testimonial to the genetic basis of behavioral differences.

One of the first such studies was done by Bastock (1956) with the yellow mutant of the fruit fly, *Drosophila melanogaster*. In competitive mating, this mutant was shown to be less successful than the wild type, partly because yellow males spent less time in the wing vibration phase of courtship.

Rothenbuhler's (1967) studies of two strains of honey bees provide another beautiful example. One strain is very susceptible to the bacterial infection called American foulbrood, while the other is resistant. This difference is largely due to a single behavioral difference in worker bees of the resistant strain that consistently remove dead larvae, thus halting the spread of the infection. When crosses between this "hygienic" strain and the susceptible "nonhygienic" strain were made, the resultant offspring were nonhygienic, indicating that the genes conferring resistance were recessive. Going further, Rothenbuhler did a series of backcrosses to the homozygous recessive resistant line; roughly a quarter of the resultant colonies were hygienic. Among the nonhygienic colonies produced, a third uncapped the cells of dead larvae but would not remove the larvae. Another third would remove the larvae, but only if the caps were artificially removed for them. The remaining third would do neither. Thus, the behavioral and genetic results were consistent in indicating that two loci were involved—one controlling uncapping, the other affecting removal—in the difference in hygienic behavior between the two strains (Fig. 1-11).

In practice, unfortunately, one rarely encounters such genetically simple behavioral phenotypes. Far more commonly a behavioral trait appears to vary along a continuum. The influence of a single gene is undetectable, for genes at many loci have small additive effects, and the environment may have a large influence upon variability in the expression of the trait. With such polygenic systems statistical approaches are necessary to estimate the roles of inheritance and environment. The principal one of these is analysis of variance. A powerful tool of behavioral genetics, the derivation and performance of analysis of variance is well described in most textbooks on statistical analysis.

**Selection experiments,** another important way to measure the degree to which a given behavioral trait is genetically determined, involve choosing only a certain part of each generation to become parents of the next generation. By such "artificial selection" the population becomes altered through time for the quantitative character under consideration, in a direction that depends upon which part of the population is chosen for succeeding generations (Fig. 1-12).

In artificial selection experiments, two measures can easily be determined empirically (Fig. 1-13). One is the *selection differential* ($S$)—how much the chosen parents deviate from the whole population in their mean value for the trait under selection. The other is the *magnitude of the response* ($R$)—how much the mean values differ between the parental and filial generations. Thus, a very useful measure called **heritability** ($h^2$) can be calculated—the degree of genetic determination of the variability that is present in the sample population, for the kind of behavior observed and for the precise method of observation used. In mathematical terms, heritability may be expressed as follows:

$$h^2 = \frac{R}{S}$$

What would happen if artificial selection were attempted on an inbred homozygous strain? In this case, none of the variation present has a

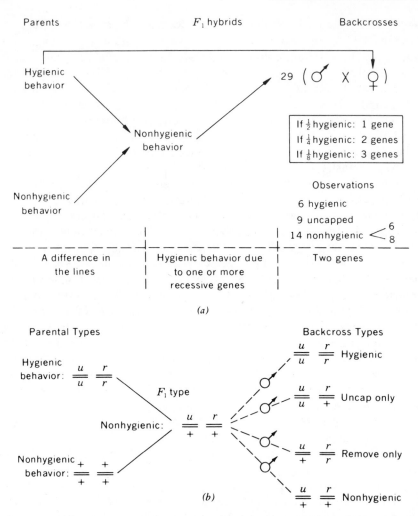

Figure 1-11  Genetics of hygienic behavior in honey bees: (a) Experiments and their results as expressed in number of colonies (not individuals). (From Rothenbuhler, 1967 in *Behavior-Genetic Analysis* by J. Hirsch. Copyright 1967. Used with permission of McGraw-Hill Book Company.) (b) Genetic hypothesis to explain responses to brood killed by American foulbrood in 63 bee colonies. (From Rothenbuhler, 1964.)

genetic basis. Thus, the population mean will not significantly change, regardless of the sort of selection being practiced; nor will the mean of the filial population differ significantly from that of the parental population, regardless of the selection differential. Therefore, $h^2$ would be 0. What if the filial population turned out to have a mean that was more or

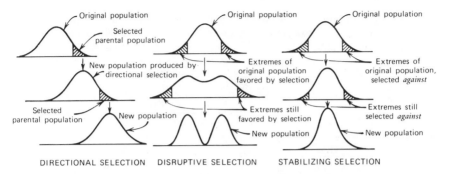

DIRECTIONAL SELECTION   DISRUPTIVE SELECTION   STABILIZING SELECTION

**Figure 1-12** Three different modes of selection yield quite different characteristic means and frequency distributions in the subsequent generations. Ordinate indicates the frequency of individuals in the population. Abscissa indicates variation for the quantitative character being considered, as expressed in some linear metric measure. Directional selection is the most common mode (see Fig. 8-6).

less identical to that of the selected sample? This would indicate that the trait was completely under genetic determination, and $h^2$ would be equal to 1. Most cases, of course, lie somewhere between these two extremes. As a general rule, most arthropod behavior is highly stereotyped and will tend to yield rather high heritability measures relative to those of the vertebrates.

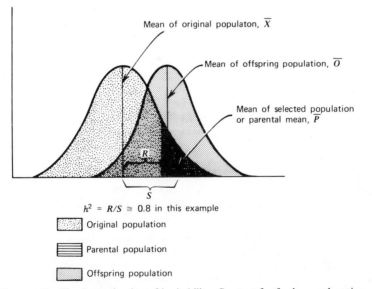

$h^2 = R/S \cong 0.8$ in this example

Original population

Parental population

Offspring population

**Figure 1-13** The determination of heritability. See text for further explanation.

**Table 1-1**  Selection for age at first oviposition in females of the milkweed bug, *Oncopeltus fasciatus*. Change in mean age at first reproduction ($\bar{\alpha}$) with selection and calculation of realized heritability from $R = h^2S$.[a] From Dingle (1974).

| Generation | $\bar{\alpha}$ | Selection differential $S$ | Response to selection $R$ | Heritability $h^2$ |
|---|---|---|---|---|
| 1 | 63.5 | | | |
| 2 | 61.3 | 6.6 | 2.2 | 0.33 |
| 3 | 50.9 | 9.6 | 10.3 | 1.08 |
| 4 | 37.8 | 7.3 | 13.1 | 1.79 |
| 5 | 35.3 | 5.3 | 2.5 | 0.47 |
| 6 | 21.7 | 10.3 | 13.6 | 1.30 |
| 7 | 16.7 | 4.4 | 5.0 | 1.14 |
| 8 | 14.9 | 2.7 | 1.8 | 0.67 |

[a] The five earliest pairs to reproduce were chosen as the parents of the subsequent generation for each generation. Environmental conditions were kept constant throughout the entire study.

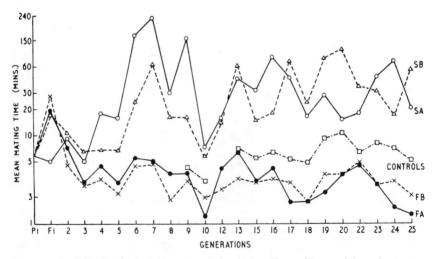

**Figure 1-14**  Selection for latency to copulation in four lines of *Drosophila melanogaster*. Mean mating time is the elapsed time from the release of flies into the container until copulation begins, plotted on a logarithmic scale. Selection was for two fast lines (FA and FB) and two slow lines (SB, SA); selection was relaxed in generations 11, 14, and 21. (From Manning, 1961.)

20

Selection experiments on milkweed bugs under laboratory conditions shifted the age at first oviposition from 63.5 days to 16.7 days in seven generations (Table 1-1). Average heritability from all generations was 1.18, which indicates that heritability for this trait must be very close to 1.0, since additive genetic variance cannot be greater than total variance. Compare this with the result of another set of selection experiments in which Manning (1961) attempted to select fruit flies for mating speed (latency to copulation) (Fig. 1-14). For seven generations, in each of which he selected the first 10 pairs in the fast lines and the last 10 pairs in the slow lines, considerable divergence was attained. However, further selection to 25 generations did not greatly increase this divergence. Furthermore, from one generation to the next the fluctuations in all the lines tended to parallel each other.

## INSECT BEHAVIOR AS A FIELD OF STUDY

Four great disciplines contribute to the study of behavior—physiology (particularly neurophysiology), ecology, ethology,* and psychology. There are no distinct boundaries, yet each has had its own developmental history and tradition of established methods and brings its own viewpoint to the subject. Somewhat facetiously, they have been distinguished by Roeder (1967) as follows: The ethologist, attempting to leave the animal as unrestricted as possible in order to study its "normal" behavior, tolerates any necessary discomforts while enclosing himself in a blind. The psychologist, attempting to reduce external variables, places the blind around the animal, thereby making it uncomfortable. The physiologist, attempting to learn what makes the animal behave, probes directly into its nervous and motor systems. The ecologist, we might add, spends his time studying how the blind changes the animal's environment.

While these different approaches have begun to show signs of fusion, with profitable sharing of techniques and concepts, it is the ethological approach that is reflected in this book. Throughout, the themes of evolution and adaptation will be used to place behavior within a larger perspective, for the behavior patterns of present-day organisms are as much their product as is their anatomy.

* The word **ethology** has a long pedigree. Based on the Greek *ethos*, which has a variety of meanings, the term has been applied to everything from stage actors who portray human characters to people who study ethics. However, as it is used today, ethology means the study of the behavior of animals in their natural habitat.

## Ethology: Historical Perspectives

Man's interest in insects is long standing, and ancient perceptions of insect behavior were often surprisingly astute. For example, a famous gold pendant from the Minoan culture some 2000 years B.C. depicts very accurately the essentials of the life cycle of a common social wasp (Fig. 1-15) and reveals a rather sophisticated understanding of these insects. And, in fact, over 3000 years ago, King Solomon the Wise recommended the study of insect behavior with those famous words, "Go to the ant, thou sluggard!"

A serious look at the historical development of animal behavior, and insect behavior as a subdiscipline within it, really begins at the end of

**Figure 1-15** One of the most outstanding examples of early Minoan goldworking, an ancient pendant depicting two paper wasps. Their legs embrace a granulated disc thought to represent the paper nest; a droplet of food or wood pulp is held in their jaws. The cagelike fixture atop the heads may be an attempt by the artist to depict the vigorous antennal beating characteristic of all encounters between individuals on a nest. The opposition of the abdominal tips may have been intended to suggest mating but is biologically inaccurate; probably it reflects simply the heraldic symmetry so common in Minoan art. Most classics texts incorrectly refer to the insects as "probably bees." (Courtesy of S. Alexious and The Archaeological Museum of Heraclion; see also Hood, 1976.)

the eighteenth century as French naturalists led the way in founding the great modern tradition of natural history. At this time, all of European society was becoming "scientific," and each important expedition began to include at least one professional naturalist. Linnaeus' method of classification and Lamarck's transformism both were gaining increasing exposure and acceptance. In entomology, the work of Reamur and his disciples began to introduce elements of scientific discipline to insect study; between 1734 and 1742, six large volumes of *Memoirs pour Servir a l'Histoire des Insectes* applied precise observation, detailed experimentation, and accurate recording to phenomena as varied as social life, parasitic habits, and leaf mining.

During the first half of the 1800s, most of the important European entomological societies were being founded. It was a time when general observation predominated over specialization. However, enthusiasm and subjectivity often affected the accuracy of behavioral observations, and interpretations were often slanted to embrace a particular philosophical creed. (In one area in particular—instinct—the terms were to be discussed for the next century.) Thus, by the 1850s an increasing body of descriptive behavioral material existed, but there was little definitive interpretation and generalization.

During the latter half of the 1800s, this broadly naturalistic tendency was joined by an expansion in specialized objective analyses. For example, many investigators became interested in the sensorial bases of behavior, particularly in color vision and chemical sensitivity. An interesting shift began to appear. Previously, behavioral traits had been depicted as shifting, will-o'-the-wisp phenomena. Now naturalists began to show that behavior often involved patterns of invariable acts, inevitably linked together in a necessary order; these patterns appeared particularly clear in sexual behavior, oviposition, and nesting behavior.

The publication of Fabre's romantically titled *Souvenirs Entomologiques* in the late 1800s popularized insect behavior in writings that achieved the heights of literary excellence. They also served to establish standards of observational patience and accuracy which subsequent workers were to continue. And while it became fashionable in the earlier part of the present century to cast doubt on his observations (Fabre was absolutely convinced of the "fixity of instinct" and never accepted the theory of evolution), critics frequently made valuable contributions in the process, helping to focus on the variability of behavior and emphasizing the necessity for a firm taxonomic foundation for behavioral studies.

Born in the principles of evolution and species formation perceived by Charles Darwin, modern insect behavior study went on to show its greatest development after the turn of the century with the emergence of the European school of ethology. In the 1920s and 1930s, when the rest of behavioral research was emphasizing the role of learning and hormones in the modification of behavior, Lorenz in Germany and Tinbergen in Holland began publishing their intensive and extensive natural behavior studies. Although dealing mostly with birds, their research also included fish and insects. As their carefully detailed work unfolded, so did the foundations of ethological theory (see Chapter 2).

How was their work different? First, it provided hints of a general order and logic in the behavior of animals. Second, it demonstrated that certain behavior patterns are just as characteristic of species as are certain morphological features. Third, because behavior was regarded as part of every organism's equipment for survival and the product of adaptive evolution, the focus of ethology was the objective study of whole patterns of animal behavior under natural conditions. Finally, ethologists emphasized the *functions* and *evolutionary history* of behavior patterns. Behavior, they said, always has a cause, and it always has one or more functions.

The emergence of insect behavior as a separate subdiscipline is relatively recent, with the past decade showing a great expansion of interest in this area. Comparative psychologists are turning increasing attention toward the "forgotten majority" of invertebrate examples as they seek verification of behavioral principles typically derived from a vertebrate perspective. Many ecologists are coming to view the behavior of individuals and populations as the basis for much of the central dogma of their discipline. More and more systematists are seeking behavioral information to aid their understanding of species and phyletic relationships.

However, it is important to remember that all of behavioral research is a relatively young, emerging science, and as such it lacks the strong theoretical framework that has come to typify more mature sciences such as chemistry. Additionally, since insects comprise over three fourths of all animals, insect behavior is a potpourri of knowledge and ignorance. Some of the most outstanding and best developed examples of behavioral phenomena—notably in the areas of communication, courtship, mimicry, and the development of sociality—are provided by the insects. But at the same time, even according to very liberal estimates, fragmentary behavioral data (host records, food plants, etc.) are recorded for no more than 5% of the described insect species; the number of species subjected to intensive investigation is far smaller yet. For a summary of the history of behavioral studies on insects, see Richard (1973) and Klopfer and Hailman (1967).

## Some Pitfalls of Behavioral Description

The detailed description of behavior has certain difficulties. Basically, they all revolve around the simple truth that our language lacks the words to accurately convey what has been observed. Almost all descriptive terms possess inherent human connotations that often imply human purpose and/or motivation. Descriptions of a cricket "happily" chirping on the hearth or a "fearful" cockroach scurrying across the floor may be commonplace in nature books for children but have no rightful place in the scientific study of behavior. We simply don't know what the sensation of happiness is to a cricket or fear to a cockroach!

Such attributing of human characteristics to other species of animals is termed **anthropomorphism**; it is unwarranted and should be constantly guarded against. But because total removal of all such terms from the vocabulary of behavior makes for very bland prose, it is not surprising to find them still appearing today. When they are used or encountered, there should be full understanding that an interpretation is being made of the motivations of the animal and that such an interpretation is highly suspect.

More insidious, and surprisingly pervasive in the field of animal behavior, is the problem of **teleology**. A form of anthropomorphism, teleology is the doctrine that the processes of nature are directed toward some discernible "goal." In effect, it endows animals with motivations similar to those which humans might show under like circumstances. Much of man's behavior is goal directed, or *purposive*; for example, when hungry one goes to the table or refrigerator in anticipation of food. However, most animal behavior is *directive,* that is, the outcome of the response is not foreseen by the organism. Thus, there is a confounding of ends and means in statements such as "bees visit flowers because they want nectar." The bee is brought to the nectar by its response to the stimulus provided by the flower; such a response is adaptive, but its outcome is not seen in advance by the bee.

Finally, because most behavior is described in terms of its end result, the very act of labeling a particular observed behavior tends to color subsequent interpretations of it. For example, labeling a chemical secreted by a female moth as a "sex attractant" is done because the end result of her secretion is that males arrive and attempt to mate with her. Yet, as Kennedy (1972) so eloquently points out, in the strictest sense the one thing these "attractants" have not been shown to do is to attract (i.e., to provide directional clues of use to an approaching insect orienting over some distance). The label "sex attractant" masks a whole congregation of behaviors by the respondent such as orientation to wind and light. Rather than focusing attention on such behaviors, the effect of the labeling has been to turn intensive research effort toward identification and synthesis of the chemicals involved.

One additional caution should be made before concluding this chapter. After several generations in the laboratory, artificial selection may alter the cultured insects' behavior in ways that adapt the insect for the more simplified environment. Therefore, behavioral data based on laboratory-reared insects may not accurately reflect the behavior of natural populations of the same species. In fact, different behavioral results obtained by investigators working on the same insect in separate laboratories may be explained as strain differences arising in cultures (in Chapter 9 we cite an example from research on *Leucophaea* cockroaches). Few behaviorists maintain careful checks on the "quality" of the laboratory-reared insects that serve as their research subjects. Chambers (1977) provides some additional insights on this subject.

## SUMMARY

Insects, by far the most abundant land arthropods, probably arose in the Devonian Period. The development of wings, the wing flexion mechanism, and complete metamorphosis have been important evolutionary advances contributing to their success.

Behavior, the ways in which an organism adjusts to and interacts with its environment, includes both maintenance and communicatory activities. Behavior tends to reflect morphologically determined phylogenetic relationships. Cockroaches provide one well-researched example; predatory solitary wasps are another.

Deducing apparent behavioral progressions must be done with extreme caution. Behavior does not fossilize well, and in the absence of a fossil record or compelling logical argument it is difficult to even indicate direction of the evolution of a trait in question. Plotting behavioral evidence against a morphologically/taxonomically determined "family tree" may help, as do the concepts of homology and evolutionary convergence. Ethoclines are relatively little studied in behavior. Sand wasps, a well-documented example of the phenomenon, also illustrate adaptive radiation.

Species-isolating devices are often most clearly evident in mating behavior. Fireflies provide examples of the use of behavioral traits to confirm and discover phylogenetic relations. Whether structure or behavior is the pacemaker of evolution is a chicken–egg question.

Tracing a behavioral component back to its genetic foundation is a difficult task compounded by multiple gene effects. Crossing experiments and selection experiments are useful methods for verifying the genetic determination of behaviors. Crossing experiments depend upon mating individuals that differ in a particular kind of behavior and then examining the behavior of their offspring; honey bee hygienics provide an exemplary study. Selection experiments involve "artificial selection," or choosing only a certain part of each generation to become parents of the next generation. Calculations of heritability give a measure of the degree of genetic determination of the variability in a given behavior present in the sample population.

Physiology, ecology, ethology, and psychology have been great influences on the study of behavior. Ethology, the study of the behavior of animals in their natural habitat, is the major focus of this book within the themes of evolution and adaptation.

Insect behavior as a discipline has its roots in the eighteenth century development of animal behavior begun by French naturalists. By the 1850s, an increasing body of descriptive behavioral material existed, but not until the 1920s and 1930s did the foundations of ethological theory begin to provide a general order and logic in the behavior of animals as ethologists such as Lorenz and Tinbergen emphasized the functions and evolutionary history of behavior patterns. Insect behavior as a separate subdiscipline is a relatively recent emergence and is expanding rapidly. As such it lacks a strong theoretical framework and is very unevenly investigated.

Pitfalls of behavioral description center around the impreciseness of human language as a descriptive tool. Anthropomorphism attributes human characteristics to other species. Teleology attributes goal direction to the processes of nature, including animal behaviors and motivations. Both yield highly suspect interpretations and should be used warily or be avoided. Because most behavior is described by its end result, labeling itself can be a pitfall. The very act of doing so tends to color and limit subsequent interpretations of the activity. Finally, the extrapolation of observations beyond the insects upon which they are based must be tempered with caution. Not only is there great variety in the insect world, but among laboratory-raised insects, in particular, unrecognized strain differences within a single species may be common.

Several additional sources are available for those wishing to augment their background in insect biology. Of the numerous texts on entomology, Wigglesworth (1964) gives an excellent overview, well illustrated and containing a heavy dose of behavior studies. Roeder (1967) and Carthy (1965) provide a physiological slant to the subject. From an evolutionary perspective, the well-balanced behavior texts of Alcock

(1975), Eibl-Eibesfeldt (1970), and Brown (1975) include a wealth of insect examples. The most recent encyclopedic overview of the field of insect behavior is that of Markl (1974); compilations of selected aspects include Haskell (1966), Marler and Hamilton (1966), and Barton Browne (1974). A delightful popular book on insect behavior is that of Evans (1966).

## SELECTED REFERENCES

Alcock, J. 1975. *Animal Behavior. An Evolutionary Approach.* Sinauer Associates, Sunderland, Mass., 547 pp.

Barber, H. S. 1951. North American fireflies of the genus *Photuris. Smithsonian Inst. Misc. Coll.* **117:** 1–58.

Barton Browne, L. 1974. *The Experimental Analysis of Insect Behaviour.* Springer-Verlag, New York, 366 pp.

Bastock, M. 1956. A gene mutation which changes a behavior pattern. *Evolution* **10:** 421–439.

Bentley, D. 1975. Single gene cricket mutations: effects on behavior, sensilla, sensory neurons, and identified inter-neurons. *Science* **187:** 760–764.

Bentley, D. and R. R. Hoy. 1972. Genetic control of the neuronal network generating cricket (*Telogryllus gryllus*) song patterns. *Anim. Behav.* **20:** 478–492.

Benzer, S. 1973. Genetic dissection of behavior. *Sci. Amer.* **229:** 24–37 (December).

Borror, D. J., D. M. DeLong, and C. A. Triplehorn. 1976. *An Introduction to the Study of Insects,* 4th ed. Holt, Rinehart and Winston, New York, 812 pp.

Brothers, D. J. and C. D. Michener. 1974. Interactions in colonies of primitively social bees. III. Ethometry of division of labor in *Lasioglossum zephyrum* (Hymenoptera: Halictidae). *J. Comp. Physiol.* **90:** 129–168.

Brown, J. L. 1975. *The Evolution of Behavior.* W. W. Norton, New York, 761 pp.

Carpenter, F. M. 1973. Geological history and evolution of insects. In *Syllabus. Introductory Entomology,* V. J. Tipton (ed.). Brigham Young Univ. Press, Provo, Utah, pp. 78–88.

Carthy, J. D. 1965. *The Behavior of Arthropods.* W. H. Freeman, San Francisco, 148 pp.

Chambers, D. L. 1977. Quality control in mass rearing. *Annu. Rev. Entomol.* **22:** 289–308.

Dingle, H. 1974. The experimental analysis of migration and life-history strategies in insects. In *The Experimental Analysis of Insect Behaviour,* L. Barton Browne (ed.). Springer-Verlag, New York, pp. 329–342.

Ehrman, L. and P. A. Parsons. 1976. *The Genetics of Behavior.* Sinauer Associates, Sunderland, Mass. 390 pp.

Eibl-Eibesfeldt, I. 1970. *Ethology. The Biology of Behavior.* Holt, Rinehart, and Winston, New York. 530 pp.

Evans, H. E. 1962. The evolution of prey-carrying mechanisms in wasps. *Evolution* **16:** 468–483.

Evans, H. E. 1966. *Life on a Little-known Planet.* E. P. Dutton, New York, 318 pp.

Evans, H. E. and R. W. Matthews. 1973. Systematics and nesting behavior of Australian *Bembix* sand wasps (Hymenoptera: Sphecidae). *Mem. Amer. Entomol. Inst.* No. 20, 386 pp.

Evans, H. E. and R. W. Matthews. 1975. The sand wasps of Australia. *Sci. Amer.* 233: 108–115 (December).

Ewing, A. W. and A. Manning. 1967. The evolution and genetics of insect behavior. *Annu. Rev. Entomol.* 12: 471–494.

Fabre, J. H. 1879–1908. *Souvenirs Entomologiques*. 10 Volumes. Delagrave, Paris. (Translations by A. T. de Mattos later published by Garden City Publ. Co., New York).

Fuller, J. L. and W. R. Thompson. 1960. *Behavioral Genetics*. Wiley, New York, 396 pp.

Halffter, G. and E. G. Matthews. 1966. The natural history of dung beetles of the subfamily Scarabaeinae (Coleoptera: Scarabaeidae). *Folia Entomol. Mexicana* 12–14: 1–312.

Haskell, P. T. (ed.). 1966. *Insect Behaviour. Roy. Entomol. Soc. Lond., Symp. No. 3*, 113 pp.

Hickey, L. J. and R. W. Hodges. 1975. Lepidopteran leaf mine from the early Eocene Wind River formation of northwestern Wyoming. *Science* 189: 718–720.

Hirsch, J. (ed.). 1967. *Behavior—Genetic Analysis*. McGraw-Hill, New York, 349 pp.

Hood, S. 1976. The Mallia gold pendant: wasps or bees? In *Tribute to an Antiquary. Essays presented to Marc Fitch by some of his friends*, F. G. Emmison and R. Stephens (eds.), Leopard's Head Press, London, pp. 59–71.

Hotta, Y. and S. Benzer. 1972. Mapping of behaviour in *Drosophila* mosaics. *Nature* 240: 527–535.

Hoy, R. R. and R. C. Paul. 1972. Genetic control of song specificity in crickets. *Science* 180: 82–83.

Kennedy, J. S. 1972. The emergence of behaviour. *J. Aust. Entomol. Soc.* 11: 168–176.

Klopfer, P. H. and J. P. Hailman. 1967. *An Introduction to Animal Behavior. Ethology's First Century*. Prentice-Hall, Englewood Cliffs, N.J., 297 pp. (especially Chapters 1 and 2).

Levi, H. W., L. R. Levi, and H. S. Zim. 1968. *A Guide to Spiders and Their Kin*. Golden Press, New York, 160 pp.

Lloyd, J. E. 1966. Studies on the flash communication system in *Photinus* fireflies. *Misc. Publ. Mus. Zool., Univ. Mich. No. 130*, 95 pp.

McClearn, G. E. and J. C. DeFries. 1973. *Introduction to Behavioral Genetics*. W. H. Freeman, San Francisco, 349 pp.

McKittrick, F. A. 1964. Evolutionary studies of cockroaches. *Cornell Univ. Agr. Exp. Stn. Mem.* 389, 197 pp.

Manning, A. 1961. Effects of artificial selection for mating speed in *Drosophila melanogaster. Anim. Behav.* 9: 82–92.

Manning, A. 1967. Genes and the evolution of insect behavior. In *Behavior-Genetic Analysis*, J. Hirsch (ed.). McGraw-Hill, New York, pp. 44–60.

Markl, H. 1974. Insect behavior: functions and mechanisms. In *The Physiology of Insecta*, Vol. III, 2nd ed. M. Rockstein (ed.), Academic Press, New York, pp. 3–148.

Marler, P. and W. J. Hamilton, III. 1966. *Mechanisms of Animal Behavior*. Wiley, New York, 453 pp.

Mayr, E. 1970. *Populations, Species and Evolution.* Harvard Univ. Press, Cambridge, Mass., 453 pp.

Parsons, P. A. 1967. *The Genetic Analysis of Behavior.* Methuen, London. 174 pp.

Richard, G. 1973. The historical development of nineteenth and twentieth century studies on the behavior of insects. In *History of Entomology,* R. F. Smith, T. E. Mittler, and C. N. Smith (eds.). Annual Reviews, Palo Alto, Calif., pp. 477–502.

Roeder, K. D. 1967. *Nerve Cells and Insect Behavior,* rev. ed. Harvard Univ. Press, Cambridge, Mass., 238 pp.

Ross, H. H. 1962. *How to Collect and Preserve Insects.* Circular 39, Illinois Natural History Survey, Urbana.

Ross, H. H. 1964. Evolution of caddisworm cases and nets. *Amer. Zool.* **4:** 209–220.

Roth, L. M. 1970. Evolution and taxonomic significance of reproduction in Blattaria. *Annu. Rev. Entomol.* **15:** 75–96.

Roth, L. M. and E. R. Willis. 1960. The biotic associations of cockroaches. *Smithsonian Inst. Misc. Coll.* **141:** 1–470.

Rothenbuhler, W. C. 1964. Behavior genetics of nest cleaning in honey bees. IV. Response of F₁ and backcross generations to disease-killed brood. *Amer. Zool.* **4:** 111–123.

Rothenbuhler, W. C. 1967. Genetic and evolutionary considerations of social behavior of honey bees and some related insects. In *Behavior-Genetic Analysis,* J. Hirsch (ed.), McGraw-Hill, New York, pp. 61–106.

Tinbergen, N. 1968. On war and peace in animals and man. An ethologist's approach to the biology of aggression. *Science* **160:** 1411–1418.

Wigglesworth, V. B. 1964. *The Life of Insects.* The New American Library, New York, 383 pp.

Wilson, E. O. and R. W. Taylor. 1964. A fossil ant colony: new evidence of social antiquity. *Psyche* **71:** 93–103.

# 2

# The Programming
# and Integration
# of Behavior

A headless male mantid can complete a sequence of mating behavior. A decapitated cricket will sing several of its song patterns when its neck connectives are stimulated. *Cecropia* moths debrained as pupae still successfully maneuver from their cocoons. How necessary is an insect brain?

Having followed a tail-wagging dance for the first time, novice honey bees can fly in the indicated direction and distance. Aphids actively discriminate between plants, and their choice of hosts can be influenced by the plants on which they were raised. Hand-raised mantids react to exactly the same stimuli the first time one releases their preying stroke as do experienced animals. How much of insect behavior is controlled by instinct?

Even headless cockroaches can learn to keep a leg flexed to avoid electrical shocks. Honey bees not only learn to avoid a food source from which they have received an electric shock but can remember the previous location of a displaced hive even 12 days later. Female *Ammophila* sand wasps return to their temporarily closed nest and, on the basis of a quick inspection, decide whether many or few caterpillars must be provided to the young larva inside. How much reasoning can insects do?

Insect behavior—obtrusive, incredible, sometimes admirable, and often awe inspiring—has always been a source of amazement and curiosity. Even with their extra legs and inexpressive faces, insects look and act enough like little people. What mechanisms underly the expression of insect behavior? The question has a long history and is still far from a good answer. Yet behavior fairly begs for some degree of physiological explanation, for ultimately behavior means making physiological decisions—when to move and when to remain still, what objects to approach and what to avoid, which muscles to contract and which to relax.

31

Two distinct but closely interconnected internal communication systems are known to mediate insect behavior. One is chemical, through specific substances (hormones, neurohumors, mediators) produced and released by specific body cells. As these chemicals are transported through the body, many cells are exposed to them, but only certain uniquely sensitive cells respond. The other internal communication system is electrochemical, through impulses traveling over the surface of specialized cells called nerve cells or **neurons**. The specificity of this latter system resides not in the differing nature of the messages, for all are due to the same type of physiochemical event, but in the nature of the pathways over which they travel.

## NERVOUS COORDINATING MECHANISMS

At any given moment the environment contains a kaleidoscopic array of stimuli only a fraction of which contain biologically relevant information. Gathering accurate, significant information while filtering out nonappropriate stimuli is a problem of primary magnitude for any animal. For this purpose sense organs comprised of specialized nerve cells have evolved. Uniquely sensitive to particular kinds of environmental energy, they simultaneously screen out all stimuli outside a selected signal range.

All receptors code their information in units called **action potentials**, which are self-regenerating standard signals that travel through other nerve cells along the length of long cytoplasmic cell projections (*axons*) through small gaps (*synapses*), the switchboards of the nervous system, where they are transmitted by the release of chemicals having specific effects upon the neuron or muscle cell across the synapse. Stimulation usually leads to the production of not one but many nerve impulses all of the same amplitude. Information about the stimulus is coded in the number and frequency with which they follow each other, within limits of the system.

In some ways the nerve cell action potentials are analogous to the muscle actions of animals. First, in both a lack of action does not mean passivity; rather, the systems are in a poised state, ready to be tripped into a release of energy. Second, the amount of energy released, whether through motor activity or through current flow across an axon membrane, is not always directly related to the magnitude of the triggering stimulus, being also a function of the internal state of the living system. Third, in cases where the poised state is relatively stable and the response is all-or-none, both require a certain measurable level of stimulus to release a reaction. This well-known measure of sensitivity is called the **threshold of response**; when a system requires a large stimulus to trigger action, it is said to have a high threshold, and vice versa. Finally, a great many behaviors, both of nervous systems and of whole animals, appear to sweep from one sensitivity extreme to the other. When an activity appears to be triggered in a manner independent of external changes but very dependent upon the animal's internal state, it is termed spontaneous, or *endogenous*.

Within the arthropod body, neurons do not occur singly or randomly but in chains of knotlike *ganglia*, which in turn are aggregated into two highly structured general systems. One is the *central nervous system* (CNS) (Fig. 2-1), which includes a brain composed of three major ganglia and coordinates the peripheral sense organs and muscle systems. The other is the *visceral* or *sympathetic system*, which controls alimentary canal movements and is closely concerned with the process of neurosecretion. For more information on the insect brain, see Howse (1975).

In what way do the brain and central nervous system function? Almost certainly not merely as simple relays, but rather as integrative machines. Part of the evidence for this comes from ablation experiments whereby anatomically distinct part of the system are surgically removed and their role deduced by comparing behaviors before and after excision. The crudest of these is to simply cut off the insect's head, an "experiment" that has been performed since far antiquity.

For many stereotyped behaviors the insect brain is unnecessary, an observation which led most biologists prior to this century to believe that insects had no brain. In fact, one criterion in Linnaeus' definition of the Insecta was the lack of a brain! A decapitated insect may even show complex behavior for many days, for unlike the vertebrates the insect has not concentrated its life support systems within the brain. Not only are crucial controls of functions such as respiration decentralized but so are many postural, locomotor, sexual, and grooming mechanisms. At the same time, however, certain crucial differences can be noted when an insect is deprived of various parts of its brain. Most notable is that the relative incidence of various types of actions changes markedly. Often, there is an uncontrolled release of competing behavior modes. For example, a mantis deprived of its protocerebrum simultaneously and continuously performs two opposed behaviors—grasping, which holds it back, and walking, which pulls it forward. This is behavioral nonsense, resulting in a hopelessly entangled and exhausted mantis. Intact mantises, of course, spend most of their time doing neither; rather, they wait motionless, often for hours, to ambush unsuspecting prey.

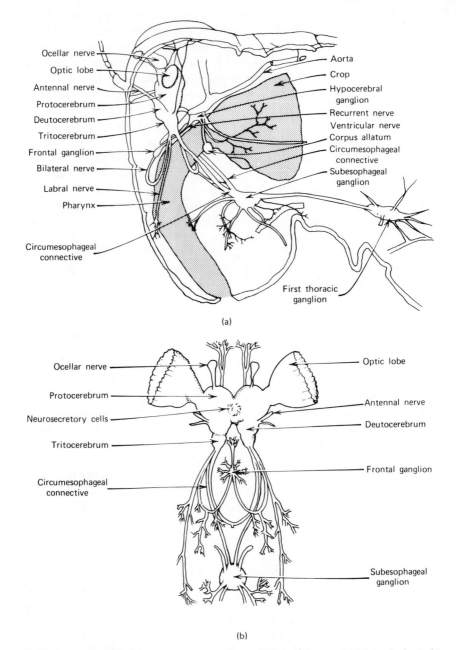

**Figure 2-1** Generalized insectan nervous system in a grasshopper: (a) lateral view, (b) frontal view. The insect brain is composed of three major regions—the protocerebrum, the deutocerebrum, and the small tritocerebrum. The protocerebrum, perhaps the most

When the subesophageal ganglion is removed along with the protocerebrum, however, the mantis becomes permanently immobile. Does the neural program for leg movement reside in the subesophageal ganglion? No, for electrical stimulation of a disconnected thoracic ganglion produces vigorous and complete limb movements. Rather, **neural inhibition** occurs—the capacity of a neuron to exert a blocking action on cells connected to it. In an intact mantis, the protocerebral lobes apparently send out endogenous inhibitory messages that differentially block parts of the excitatory activity being generated by the subesophageal ganglion. The excitatory messages that are allowed to pass are transmitted to the thoracic ganglia where grasping or locomotion is initiated.

But what of the copulatory behavior of the decapitated male mantis? When his head is destroyed, as often occurs during courtship with the predatory female (see Chapter 9, Fig. 9-8), the mantis walks in a circle while vigorously performing continuous copulatory movements. Here the chain of ventral ganglia themselves possess an endogenous activity which is usually inhibited by the subesophageal ganglion—a classic example of *central inhibition*. Such inhibition is a fundamental property of nervous systems. If a nervous system were unable to inhibit those circuits competing with the one responsible for a desired behavior, the result would be behavioral chaos.

### Simple Reflex and Repeated Motor Patterns

Many rapid autonomous behavior patterns in insects depend upon relatively simple neural circuitry. One category of such patterns are the **reflexes**. These are familiar to us; a gentle tap on one's knee elicits a knee jerk reflex, for example. Movement is the commonest form of response to a stimulus. It is also the least complex, for the input and output have a one-to-one relationship. A knee jerk, for example, does

---

complex part of the insect brain, directs neural traffic at the crossroads between sensory input (predominantly from the antennae and eyes) and motor output. This mass also contains medially situated neurosecretory cells that extend to the corpora cardiaca which, together with the corpora allata, form an important part of the insect endocrine system. The deutocerebrum is concerned primarily with nerves associated with the antennae. The anteriorly located tritocerebrum contains nerves connecting with the frontal ganglion and labrum; from its posterior, the circumesophageal connectives connect the brain to the subesophageal ganglion, a complex structure supplying nerves to mouthparts, neck, and salivary glands. Posterior to it, a chain of ventrally situated segmental ganglia furnishes nerves to the peripheral sense organs and muscle systems of the thorax and abdomen. Neurosecretory cells also occur in large numbers in the ventral ganglia where they form a bridge between the two internal communication systems, nervous and endocrine. (Modified after Snodgrass, 1935.)

not repeat itself unless the knee is tapped repeatedly. Because of their relative simplicity, reflexes provide some of the most lucid examples of the way in which behavioral stereotypy depends on properties of nervous systems.

A related set of simple behaviors are those involved in various rhythmically repeated motor patterns such as insect songs, flight, and walking. How are these physiologically generated, maintained, and coordinated? Historically, one explanation, called the **cyclic-reflex** hypothesis, stated feedback from the act itself was sufficient to cause the act to be repeated. Thus, such rhythmic actions would continue in a repetitive circular loop (like perpetual motion) until inhibited by other reflex paths or by the brain. Since the 1960s, however, accumulating evidence suggests that many patterns are generated in the central nervous system—the **central pattern generator** hypothesis.

To date, perhaps the most elegant experimental results supporting the latter hypothesis have been those of Wilson (1964, 1968) on locusts. Using tethered locusts, which will perform normal flight movements in a wind tunnel, Wilson demonstrated that the central nervous system, without any sensory input from wing sense organs, can generate a pattern of motor neuron output that closely resembles the pattern produced in normal flight (Fig. 2-2). When the winds were kept still, the basic pattern of flight motor nerve discharge remained unaltered. Even after he severed the nerve fibers from the stretch receptors which register wing movement, the pattern persisted. Clearly, the locust is not timing its wing beat through sensory cues created by wing movement. Through additional experiments Wilson also dispensed with timing cues from other moving body parts.

The concept of central coordination does not exclude peripheral influences, however. Though the sensory and motor systems are not linked in the locust in terms of timing, the average *frequency* of wing beat is correlated with the discharge rate in the receptor nerves and slows when receptor nerves are cut. (Similarly, cockroaches deprived of sensory feedback maintain a typical walking rhythm, but their movements are in slow motion.) The safest generalization at present is that both mechanisms are probably involved in a great many behaviors.

Perhaps the most striking characteristic of many simple reflexlike actions of insects is their speed. For example, the complete strike of a praying mantis may require but a twentieth of a second. To an escaping insect a few thousandths of a second may be the difference between the quick and the dead. How are cockroaches, for example, so proficient at reacting to our presence, even from behind?

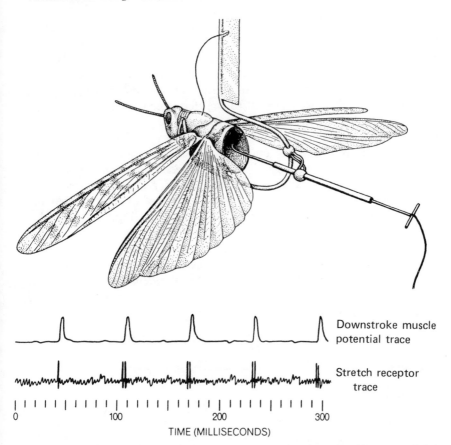

**Figure 2-2** Sensory discharges in nerves from the wing and wing hinge recorded by manipulating wires into a locust's largely eviscerated thoracic cavity. Downstroke muscle potentials repeat at the wingbeat frequency, while stretch receptors fire two to three times per wingbeat. (Slightly modified from "The Flight-Control System of the Locust" by Donald M. Wilson, Copyright © 1968 by Scientific American, Inc. All rights reserved.)

CASE STUDY: COCKROACH STARTLE RESPONSE

Behind a cockroach's abdomen are a pair of antennaelike filaments (cerci) containing deflection-sensitive hair receptors. Nerves connect these with the ventral nerve cord where they synapse with a number of special **giant neurons** with axons about 30 microns in diameter rather than the "normal" axon diameter of less than 5 microns. Two lines of indirect evidence suggest that these giant neurons, which run for great lengths between sensory neurons and motor neurons, are developments for rapid evasive behavior. One is that giant neurons occur in a wide variety of organisms—from fish and lampreys to cephalopods and crayfish—always in connection with mechanisms of withdrawal and escape. The other is that nerve impulse transmission speed has been shown to be roughly proportional to the diameter of the fiber traveled along.

The relation of neurology and evasive behavior has received its most thorough laboratory study through a series of experiments by Roeder (1967). In documenting the relation between one series of interconnected nerve cells and one behavioral response, Roeder also has provided a useful model for investigations into the physiology of insect behavior. First, through a simple but ingenious technique (Fig. 2-3), Roeder established the general time interval within which the overall startle-response system was operating. Then, by successively inserting electrodes into various positions along the cockroach nervous system, he recorded the spike potentials registered when the cerci were stimulated. Through many hours of careful work, a picture began to emerge of the neural events between stimulus and response.

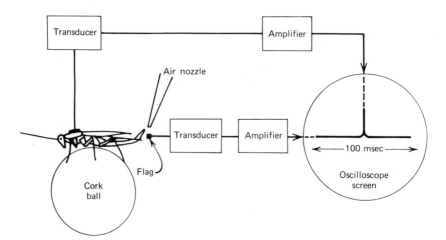

**Figure 2-3** Apparatus used in the experimental study of the cockroach startle response. Glued with wax to a stick, the cockroach is suspended so its legs grasp an unattached cork ball. Near the cockroach's cerci are an air nozzle and a small paper flag. Both flag and attachment stick are connected to an oscilloscope. As an air puff simultaneously strikes the cerci and flag, a deflection registers upon the oscilloscope screen. As the cockroach reacts, the cork ball begins to spin, causing a second deflection. Photographic records of the screen indicate a startle-response time of 28 to 90 milliseconds, with an average of 54 milliseconds. (Redrawn after Roeder, 1967.)

When a mechanical stimulus (such as a puff of air) bends hairs on the cerci, sense cells at their bases are excited. The sense cell axons, about 150 in all, unite to form the cercal nerve, which enters the terminal (sixth) abdominal ganglion to synapse with about four ascending giant fibers and probably several smaller axons. Since stimulation of the cerci excites many sense cells at once, a volley of action potentials arrives at the synapse more or less simultaneously. This source-related additive effect, called *spatial summation*, generates an impulse in a giant fiber. (Actually some discrimination is possible at this point since the giant fibers will not discharge if the summed cercal nerve impulses received at the synaptic region are below the threshold level.)

Sweeping anteriorly along the giant axons in the abdominal nerve cord, impulses eventually reach the metathoracic ganglion where they presumably synapse with motor neurons running to the hind leg muscles.* Interestingly, Roeder found that impulses would not be propagated along a motor axon until at least two, often three to four, volleys of impulses had arrived at this synapse in a time-related additive effect (*temporal summation*). When Roeder continued to send more volleys of impulses along the giant axons, however, they quickly ceased to respond to the continued stimulation. On behavioral grounds, this was not unexpected—in a natural situation, continual directives to begin running are useless and redundant information to a cockroach already fleeing! An additional observation was that once impulses in the motor fibers were generated, they continued to repeat for several seconds after all input from the giant fibers had ceased. Again this was not surprising, since an important adaptive aspect of evasive behavior is that movement continues, or "overshoots," for some time after contact with the stimulus thereby serving to take the animal well out of harm's way.

Using only the shortest possible durations for each neural transaction and summing the times required for the separate parts, Roeder calculated the minimum total elapsed time required to set the cockroach in motion after disturbance. It came to about 20 milliseconds, a figure which, happily, was less than but not too different from the shortest behaviorally measured response time of 28 milliseconds.

The giant axon and its synapses thus appear to form the hub of a very efficient response system. Because of spatial summation, these cells relay information about only very large changes in sensory input (hence

* Recently, it has been found that selectively stimulated giant axons do not elicit the escape movements. Rather, the slower-conducting smaller neurons appear to actually mediate the startle response. The precise function of the giant fibers is not certain, but a possible inhibitory role is implicated and excitation of antennal muscles has been demonstrated as well (see Huber, 1974).

cockroaches do not dash madly off in response to stimuli such as kitchen drafts, which would cause only a few cercal setae to be deflected and fire their sense cells). Discrimination seems to occur at the synapses, which act as switches for the system, requiring input above a certain threshold to turn on but failing to operate after repeated stimulation. Thus, the startle-response system is a simple one. The cockroach needs to make only one of two choices: yes–go or no–stay. Once the decision is made to go, additional information is unnecessary.

## The Release of Stored Programs

Sand wasps, with no contact with members of a previous generation, perform complex nest-burrowing behaviors. Caddisflies reared in isolation spin perfect cases. Male crickets reared in solitary confinement still sing species-typical aggressive and rivalry songs when confronted with another male. In many ways, the behavior of insects gives the impression of the acting out of a prewritten script. Faced with a given situation, even the novice behaves appropriately.

The concept of "instinct," so emotionally controversial in vertebrate behavioral studies, has been more easily accepted when applied to insect behavior. As numerous studies have shown, even in insects that have never been in contact with conspecifics, all individuals of the species still exhibit many species-specific motor patterns in exactly the same form. Such **fixed action patterns** (FAPs), first described and elucidated by Lorenz (1970) and Tinbergen (1951) in the 1930s, seem to be controlled centrally. Once initiated, they require few or no external stimuli or additional sensory cues for their maintenance or completion. For example, many insects, after certain body parts are removed, will proceed to clean the missing appendages as though they were still there. Thus, FAPs have been likened to a piece of memorized music played by a pianist without hearing, seeing, or feeling tactile sensation from his hands.

Fixed action patterns can be evoked by a variety of stimuli, sometimes from different sensory modes, and they may show variability in orientation. Their intensity, completeness, and repetition rate may vary. However, once evoked, their basic structure, that is, the temporal and spatial relation between the elementary muscle contractions comprising the behavioral performance, is stereotyped. Like morphological characters, FAPs are subject to selective pressures and have a genetic basis.

## The Control of Performance

What triggers the release of a fixed action pattern? In general terms, the sensory-neural mechanism involved has to be exposed to some very specific form of stimulation in order to set the pattern into motion. In some cases, it appears that the animal responds to the total stimulus configuration, or *Gestalt*. However, in most cases only certain limited and specific aspects of the stimulus evoke a fixed action pattern. For example, the attraction of a male mosquito to conspecific females may be evoked by a particular sound frequency (see Chapter 7), which we recognize as the females's distinctive hum; or that of a male moth, by a specific odor (see Chapter 5). Artificially produced sounds and synthetic chemicals will produce exactly the same behaviors, in the complete absence of the respective females. Such quite specific stimuli or stimulus configurations, which serve as signals that elicit a fixed motor pattern under normal conditions, are termed **releasers** or **sign stimuli**. Their adaptive significance probably lies in their ability to maximize an animal's recognition of biologically important stimuli while minimizing the amount of neural circuitry required.

A releaser need not in itself have any inherent relevance to survival. Rather, it may be some property of a particular *situation* that has biological importance and a property that is shared by all other similar situations. For example, skatol and ammonia release feeding behavior in both dung beetles and blowflies; these substances are produced by decaying matter, but neither are by themselves of any nutritive value nor owe their existence to dung beetles or blowflies. This type of releaser has been termed a **token stimulus** (see also Chapter 4).

In passing, it should be noted that the naturally occurring sign stimulus for a behavior may not necessarily be the optimally effective one. Male *Argynnis* butterflies, for example, are more strongly attracted to solid orange than to the orange-black pattern of the female (see Fig. 6-12). A male fly's sexual jump is optimally released by an object two to three times the female's size. Such **supernormal sign stimuli**, signals which are even more effective in eliciting a given behavior than the natural signal, occur as a result of evolutionary compromise. For example, in the sexual approach of *Heliconius* butterfly males, use of experimental models has shown that pure red wings upon the female would be the optimal attractant. However, since the coloration at the same time has mimetic functions, natural selection has dictated that the red signal be limited to a stripe.

A simple and useful heuristic concept has arisen easily from discussion of fixed action patterns and releasers, namely, that for each FAP there exists a sensory-neural release-controlling mechanism, called the **innate releasing mechanism**, or IRM. This mechanism evaluates the afferent signals of one or several sense organs and triggers (or fails to trigger) a reaction. In complex behaviors, the entire sequence of actions depends, at every step, upon the appropriate external stimuli and internal physiological state before that particular subset of stereotyped motor patterns can be released. If any link in the chain is broken, the genetically preprogrammed sequence of behaviors does not continue and hence cannot run to completion. At each step, the threshold for activating the innate releasing mechanism may be highly variable and may respond to both internal and external influences. In common experience, we say an organism shows varied "motivation" to perform the act. A highly motivated individual may be so ready to act that it will perform a fixed action pattern when confronted by a stimulus bearing but slight resemblance to the typical releaser. For example, if the sand wasp *Ammophila* is thwarted too long from pulling a paralyzed caterpillar into her nest, she will begin to retrieve substitutes. A nonrewarded individual or a satiated one, on the other hand, is less motivated. Male *Nasonia* wasps, for example, will mate continually if virgin females are available; when confronted by only nonreceptive (previously mated) females, however, their courting gradually wanes.

The IRM is, by and large, a "black box" sort of concept, for it considers only input–output rather than probing the heart of the "box" to examine the physiological components behind the control of fixed action patterns. Such an analysis has been undertaken in only a few instances. One of the most elegant and complete has been the investigation of the releasing mechanisms for the evasive behavior of noctuid moths by Roeder (1965, 1966, 1967).

It is easy to show that some moths respond to high-pitched sounds. A variety of man-made sounds are effective. As a simple experiment, suddenly jingle a bunch of keys near a group of moths flying around a street light or window screen. The response of the moths will at first seem chaotic. Some nearly fall to the ground, while the flight of others becomes quickened and more erratic. Some that were fluttering may become motionless, while others, previously motionless, may take flight. What pattern can there possibly be to all of this? To answer this question, one must appreciate the importance of a principal predator of moths, namely, bats.

Bats and moths have been playing the proverbial cat-and-mouse game for millions of years. Moths are one of the main sources of food for certain families of bats. Yet the game is clearly a balanced one—while all bats in these families probably locate and capture some moths, some moths locate and evade all bats. Knowledge of the bat's sonar system was provided by Griffin (1958), whose precise and ingenious experimentation revealed a bat echolocation system capable of indicating size, distance, location, and considerable detail about its surroundings, down to items as tiny as insects smaller than midges.

Except in the case of certain "singing" insects, one does not usually think of insects as possessing ears. But members of certain families of moths, especially the Noctuidae, possess true tympanic organs, located in the thorax just below the second pair of wings (Fig. 2-4). From the physiologist's standpoint, a simpler system for experimental analysis would be difficult to find, for each of the moth's tympanic organs consists of only two acoustic sense cells, each with a different threshold, coupled to a thin tympanic membrane. The more sensitive acoustic cell is known as A1, the less sensitive, as A2. Although unable to discrimi-

**Figure 2-4** Profile of a noctuid moth, with the wings raised to reveal the external opening of the left ear (arrow). (Courtesy of H. R. Agee.)

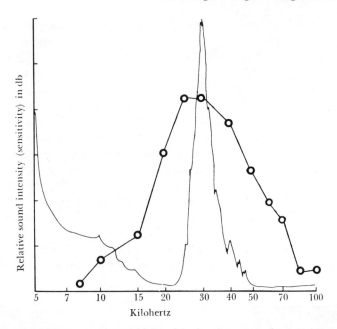

**Figure 2-5**   The relative acoustic sensitivity of the hearing organ of a noctuid moth, *Feltia subgothica,* at various frequencies (open circles), contrasted with summed intensites of all natural environmental noise recorded in the moth's environment at night (solid line). Sounds below 15 kHz were mostly of insect origin. Those between 23 and 50 kHz came from passing bats. It is apparent that moth ears are maximally sensitive to sounds in the latter range and relatively insensitive to other sounds. (From Roeder, 1970. Reprinted by permission, *American Scientist,* journal of Sigma Xi, The Scientific Research Society of North America.)

nate between tones of different pitch, the tympanic organs are tuned for the reception of ultrasound and have revealed themselves to be especially sensitive to the calls of bats, as shown in Figure 2-5.

Case Study: How Moths Hear Bats

The first verification that moth tympanic nerves respond to bat chirps was obtained quite by accident at Tufts University, when a hibernating bat was brought out of a refrigerator into the laboratory near a moth tympanic nerve preparation linked to an oscilloscope and microphone. Unexpectedly the bat recovered from the cold enough to shriek, bite the investigator's hand, and escape into the room, where it flew "silently" around near the ceiling. Throughout this flight, the tympanic nerve proceeded to deliver a rapid series of short spike bursts.

Excited and encouraged at this, Roeder and his students eagerly lugged some 300 pounds of electronic gear out onto a grassy hillside where bats were known to feed in the evening and set up a moth for tympanic recording. Very soon, they found themselves able to decode the movements of bats from the nature of the moth's tympanic nerve responses. Rigging up a floodlight so they could actually see the bats (since they were unable to hear them independent of the moth's ears), they quickly learned that the range of the moth ear was much greater than they had expected. Later experiments, in fact, showed it to be in the nature of 100–200 feet when bats were flying toward the moth at an altitude of about 20 feet. In contrast, a bat is unable to track a moth at ranges greater than about 10 feet. The bat, however, has the advantage of much greater speed.

When Roeder turned his attention to the moth's flight behavior, the complexity of the natural situation seemed almost to defy study—"a dizzy dogfight" was his impression, "extrapolation of a string of acoustic dots in time. . .pitted against unpredictability; power and speed against maneuverability." It might be compared to combat between a fighter plane and a helicopter, the bat flying a time intercept course while the moth undertakes selective evasive tactics on the basis of the bat's range and speed of approach. To reduce this complexity, Roeder replaced the bat with an artificial one: a stationary multidirectional electronic transmitter of ultrasonic pulses, linked to a camera system and perched 16 feet in the air. When exposed to these electronic bat cries, moths showed the same bewildering variety of reactions as they did to real bats. After many hours of observation and over a thousand photographically recorded moth tracks, Roeder determined that the nature of the evasive tactics of the naturally flying moths was related to the distance between the moth and loudspeaker (Fig. 2-6). Moths that flew close to the loudspeaker and encountered ultrasound pulsed at 10–30 times a second responded with a variety of maneuvers, usually ending in a dive. Those cruising at greater distances characteristically turned and flew directly away from the sound source. The adaptive value of this response difference seems obvious. There would be little survival advantage to the moth in attempting to flee a bat at close range or in making erratic turns and twists while the predator was still distant, but every advantage in erratic behavior when the bat was close. Not surprisingly, straight directed fleeing occurred only in response to low-intensity sounds; when the loudspeaker signals were made progressively weaker, the distance at which directed fleeing was released was correspondingly reduced.

Interestingly, moths cruising at about the same height as the loud-

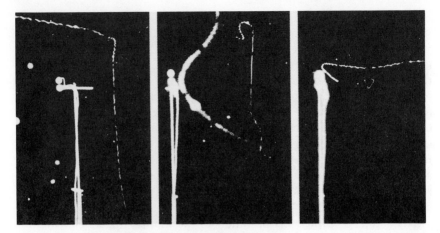

**Figure 2-6** Flight tracks of free-flying moths in response to artificial bat cries from a miniature loudspeaker mounted on a thin tower. The white streak represents the moth's flight track; undulations are due to wing movements. Isolated white spots and small blurs are due to other insects. Depending upon stimulus intensity, moths respond by power dives (left) or passive drops (center) from intense signals. In response to faint ultrasound moths fly directly away (right). (Photograph by F. Webster; see Roeder, 1965.)

speaker turned and flew away in the horizontal plane; those above it were observed to fly directly upward, or redirected their flight by making a sharp turn before flying straight away. How can a moth equipped with only four sense cells orient and steer itself with respect to a sound that comes from various angles? As background for an answer, one must remember that free-flying moths typically flap their wings many times per second. Unlike an airplane, they rarely fly for long in the same direction on an even keel. This wing movement has profound effects on the acoustic sensitivity and directionality of the moth's ears, for the position of the ear below the wing means that on the downstroke most sound will be screened from reaching the ear, while at full upstroke the ear is accessible to sound from all directions.

To investigate the effects of wing position more fully, Roeder and his associates devised an elaborate apparatus (Fig. 2-7) designed to measure the acoustic sensitivity of one ear to sound coming from all points in an imaginary sphere surrounding a moth fixed in different flight positions. Obtaining these recordings was a long and arduous task, and Roeder was fully aware that the surgical insult of implanting electrodes, plus the restriction of wing movements that was necessary, might affect the results. However, recordings of the responses of the A fibers, when analyzed, revealed two types of acoustic assymetry between the two

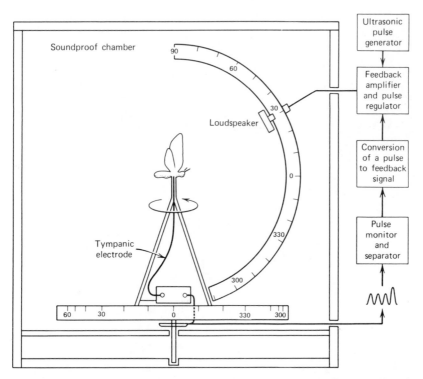

**Figure 2-7** Diagram of apparatus used by Roeder and colleagues for measuring the directional sensitivity of a moth's ear. With wings held in a fixed position, a moth is impaled on a revolving tower which serves as an indifferent electrode. An electrode is inserted into the tympanic nerve, and both are connected to a continuous recorder. A loudspeaker positioned at various angles to the moth emits ultrasonic pulses the intensity of which is continuously regulated by feedback from the moth's nerve spike response as the moth is rotated 360 degrees. (Redrawn after Roeder, 1967.)

ears. One was a left–right assymetry. The right ear detected sounds most effectively when they originated on the right half of the sphere, and best of all when the sound source was roughly at right angles to the body axis with the wings at the top of their upstroke. At the same time, hearing in the left ear was at its minimum. Alternating with this was a dorsal–ventral assymetry, for the flapping of the wings had the effect of damping the sound out nearly altogether when the wings were in the down position. Thus, the left–right differences in sound intensity would alternate with the temporary disappearance and reoccurrence of sound occurring some 10–40 times per second in synchrony with the rate of wing flapping.

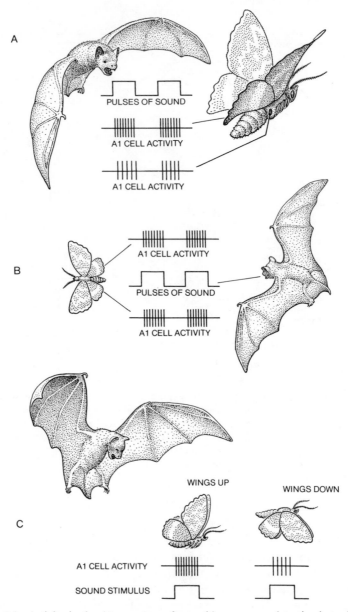

A

PULSES OF SOUND

A1 CELL ACTIVITY

A1 CELL ACTIVITY

B

A1 CELL ACTIVITY

PULSES OF SOUND

A1 CELL ACTIVITY

C

WINGS UP

WINGS DOWN

A1 CELL ACTIVITY

SOUND STIMULUS

**Figure 2-8** Activity in the A1 receptors of a moth's ears upon detecting bat cries from different points in space (not drawn to scale): (A) The bat is to one side of the moth; the receptor on the side closest to the predator fires more rapidly than the shielded receptor. (B) The bat is directly behind the moth; both A1 fibers fire at the same time and rate. (C) The bat is above the moth; activity in the A1 receptors fluctuates in synchrony with its wing beat. (From Alcock, 1975, based on Roeder, 1967.)

How do these facts relate to the ability of a moth to escape a hungry bat? A left–right difference in A-fiber discharge when the wings are up probably provides the moth with a rough horizontal bearing on the position of a bat with respect to its own line of flight. The absence of a left–right differential discharge and the presence of similar levels of on and off from both ears might inform the moth that the bat was above it. If neither variation occurred at the regular wing beat frequency, it would mean that the bat was below or behind the moth (Fig. 2-8).

It would seem that a single sense cell in each ear would be capable of transmitting sufficient information to inform the moth of the bearing of a cruising bat. Yet, as mentioned earlier, moth ears each have two acoustic sense cells. Why? The key seems to lie in their differential sensitivities. Roeder's comparative measurements of the response of the A1 and A2 cells to a range of intensities of ultrasound revealed that A1 responded first, over a range of low to moderate intensities. Cell A2 began to fire only at moderate intensities and fired even faster at high sound intensity, a range within which A1 was saturated and therefore incapable of further increase. Thus, by operating in piggyback fashion, A1 and A2 appear to provide a combined signal from which intensity differences can be discriminated over a range wider than either alone could accomplish. In effect, their combined signal informs the moth how far away the bat is, which ultimately decides the form of the moth's evasive behavior. Thus, moths hear and decode several parameters of bat sounds that are of importance in survival. A popular account of this study is given by Roeder (1965) and an interesting sequel is Roeder (1976).

## The Hierarchical Organization of Behavior

To this point, we have considered behavior as though it consisted only of different elementary performance units, all of equal status, which take polite turns being in charge of the insect body. However, the vast majority of motivational factors (internal and external) initially generate quite unspecific behavior, sometimes simply locomotion. Thereafter, a series of increasingly specific "appetitive" behaviors extend, each with its own more specific releaser, ultimately culminating in those stimuli which can trigger the terminal "consummatory act." In the chainlike reciprocal signal exchange that occurs between the two sexes during courtship (see Fig. 6-13), the consummating act of copulation can occur only at the end of a hierarchical sequence of several releasers and FAPs. The prey-catching behavior of the bee wolf (see Chapter 4) and the provisioning behavior of *Ammophila pubescens* (see p. 56) illustrate additional examples of the hierarchial sequences of behaviors.

Our model has been too simplistic also in another way. Motivations for several types of behavior exist in an animal concurrently, and in the environment there often exist sign stimuli for quite different and sometimes even contradictory forms of behavior. However, an animal can normally do only one thing at a time. Thus, it is not surprising to find that some conflicts occur; what is surprising is that there are not more. It appears that under normal circumstances behavior outcome is dependent upon the relative strengths and effectiveness of the different motivational factors and sign stimuli involved. Moreover, the behavior that is activated apparently suppresses other noncompatible behaviors either wholly or partially. For example, in aphids flight and settling behaviors inhibit each other. When honey bees are exposed to releasers for activation and inhibition of their communicative dances simultaneously, trembling dances may occur. In a great many insects, escape behavior is inhibited during copulation.

When actual conflict does occur, apparently nonfunctional stereotyped actions called **displacement activities** sometimes occur. Most recognized cases have involved cleaning or preening behaviors. A dragonfly nymph whose prey is suddenly withdrawn, for example, may begin apparently nonfunctional grooming behavior. Bees faced with the antagonistic behavioral motivations to stay at a food source and to leave it will begin grooming, in a displacement behavior independent of actual "dirtiness." Dusting the bee with pollen or flour does not change the frequency of this behavior; rather, it merely directs the preening movements toward the dustier body parts. Interestingly, there are no known cases of displacement activities being ritualized as parts of more complex behaviors, a common and conspicuous event in vertebrate behavioral evolution.

We have briefly considered the motivational and stimulus parameters needed to *trigger* a behavioral response. Organization, however, also implies control through coordination and termination of behavioral sequences. In some simple cases this seems unimportant; in the cockroach startle reaction, for example, the "yes–go" response was totally unaffected by subsequent sensory input. In "open" systems such as these, there is no provision or need for correction subsequent to the initial stimulus. The majority of ethologically interesting behaviors, however, are "closed"; environmental and/or internal responses continuously influence the animal's behavior through feedback loops.

Critical to the explanation of such feedback loops is the requirement for some type of monitor acting in a manner analogous to a thermostat. A classic example is provided by the feeding behavior of the blowfly (Gelperin, 1972; see Fig. 4-23). Although releasing stimuli for feeding may be continuously present, central inhibition and chemoreceptor adaptation cause the fly to interrupt feeding sporadically. However, food uptake is not finally ended without inhibitory stimuli arising from stretch receptors in the gut wall, which serve as internal monitors that fire as the gut fills. Action potentials carried to the brain via the recurrent nerve trigger the eventual motor response, namely, retraction of the fly's proboscis from the food. Cutting the recurrent nerve eliminates the negative feedback from the stretch receptors, with the result that the fly literally explodes from overeating!

The role of feedback mechanisms in behavior is treated in detail by McFarland (1971).

## The Mental Capacity of Insects

Dragonfly larvae of the genus *Anax* raised in aquaria soon come to associate the sight of their caretaker with food; when he appears, they may even snap in anticipation before food actually appears. A mantis, although it originally would attack, will learn not to strike at an object after it has received an electric shock or bad taste from this object. Such associations between a signal and an experience are actually not uncommon among insects. Many learn to associate food, nest location, or imminent unpleasantness with other stimuli and react accordingly.

### *Memory and Learning*

Learning may be simply defined as any enduring or relatively permanent change in behavior that occurs as a result of experience or practice. Memory—the capacity to store information—is a prerequisite, resulting in a linkage between stimulus and response that would not have occurred without the previous experience with that stimulus. The commonest form of insect learning is **habituation**; that is, the gradual lessening of responsiveness to a stimulus as experience finds it to be harmless or at least unavoidable. For example, a variety of insects such as ants and mantids can be "tamed"; if handled frequently enough, they will come to respond calmly to being picked up and moved about. Habituation should be distinguished from neural adaptation and sensory or motor fatigue, which are very short-term failures in response. Habituation, a more enduring lack of response, is not adaptation or fatigue but rather a mechanism to reduce the instances in which a certain innate releasing mechanism (IRM) is needlessly activated.

Also very common in insects is **associative learning**, that is, the ability to form associations between previously meaningless stimuli and reinforcements such as reward or punishment. This type of conditioning has been widely exploited in experiments designed to delimit the sensory capacities of insects. A classic example is von Frisch's experiments on the language of honey bees (see Chapter 7). In insects, as in most animals, associative learning is undoubtedly an important part of individ-

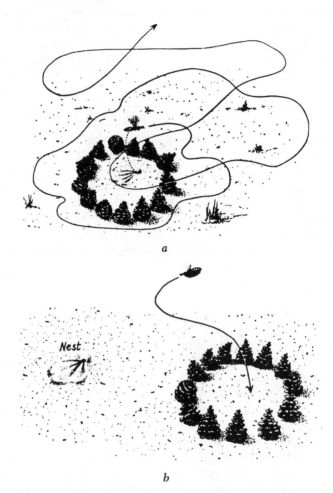

*a*

*b*

**Figure 2-9** Landmark learning by the bee wolf, *Philanthus triangulum*. After the nest entrance is ringed with fir cones, the wasp learns to associate her nest with this distinctive landmark through an orientation flight (a). While the bee wolf is away hunting, the ring of cones is displaced. Upon return the prey-laden female flies to the center of the fir cones (b). (From Tinbergen, 1951; see also Tinbergen, 1972.)

ual accommodation to a changing environment, a method by which an IRM can be fitted more exactly to the environmental situation.

A third basic type of insect learning is **latent learning**, sometimes termed "exploratory learning." Bees and wasps, for example, learn the location of their nests through landmarks recognized and remembered from a previous orientation flight (Fig. 2-9). Latent learning differs from associative learning in that it occurs without apparent reinforcement. The motivation for latent learning is undefined, but by learning the characteristics of its surroundings an insect increases the chances of its own survival.

The area over which an animal normally wanders in search of food, shelter, and/or mates is called its **home range**; within it, conspecifics may be tolerated and the space thus shared. Among insects many species long-lived as adults have a well-defined home range, such as the foraging area of an ant or termite colony. Learned home ranges are characteristic of tropical *Heliconius* butterflies, within which they memorize the location of their food flowers and larval host plants; adults tend to return nightly to a particular communal roosting place for up to six months. Similarly, tropical euglossine bees fly regular plant "trap lines" to which they return day after day to collect pollen and nectar from newly opening flowers (Fig. 2-10).

While insects are undoubtedly capable of other types of learning, behavioral knowledge about insect learning is still embarrassingly meager. Almost all behavioral learning studies have been performed with a mere handful of insects from four of the 28 or so insect orders. These animals—cockroaches, grain beetles, fruit flies, blowflies, ants, wasps, and bees—have been chosen not because of their representative nature but primarily for convenience and availability.

### Insect Intelligence

Honey bee workers can learn signals quickly and with apparent ease in every known sensory modality. Multiple tasks pose few problems; they can link up to five different visual signals with correct turns in a maze. Nor do sequential tasks pose great difficulty, for bees may learn to visit a single location up to six different feeding times during the day or four different places at four different times. Some ants, such as *Formica pallidifulva* workers, can learn a six-point maze at a rate only two to three times slower than that of a relatively advanced vertebrate, the laboratory rat. Female *Ammophila* wasps that have made an inspection visit to a temporarily closed nest burrow can still remember and

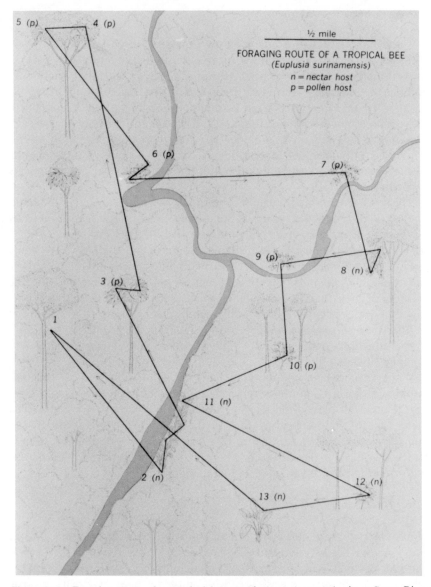

**Figure 2-10** Foraging route of a tropical bee, *Euplusia surinamensis,* in a Costa Rican forest. The female's nest (1) is constructed 130 feet above the ground under loose tree bark. Each day she flies this foraging route or trapline which includes ground plants (2, 11, 13), shrubs (6, 7, 9), understory trees (10, 12), vines (3, 5, 8), and an epiphytic shrub (4) to collect nectar (*n*) and pollen (*p*) from newly open flowers. Each plant visited by these bees produces only one to a few new flowers each day, but flowering may continue for up to six months. That the bees memorize the trapline was shown by the fact that marked females returned to the plants daily even though the flowers were artificially removed prior to the bee's visit. (From Janzen, 1974.)

properly perform the action determined in that single visit, even 15 hours later.

Faced with a myriad of such examples, it is easy to succumb to a sense of wonder and conclude that some insects are comparable to vertebrates in intelligence. (It is small wonder that many fiction writers have done so.) However, these are but fragments of information from the insect's behavioral repetoire, and "intelligence" in this sense of the word is a meaningless measure. In fact, in this sense animals do not even possess varying degrees of "intelligence"; rather, they possess specific abilities to learn selected things.

For any organism there are certain types of tasks that can be mastered and others that cannot. This is no less true with insects. Honey bees, which can learn quickly to orient with respect to attractive odors, cannot do so at all with respect to repellents. Thus, one is often struck by an apparent paradox—the association, side by side, of a seemingly startling propensity for learning and very inflexible behaviors. For example, the predatory wasp *Liris nigra* (see Fig. 8-29) is no less able than others of its kind to utilize latent learning to relocate nests after foraging trips. Yet in prey capture, which normally involves stinging a cricket in four ganglia ending with the subesophageal, the wasp is intensely "frustrated" when confronted with a decapitated prey. She will sometimes spend more than an hour vainly searching the cricket for the site of the missing ganglion (Steiner, 1962).

In a more limited and useful sense, intelligence means the ability to perform *rational* operations, that is, to generalize learned information by transferring it from one set of circumstances to another. How do insects measure up in this regard? Not very well. Their ability to transfer memories to assist in the learning of new situations is nearly or totally absent. Nor are they capable of *insight*, that is, reorganizing their memories to construct a new response in the face of a novel problem. The great naturalist J. H. Fabre demonstrated this about a hundred years ago with another hunting wasp that briefly drops her cricket near the hole to enter her burrow before reappearing to drag it inside. While she was inside the burrow, Fabre moved the cricket a short distance away. With insight, a wasp might be expected to recognize that her prey had merely been moved and that her burrow, having been inspected, was ready for stocking. Instead, when the wasp reappeared, she returned the cricket to its "proper" place at the edge of the hole once more, then descended again, alone. Fabre moved the cricket; again, the wasp reappeared to reposition it and reenter her hole alone. Fabre reelicited this response 40 times before he lost patience!

In other situations where insight or transfer learning might be expected, insects often substitute still another reaction. Rather than showing insight, they may merely disregard conflicting or novel informa- tion. If a familiar sign stimulus appears out of proper chronological sequence, for example, it may simply be ignored. The digger wasp, *Ammophila pubescens*, illustrates this reaction well.


CASE STUDY: THE DIGGER WASP, *Ammophila pubescens*

Unlike most other digger wasps, *Ammophila pubescens* provisions up to 15 nests at a time, not only remembering the locations of each but providing each with the proper number of caterpillars. Because the nests were begun at different times, the young inside may range from a newly hatched larva capable of consuming but a single caterpillar a day to an older offspring requiring three to seven caterpillars, or a full grown larva which needs not more food but sealing off of its burrow so as to allow it to mature underground unmolested.

Watching the wasp closely, the Dutch entomologist G. P. Baerends (1941) noted that the female begins each day with inspection visits. Empty-handed, she flies from nest to nest, inspecting the contents of each. Might this behavior provide a clue to her ability to do so many complex things at once? To answer this question, Baerends located a series of nests being attended by a single female and carefully replaced the nests with plaster of Paris casts (Fig. 2-11), which he could open and inspect at will. Since the female accepted these as authentic burrows, he was in a position to vary cell contents at will—an ideal experimental setup.

In the following days Baerends spent many hours performing a variety of manipulations upon the nest contents, substituting larvae of various sizes for each other and adding and removing caterpillars the female had brought. Very quickly it became apparent that the single morning inspection visit was setting the wasp's behavior for the rest of the day. When the mother wasp encountered a small larva during the inspection visit, that cell would receive one to three caterpillars at some point during the day. Upon finding a large wasp grub during her inspection, the mother would later bring three to seven caterpillars. Finding a fully grown one spinning its cocoon, she would close the nest and begin another.

No matter what actions Baerends took subsequent to the wasp's inspection visit, the pattern was unaltered. Substituting a larger larva after the inspection visit made absolutely no difference, nor did it matter whether he added prey or removed some that she had already brought.

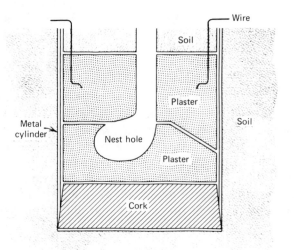

**Figure 2-11** The artificial *Ammophila* nest used to study the influence of cell contents on behavior. A metal cylinder with a cork bottom surrounds the two-part plaster of Paris nest; the upper portion of the nest can be lifted off by grasping the wire hooks. (After Baerends, 1941.)

If her inspection visit indicated permanent closure, this she would do, even if the larva had subsequently been removed so she now was sealing a completely empty nest!

Thus, provisioning behavior in *Ammophila* is paced to meet the needs of each growing larva in a way that is both sophisticated and restricted. Each step in the unfolding behavior patterns of the day is guided by the sign stimuli present in the single brief examination of nests, and after this each appropriate motor pattern is performed in a genetically determined sequence.

In concluding this section it is profitable to ask under what conditions a heavy reliance upon learning or instinct might be selectively advantageous. It appears that, in general, a high reliance upon learning ability is adaptive when an animal is relatively long-lived and/or faced with a good deal of uncertainty or variability regarding aspects of its environment that are biologically significant (see Fig. 9-20). In contrast, the advantage of a high reliance upon innate behavior patterns lies in their reliability. When a particular environmental cue can be linked dependably to a biologically appropriate response, innate fixed action patterns are certain to be successful responses (see Fig. 2-15). Selection will also favor innate behavior when the cost of an initial mistake is high, such as would be true in the case of the cockroach and moth escape reactions. Innate behavior presumably also permits economy in the nervous system (Dethier, 1971).

Once, the "nature–nurture" controversy was vigorously debated; instinct and learning were pictured as diametric opposites in command of the behavior of different kinds of animals. However, the controversy really is a spurious one, and few behaviorists nowadays seek to determine whether a particular behavior response is learned or instinctive (see Konishi, 1971). Behavior is rarely determined either solely by the type of outside events impinging on the individual or by inborn heredity alone, but rather by interaction of the two. A more useful question concerns the degree of stereotypy of a given behavioral act or what aspects of the behavior can be attributed to genetic versus environmental input.

## HORMONAL COORDINATING MECHANISMS

It is well known that insects rely heavily upon "circulating chemistry," an endocrine system closely allied with and complementary to the nervous system (see Fig. 2-1), producing blood-borne hormones serving to coordinate longer-term activities. In addition to specialized endocrine glands, insects also have clusters of neurosecretory cells in the brain and throughout the central nervous system. Various body organs, especially those associated with reproduction (such as ovaries, testes, spermatheca, etc.), also are known to have a secondary endocrine function.

A major hormonal function, the control of growth and development, is thoroughly summarized by Wigglesworth (1970). The other hormonal effects most studied to date have been connected with sexual maturation and reproduction (see Chapter 9), but internal hormone secretions have been implicated in nearly every aspect of insect life history, including important controls on migration, orientation, and periodic behaviors as well as activation of adult behavior following eclosion.

Behaviorally, hormones often act as "primers" setting the "mood" or internal motivation of an insect to perform a particular behavioral act. Evidence for this role comes from three types of observations: (1) correlation of the onset of hormone secretion with the appearance of a specific behavior; (2) precocious induction of a specific behavior by hormone application or implantation of active glands; and (3) disappearance of a particular behavior after removal of the endocrine source of a particular hormone. Alone, none of these constitutes a proof-positive test for hormonal involvement, however. Stronger evidence is obtained when removal of a particular endocrine center abolishes a specific behavior which is then restorable through gland implantation or hormone application. For example, removal of the corpora allata (an operation termed **allatectomy**) of last instar or newly emerged females of *Gomphocerippus* grasshoppers completely prevents the onset of normal sexual receptivity. Females so treated respond to courting males by kicking and performing escape reactions, which are primary defense behaviors. However, implantation of actively secreting corpora allata into such females changes their behavior toward males to a state of "copulatory readiness." In addition, allatectomy of previously sexually receptive females leads to loss of this receptivity and reappearance of primary defense behavior within six days. Thus, one can state with fair confidence that for these grasshoppers the internal "mood" toward mating is under direct control of the corpora allata.

Unfortunately, not all hormonal influences on behavior are as clear-cut as the above example. In periodic behaviors, considered in the following section, hormonal involvement has long been implicated but the evidence is more controversial.

## Hormones and Biological Rhythms

After 17 years of silence underground as nymphs, great numbers of periodical cicadas emerge to fill the air with their raucous noise. The snowy tree cricket sings in such a rhythmic tempo that one can ascertain the temperature on a summer's eve by listening to its chirp with a watch in hand. Silverfish scurry about each night, resting in cracks and crevices during the light of day. Much of what insects do is rhythmic, from heartbeats and songs to cycles of sleep and wakefulness, to reproductive cycles measured in weeks, months, or even years. All such biological rhythms, however, appear to fall into four broad categories—cellular, physiological, gated, and reiterative. Of these only the latter two have particular relevance to the programming of behavior.

A great many behavioral rhythms occur with a regular repeated periodicity in the life of a single individual. Feeding and locomotory cycles are common examples. Such **reiterative behaviors** may cycle about a relatively long period, such as the lunar periodicity of nesting behavior in the nocturnal sweat bee, *Sphecodogastra texana* (Fig. 2-12). Typically, however, reiterative behavioral rhythms have a periodicity of about 24 hours and thus are termed **circadian rhythms**, derived from the Latin *circa*, about, and *diem*, day. For example, anyone who has cohabited with cockroaches has noticed their characteristic pattern of daily activity which begins shortly after dusk (Fig. 2-13).

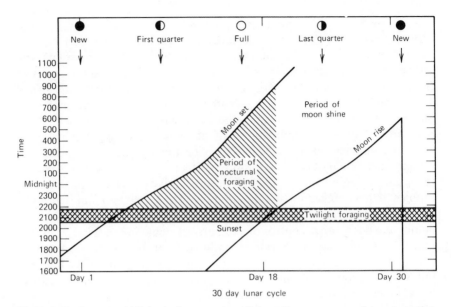

**Figure 2-12** Lunar periodicity in the sweat bee, *Sphecodogastra texana*. In good weather adult female activity always began about sunset. Concentrated pollen collection and nest cell construction were observed only during that part of the lunar cycle where moonlight was continuous with the twilight, thus permitting extended nocturnal foraging. When moon rise occurred *after* the close of twilight (ca. 9:30 P.M.), the bees closed their nests at the end of twilight and made very few pollen collections from their host plant (the evening primrose, *Oenothera rhombipetala*) although pollen was abundant. Brood development in excavated nests correlated with the observed foraging activity. (Based on Kerfoot, 1967.)

On the other hand, certain other behavioral events, such as the cicada's emergence, occur only once in any individual's life, but appear as overt rhythms when one views the synchronous population. These are termed **gated rhythms**, for at the appropriate stage of the life cycle it is almost as though some mechanism were opening a behavioral "gate" at the appointed time. All those individuals streaming through are allowed to begin a given behavior, but an individual that misses a given gate opening time must wait for the next one. (When the rhythm centers about the 24-hour day, it may also be called circadian; contradictory as this may seem at first, analysis of gating in the context of the population shows clearly that the underlying control of the individual is circadian, even though the behavior is "one shot.") The pupal eclosion cycles of many *Drosophila* species, for example, have been shown to have an innate, endogenous circadian rhythm. The discovery of "clock mutants"

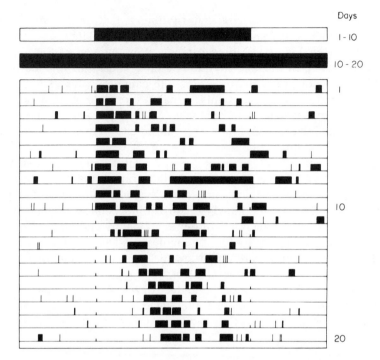

Days

1 - 10

10 - 20

1

10

20

**Figure 2-13** Locomotor activity of a cockroach, *Leucophaea maderae,* as recorded by an actograph. Each horizontal line represents 24 hours; the vertical black bars indicate periods of activity. Successive days are arranged from top (day 1) to bottom (day 20). For 10 days the animal was maintained in a 12-hour light–12-hour dark cycle. Beginning on the eleventh day the animal was kept in total darkness; activity continued to show periodicity but exhibited a free-running drift away from solar time, a characteristic of all endogenously timed daily rhythms. (From Brady, 1974; after Roberts, 1962.)

in *D. melanogaster* confirms that the periodicity is innate and genetically coded (Saunders, 1974).

Circadian rhythms all share four important characteristics. First, their cycles average about 24 hours but are never exactly so. Thus, they will tend to "drift" if not kept in line by certain external environmental cues. Related to this, their second characteristic is that they are synchronized to, or *entrained* by, a cyclical environmental cue called a **Zeitgeber**. Most circadian rhythms studied to date have used light–dark transitions as their *Zeitgeber*, but other cues may also prove important. Some insects held in continuous darkness, for example, distinguish between long and short days in a temperature cycle.

A third characteristic of circadian rhythms is that, in the absence of a *Zeitgeber*, they will *free run* in cycles of approximately 24 hours for some time (see Fig. 2-13). Free running after removal of the apparent *Zeitgeber* indicates that the rhythm is endogenously maintained but, of course, does not prove that it is, for one cannot always exclude other environmental cues with certainty. Finally, circadian rhythms are temperature compensated. Unlike most other physiological processes, within normal biological limits the periodicity remains stable under changing temperature conditions.

In discussions of biological rhythms, the term "clock" is commonly used. However, it should be noted that, rather than referring to a specific physiological mechanism, "clock" is only a convenient catch-all term used to describe the largely unknown biochemical systems driving the rhythmic cycles, which may differ significantly and fundamentally in different species. It appears that timing devices have evolved more than once, and their ecologically important features result from convergence not common ancestry (Saunders, 1976).

The rhythmic life history of the midge, *Clunio marinus*, which spends its larval and pupal life among red algae between the tidemarks of certain European beaches is instructive. Adults emerge only at low tide

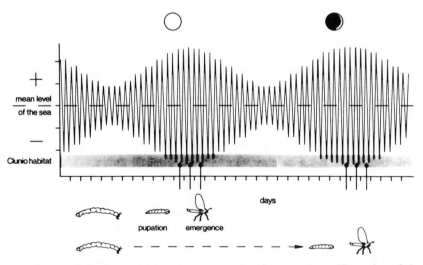

**Figure 2-14** Semilunar periodicity of emergence in *Clunio marinus*. Illustration of the temporal correlation between moon, tides, exposure of the *Clunio* habitat (gray strip), and emergence (arrows). (From Neumann, 1976. Reproduced, with permission, from the Annual Review of Entomology, volume 21. Copyright © 1976 by Annual Reviews Inc. All rights reserved.)

in a behavior which must be carefully timed, for adults live only 2 hours and must mate and oviposit before the tide advances back over their breeding ground (Fig. 2-14). A number of genetically distinct geographical races exist, each with its own emergence rhythm that is very precisely adapted to its local tidal conditions. When northern and southern races are compared, an interesting difference is found. In southern races the daily periodicity is circadian and free runs in constant light. In the Arctic, however, emergence is strictly tidally controlled and stops in constant light or dark; in this case, an "hourglass" appears to control emergence, measuring 10 hours from the time of first exposure to the previous ebb tide (Neumann, 1976).

In some cases, hormones apparently alter behavior by causing specific changes in peripheral organs. However, in most cases the influence of hormones on behavior is through direct action on the central nervous system. How do hormones act at the level of the neuron? Ideally, one attempting to find an answer to this question should seek an organism with a relatively simple nervous system, select neurons important to a clear-cut behavioral alteration, and then follow the changes occurring in those neurons during the course of hormone action. At present, the research which has come closest to meeting these criteria and reaching these goals is that of Truman (1973) studying the switch-over from pupal to adult behavior that occurs at the time of adult emergence in giant silkmoths (family Saturniidae, not to be confused with *the* silkworm moth, *Bombyx mori*).

## CASE STUDY: THE INITIATION OF ADULT BEHAVIOR IN SATURNIID MOTHS

In moths, as in most insects, **eclosion**—the emergence of the adult insect from the old pupal skin—occurs only during a specific period of the day; individuals that do not emerge during that particular time cannot eclose until the proper time on the following day. However, simple escape from the pupal skin does not herald the arrival of a fully functional adult. The *pharate* adult, which has completed its development and is waiting for its gate, displays little in the way of adult behavior and only immature rotary movements of the abdomen. The problem is not lack of appropriate neural machinery; the pharate moth possesses a fully developed adult nervous system. Nor is the lack of adult behavior merely due to the insect's restraint; peeling the pupal cuticle away does not mature the behavioral repertoire before its normal emergence gate. Complex motor patterns such as flight or walking, on the rare occasions when they can be elicited at all, are uncoordinated and abortive. Even simple reflexes, such as the "righting reflex" when overturned, are missing.

At the arrival of the eclosion gate, however, the behavior of a "peeled" moth changes strikingly. In perfect pantomime, the moth sheds its phantom pupal skin and escapes from its nonexistent cocoon. At the end of the performance, it inflates its wings and assumes the full repertoire of adult behavior.

How, asked Truman, is such adult behavior switched on? Obviously, as the peeled moth illustrated so eloquently, the control must be centrally mediated and closely linked to the timing of eclosion. Hoping to locate the site of the eclosion "clock," Truman performed a series of surgical ablation studies on various parts of the pupal nervous system. When the brain itself was removed shortly after the onset of adult development, he found that development proceeded normally and the resulting moths went on to shed their pupal skins. However, their emergence was quite abnormal; some normal behaviors were entirely omitted and others were out of sequence. Even more striking, the eclosion was no longer gated. Moths emerged at all sorts of odd hours randomly distributed throughout the day and night.

Was all this behavioral confusion simply due to removal of important neural centers? Truman tackled this suggestion by implanting "loose" brains into the abdomens of debrained pupae. In these "loose brain" moths, neural connections between the brain and nervous system were never established. Yet the resultant moths emerged at the proper gating time and displayed proper emergence behaviors! Clearly, the brain was acting humorally, that is, through crucial neuroendocrine centers. An eclosion hormone appeared likely, for injecting brain homogenates into moths prior to their normal eclosion time provoked their precocious emergence.

But what was the *actual* role of the brain? For example, did the brain include the photoreceptor, the "clock" measuring the time after lights-on or lights-off, or both? Truman knew that different species of giant silkmoths had quite different eclosion gating times. Therefore, he performed brain transplants between these species. The results (Fig. 2-15) clearly confirmed his suspicion that the brain contained the gating "clock." By interchanging the brain, he could interchange the time of emergence.

Why did he suspect that the photoreceptors might also be in the brain itself? Simply because removing a moth's compound eyes did not halt its response to light–dark cycles. Choosing 20 debrained pupae, Truman implanted brains into the head region of half of them and into the abdomen of the other half. Then he plugged them all into holes in a

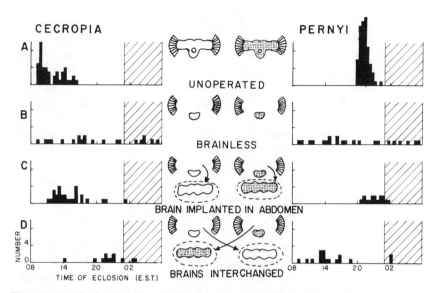

**Figure 2-15** The hormonal "gating" role of the brain in silkmoth (*Hyalophora cecropia* and *Antheraea pernyi*) eclosion. Abscissae are 24-hour periods having a 17-hour light phase and a 7-hour dark phase (cross hatched). Under these conditions intact moth pupae of the two species have distinctly different peaks or "gates" (A). Brain removal results in arhythmic emergence (B). Replacing the same brain into the pupa's abdomen reestablishes the gating rhythm (C), while brain exchange between the two species resulted in the recipient pupae taking the gate time of the donor brains (D). (Courtesy of J. W. Truman; from Truman and Riddiford, 1970. Copyright 1970 by the American Association for the Advancement of Science.)

partition and placed them into a chamber so as to expose their anterior and posterior halves to different photoperiod regimes. In each case the subsequent eclosion of the host depended only upon the photoperiod to which the brain was exposed.

## INTERNAL COMMUNICATION—THE INCREASE OF INFORMATION

A rabbit possesses something on the order of $10^8$ olfactory receptor cells. A representative caterpillar has only 48. Both herbivores, however, successfully find and discriminate between potential food plants. What features of the comparatively limited insectan receptor system serve to maximize the amount of biologically significant information transmitted to higher levels in the CNS?

It is clear that certain types of receptors are "tuned" to very specific stimuli or, more correctly, to only particular aspects of a stimulus. For example, the acoustic receptor cells of the tympanum of the noctuid moth respond only to ultrasonic pulses falling within a certain broad frequency range (see Fig. 2-5). All other frequency ranges are excluded or filtered out by these acoustic receptors. A sound, however, has several features which could convey information, including its intensity, pulse duration, the interval between pulses, and so on, and the receptor is but one point along the chain of neurons leading to the brain where selective decoding of stimuli might occur. Thus, it is possible to envision a hierarchically organized neural information processing system through which the various stimulus properties are "filtered" leading ultimately to the release of a particular behavioral response.

In this concept of multiple stages of **stimulus filtering**, each stimulus can be regarded as a sort of "key" fitting one group of "doors" en route to the brain. Thus, in the case of the ultrasonic bat cry, receptor cells ("doors") selectively respond to a particular range of frequencies ("keys"). Once the receptor cells transmit an impulse, subsequent interneurons along the ventral nerve cord are no longer "interested" in the particular frequency of the sound; rather, other sound properties such as pulse intensity, pulse duration, or interval between successive pulses are now monitored. For example, Roeder has identified a "pulse marker" interneuron in the moth that responds to three or four sensory impulses separated by short intervals by firing just once. However, it makes no distinction as to duration of a pulse, that is, 0.5 and 500 microseconds are to it the same. But the world of neural interconnections in the insects' CNS remains largely a black box for neurophysiologists. Let us therefore examine other sense receptors more readily amenable to physiological tinkering.

Whereas the moth acoustic receptor appears to fire only in response to ultrasonic sounds, remaining silent in the presence of all other frequencies, chemoreceptor recordings from a variety of insects have shown that most single chemoreceptors respond to several kinds of chemicals. However, there is considerable variation in the *pattern* of sensitivity. Response to a "preferred" chemical may be in the form of an even series of impulses. Noxious compounds, in contrast, may evoke responses in the form of irregular bursts or volleys of spikes which can often be correlated with avoidance behaviors by the insect. Thus, each different chemoreceptor might be envisioned as having an individually characteristic sensitivity pattern to a given array of compounds. The CNS could, by summing the responses of the different chemoreceptors, obtain characteristic total response profiles each uniquely representative of a particular chemical compound.

An additional informational dimension would be acquired if instead of being silent in the unstimulated (resting) state, chemoreceptor cells maintained some constant spontaneous baseline level of firing activity.

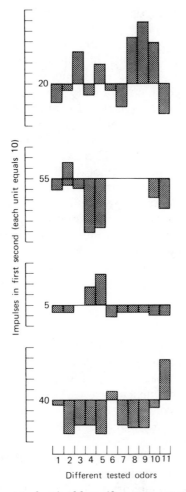

**Figure 2-16**  Activity spectra of each of four olfactory receptors of the tobacco hornworm (*Manduca sexta*) in response to 11 different odors. Each receptor's sensitivity takes the form of an increase or decrease over its unique spontaneous firing rate. (Adapted from Dethier, 1971.)

Then a particular stimulus could be recorded as either an increase *or decrease* in firing over the spontaneous rate. A coupling of this ability with the varied spectral sensitivities of the different receptors would provide a rather sophisticated information encoding system. Interestingly, the olfactory sense of caterpillars seems to operate in just this fashion, resulting in different response profiles for each receptor (Fig. 2-16). Since the caterpillar receives information from all chemoreceptors simultaneously, discriminations between numerous natural plant odors, even in excess of those likely to be encountered in nature, are easily within the caterpillar's sensory powers using only a few receptors. It is likely that the relative sensitivity of different receptors also regularly fluctuates, since it is known, for example, that many insects are rhythmic in their degree of susceptibility to insecticides.

Use of complementary chemical and neuroelectrical internal communication systems has obvious advantages. To mediate all behavioral changes solely through neural means would demand an extensive series of excitatory neural networks. This would require a corresponding increase in the size and complexity of the CNS, a luxury that insects can ill afford. In addition, while some behaviors such as alarm and escape require the split-second response that only nervous mechanisms can provide, others such as mating and oviposition often can more reliably be triggered through summation of information over time. And, because hormones have access to all neurons, they are well suited for modulating the overall responsiveness of the nervous system.

## SUMMARY

It is evident, albeit practically impossible, that to truly understand the behavior of a given organism, one should really examine thoroughly their nature from *all* aspects—how information is perceived (and excluded) by receptor systems, how effectors (muscles, glands, organs, etc.) are activated and controlled to produce behavior, and how integrative systems (nervous and endocrine) relate effector output to receptor input and central programs.

Insects do not normally display a particular behavior continuously or more than one behavior simultaneously. The apparent explanation for this is that endogenous inhibitory and excitatory controls originating in the brain ensure that only one of a variety of possible behavioral options is promoted at a given time. An insect's "mood" and "drive" or motivation can greatly influence its actions. Thus, the internal chemical milieu and previous experience affect behavioral responses. For example, a male cricket's response to cercal stimulation varies depending on its reproductive status (Table 2-1).

The nervous and endocrine systems interact in cooperative fashion to determine the behavioral and physiological responsiveness of the insect to external and internal stimuli during its life cycle. The two systems are closely connected structurally and functionally, with the neurosecretory cells in particular believed to be a link between them. In general, the endocrine system is primarily involved in regulating behaviors associated with development or relatively predictable periodic activities. Nervous controls apply especially to variable short-term behaviors where fast action is often required. Neurons receive and process input of a single or of different sensory modalities and possess filtering abilities for selecting particular information. Complex behaviors or fixed action patterns often reflect centrally programmed commands that are triggered by appropriate sensory stimuli (releasers) and modulated by peripheral feedback. As the behavior is accomplished this information is continuously interpolated into the motor output via feedback loops.

Roeder has figuratively said that neurophysiology and ethology are separated only by the skin of an animal—neurophysiology is concerned with the nervous system's action "inside the skin," while ethology deals with the nervous system "outside the skin." In a sense, this is a good analogy. Someday soon we may see these two fields truly merge, for we are witnessing new levels of understanding of the neural and endocrine control of behavior, and in the final analysis both disciplines face almost identical questions.

In this chapter we have considered only selected aspects of the physiological bases that underly insect behavior. In so doing, we have tried to lay out the general framework within which insect behavior operates and to elucidate some of the principles behind the programming

**Table 2-1** Reactions to cercal stimulation in male *Gryllus campestris* during three differing states of sexual responsiveness.[a]

| Unresponsive state (usually without a spermatophore) | Responsive state (usually with a spermatophore) | Postcopulatory state (without a spermatophore) |
|---|---|---|
| Evasion | Calling and courtship songs | Locomotion |
| Locomotion | Copulatory behavior | Aggression |
| Aggression | | Rivalry song |
| Rivalry song | | |

[a] After Huber (1974).

and integration of behavioral processes. For those who wish to delve deeper, an excellent and delightful introduction is available in Roeder's (1967) little book, *Nerve Cells and Insect Behavior*. A thorough review of the emerging field of neuroethology is given by Hoyle (1970), and an extensive update on neural integration is provided by Huber (1974) and Treherne (1974). The interrelationships between hormones and behavior are capably summarized by Truman and Riddiford (1974), while behavioral rhythmicity is reviewed by Brady (1974) and Saunders (1974). Alloway (1972) and Corning et al. (1973) provide an introduction to the growing literature on insect learning. Markl (1974) provides additional examples and references on most topics discussed in this chapter.

## SELECTED REFERENCES

Alcock, J. 1975. *Animal Behavior. An Evolutionary Approach*. Sinauer Associates, Sunderland, Mass. 547 pp.

Alloway, T. M. 1972. Learning and memory in insects. *Annu. Rev. Entomol*. **17**: 43–56.

Baerends, G. P. 1941. Fortpflanzungsverhalten und Orientierung der Grabwespe *Ammophila campestris*. *Tijdschr. Entomol*. **84**: 68–275.

Brady, J. 1974. The physiology of insect circadian rhythms. *Adv. Insect Physiol*. **11**: 1–115.

Corning, W. C., J. A. Dyal, and A. O. D. Willows (eds.). 1973. *Invertebrate Learning*, Vol. 2, *Arthropods and Gastropod Mollusks*. Plenum Press, New York, 284 pp. (see especially Chapters 8 and 9).

Dethier, V. G. 1971. A surfeit of stimuli: a paucity of receptors. *Amer. Sci*. **59**: 706–715.

Gelperin, A. 1972. Neural control systems underlying insect feeding behavior. *Amer. Zool*. **12**: 489–496.

Griffin, D. R. 1958. *Listening in the Dark*. Yale Univ. Press, New Haven, Conn., 413 pp.

Highnam, K. C. 1964. Hormones and behaviour in arthropods. *Viewpoints Biol*. **3**: 219–255.

Howse, P. E. 1975. Brain structure and behavior in insects. *Annu. Rev. Entomol*. **20**: 359–379.

Hoyle, G. 1970. Cellular mechanisms underlying behavior—neuroethology. *Adv. Insect Physiol*. **7**: 349–444.

Huber, F. 1962. Central nervous control of sound production in crickets and some speculations on its evolution. *Evolution* **16**: 439–442.

Huber, F. 1974. Neural integration (central nervous system). In *The Physiology of Insecta*, Vol. IV, 2nd ed., M. Rockstein (ed.). Academic Press, New York, pp. 3–100.

Janzen, D. H., 1974. The deflowering of Central America. *Nat. Hist*. **83**: 48–53.

Kennedy, J. S. 1967. Behavior as physiology. In *Insects and Physiology*, J. W. L. Beament and J. E. Treherne (eds.). Oliver and Boyd, London, pp. 249–265.

Kerfoot, W. B. 1967. The lunar periodicity of *Sphecodogastra texana*, a nocturnal bee. *Anim. Behav*. **15**: 478–485.

Konishi, M. 1971. Ethology and neurobiology. *Amer. Sci*. **59**: 56–63.

Lorenz, K. Z. 1970, 1971. *Studies on Animal and Human Behavior,* Vols. 1 and 2. Harvard Univ. Press, Cambridge, Mass., 403 pp, 366 pp.

Markl, H. 1974. Insect behavior: functions and mechanisms. In *The Physiology of Insecta,* Vol. III, 2nd ed., M. Rockstein (ed.). Academic Press, New York, pp. 3–148.

McFarland, D. J. 1971. *Feedback Mechanisms in Animal Behaviour.* Academic Press, New York, 279 pp.

Neumann, D. 1976. Adaptations of chironomids to intertidal environments. *Annu. Rev. Entomol.* **21:** 387–414.

Roberts, S. K. de F. 1962. Circadian activity rhythms in cockroaches II. Entrainment and phase shifting. *J. Cell. Comp. Physiol.* **55:** 99–110.

Roeder, K. D. 1965. Moths and ultrasound. *Sci. Amer.* **212:** 94–102 (April).

Roeder, K. D. 1966. Auditory system of noctuid moths. *Science* **154:** 1515–1520.

Roeder, K. D. 1967. *Nerve Cells and Insect Behavior,* rev. ed. Harvard Univ. Press, Cambridge, Mass., 238 pp.

Roeder, K. D. 1970. Episodes in insect brains. *Amer. Sci.* **58:** 378–389.

Roeder, K. D. 1974. Some neuronal mechanisms of simple behavior. *Adv. Stud. Behav.* **5:** 1–46.

Roeder, K. D. 1976. Joys and frustrations of doing research. *Persp. Biol. Med.* **19:** 231–245.

Saunders, D. S. 1974. Circadian rhythms and photoperiodism in insects. In *The Physiology of Insecta,* Vol. II, 2nd ed., M. Rockstein (ed.). Academic Press, New York, pp. 461–533.

Saunders, D. S. 1976. The biological clock of insects. *Sci. Amer.* **234:** 114–121 (February).

Snodgrass, R. E. 1935. *Principles of Insect Morphology.* McGraw-Hill, New York, 667 pp.

Steiner, A. L. 1962. Étude du comportement prédateur d'un Hyménoptère Sphégien: *Liris nigra* V. dL. (= N.p. Panz.). *Ann. Sci. Nat. (Zool.) Fr. 12e Ser. (IV),* 126 pp.

Tinbergen, N. 1951. *The Study of Instinct.* Clarendon Press of the Oxford Univ. Press, London, 228 pp.

Tinbergen, N. 1972. *The Animal in its World. Explorations of an Ethologist, 1932–1972.* Vol. 1, *Field Studies.* Harvard Univ. Press, Cambridge, Mass., 343 pp. (especially pp. 103–145.)

Treherne, J. E. (ed.). 1974. *Insect Neurobiology.* Elsevier, New York, 450 pp. (see chapter by Peter Miller).

Truman, J. W. 1973. How moths "turn on": a study of the action of hormones on the nervous system. *Amer. Sci.* **61:** 700–706.

Truman, J. W. and L. M. Riddiford. 1970. Neuroendocrine control of ecdysis in silkmoths. *Science* **167:** 1624–1626.

Truman, J. W. and L. M. Riddiford. 1974. Hormonal mechanisms underlying insect behavior. *Adv. Insect Physiol.* **11:** 297–352.

Wigglesworth, V. B. 1970. *Insect Hormones.* University Reviews in Biology. Oliver and Boyd, Edinburgh, 159 pp.

Wilson, D. M. 1964. The origin of the flight motor command in grasshoppers. In *Neural Theory and Modeling,* R. F. Reiss (ed.). Stanford Univ. Press, Stanford, Calif., pp. 331–345.

Wilson, D. M. 1968. The flight-control system of the locust. *Sci. Amer.* **218:** 83–90 (May).

# 3
# Spatial
# Adjustment

The burrows of most wood-boring cerambycid beetle larvae are oriented extremely irregularly; why? Bumble bees fly about at temperatures so low that most other insects are inactive; how? Migrating locusts appear to swarm single-mindedly toward a fixed goal. Do they? By what means? Though the scale of their movement may vary widely, the ability to change position within the environment is essential to the survival of nearly all insects. Escape from predators, food gathering, mate location, adjustment to environmental variables such as temperature and humidity—these and other important behaviors all depend upon an insect's ability to adjust its spatial relations.

Originally terrestrial organisms, insects subsequently invaded both aerial and aquatic environments. Some of their members have become the only invertebrates capable of flight. Before considering some ways in which insects meet their very specific spatial problems, it may be useful to consider briefly the ways in which insects move within their environment.

## LOCOMOTION

A tiny flea's jump may be 13 inches long; a blood-sucking bug, *Rhodnius*, may move about with a meal 10 to 12 times its weight, corresponding to a man's drink of 200 gallons weighing a ton. As insects move about, many perform activities that appear extraordinarily impressive by human standards. Are they endowed with comparatively tremendous muscular power or a different sort of muscle from those we possess? Not really; physiological studies have shown that insect muscles are quite similar in almost all respects to our own, although the insect may possess many more individual muscles than a man.

Many of the strange powers insects appear to have—as well as many of the problems they face—arise from a simple relationship between surface and mass. For any object, as the size or mass diminishes the relative amount of surface increases. (The volume of a sphere is $\frac{4}{3}\pi r^3$, where $r$ is the radius of the sphere; the surface of a sphere, however, is $4\pi r^2$.) For an organism as small as an insect, this relationship has a marked effect on muscle power. The power of a muscle is proportional to the *area* of its cross section, whereas the mass it has to move is proportional to *volume*. We are amazed by the long jump of the flea, which proportionately carried out by a man would cover 800 feet. However, we must realize that a flea the size of a man would have relatively much more mass per unit cross-sectional area of muscle than a normal-sized flea (Wigglesworth, 1964).

## Terrestrial Locomotion

Although a sluggish few use their legs primarily to cling to surfaces, walking and running are common behaviors for the adults of nearly all flying and nonflying insects, and for many nymphal and larval forms as well. Many are quite rapid and agile. Cockroaches, for example, have been clocked at speeds of nearly 3 miles per hour—a remarkably high speed in relation to their body size.

The power for most terrestrial locomotion comes from the thoracic legs, which move in various sequences at different speeds so that stability is always maintained. Coordination of these patterns, understandably crucial, is mediated both through central mechanisms and segmental reflexes. Two general principles appear to underly the walking sequence. First, no leg is raised until the leg behind it is in a supporting position. Second, the movements of the two legs of a segment alternate. A pattern of alternating triangles of support is commonly observed; with never less than three legs on the ground, an insect can stop at any point without losing stability. Stability is also enhanced by the fact that the insect body is slung between the legs in such a way that the center of gravity is low.

Many immature insects move in a manner similar to adults. However, insect thoracic legs can function only when the external skeleton is relatively rigid, so soft-bodied larval forms generally employ somewhat different methods. The crawling or creeping of caterpillars (Fig. 3-1) and sawfly larvae is carried out mainly by fleshy **prolegs**, false "legs" on the abdomen. The pressure of the body hemolymph is essential; puncture of the body wall soon renders them incapable of movement.

The ability to leap or jump appears to have repeatedly and independently evolved in insects of all sizes. But although a great many insects are capable of propelling themselves through the air without wings (Fig.

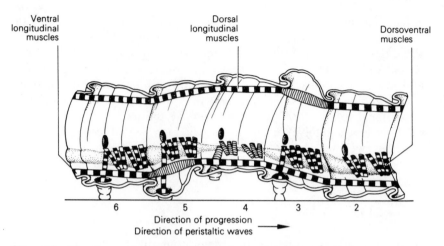

Ventral
longitudinal
muscles

Dorsal
longitudinal
muscles

Dorsoventral
muscles

6      5      4      3      2

Direction of progression
Direction of peristaltic waves ⟶

**Figure 3-1**  Movements of a caterpillar in diagrammatic section. Serial waves of contractions pass along the body from back to front, lifting body segments and prolegs as they travel. At each point in time, at least three segments are in different stages of contraction, a process that calls for a high degree of nervous coordination. (From Romoser, 1973, *The Science of Entomology,* Macmillan Publishing Co.)

3-2), only in certain groups has this become a pronounced specialization. Usually, jumping is a form of escape reaction, but some insects, particularly Orthoptera, commonly use short hops as a normal means of locomotion.

Insect gait has prompted considerable behavioral study; the reviews of Wilson (1966) and Hughes and Mill (1974) consider patterns of insect walking and their possible modes of control.

**Aquatic Locomotion**

The extend of aquatic activity of an insect is obviously affected by its respiratory habits. Some with gills or other aquatic respiratory adaptations can live permanently submerged. Others must stay on the surface or come to it frequently to renew their air supply; in addition, their store of air tends to give them buoyancy. Thus, aquatic insects have evolved two general sorts of locomotory adaptations—those enabling them to propel themselves upon or up to the top of the water and those by which they "swim" beneath the water surface.

Surface dwellers take great advantage of the relationship between their body size and the physical properties of water at temperatures and pressures characteristic of their environment. Specifically, under these

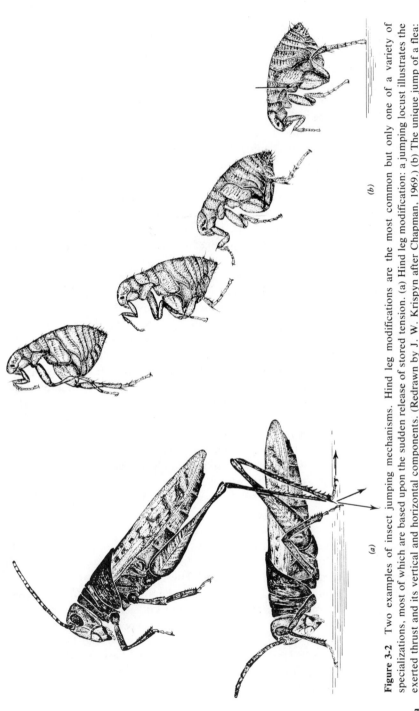

*(a)*

*(b)*

**Figure 3-2** Two examples of insect jumping mechanisms. Hind leg modifications are the most common but only one of a variety of specializations, most of which are based upon the sudden release of stored tension. (a) Hind leg modification: a jumping locust illustrates the exerted thrust and its vertical and horizontal components. (Redrawn by J. W. Krispyn after Chapman, 1969.) (b) The unique jump of a flea: crouching before takeoff serves to "cock" two cuticular catches and compress the elasticlike resilin of the pleural arch (arrow). Simultaneous relaxation of the femur-raising muscles and ventral longitudinal muscles release the stored energy, thrusting the hind trochanters against the substrate. (Redrawn by J. W. Krispyn after Rothschild et al., 1973.)

conditions water tends to have a relatively high surface tension, so that the waxy water-repellent surface of the insect cuticle is sufficient to support many small surface dwellers as though upon a thin elastic membrane. Many insects also secrete additional waxy material upon their tarsi, allowing them to walk or row across the water film without breaking its surface.

Other surface dweller adaptations, while less common, may be even more effective. For example, *Stenus* beetles live on grasses along mountain streams and often accidentally tumble into the water. Here they are able to walk upon the surface, but only slowly. In response to apparent danger, however, they release an anal gland secretion which lowers the surface tension of the water behind them. Drawn forward by the higher surface tension in front, the beetles propel themselves along at speeds of 45 to 70 cm/second, moving their abdomens from side to side to direct their movements.

Among insects that live beneath the water surface, locomotion methods vary greatly. Bottom-dwelling insects such as larval Odonata and Trichoptera walk over the substrate just as terrestrial insects do, though the larval case of some caddisflies can be quite a hindrance to movement. Others move by various types of body undulation or by the propulsion produced by jets of water. Dragonfly larvae, for example, force water rapidly out of the rectal chamber so that the body is driven forward. Most free-swimming insects, however, paddle with their hind legs, sometimes together with the middle legs. Efficency is often increased by devices such as hairs or cuticular blades and/or modification in the morphology and relative size of the legs. In contrast to the general rule in terrestrial locomotion, in swimming the two legs of a segment sometimes work together like oars. Nachtigall (1974b) provides a review of aquatic locomotion in insects.

**Aerial Locomotion**

Insects alone among the invertebrates possess the ability to fly, and flight is one of the most important reasons for their success. How and when did this remarkable ability arise? The fossil record on this subject provides some surprisingly complete answers. Throughout the fossil story wing venation has remained relatively consistent, suggesting that insect wings evolved only once, as sideways expansions of the upper part of the thorax of rather large, active insects during the Devonian or Mississippian periods. Presumably, these expansions served first only for gliding, with wing inclination control and actual flapping movements gradually being added later.

The first flapping wings extended permanently from the insect's sides, as in dragonflies, or aloft over the thorax, as in mayflies. Elevator and depressor muscles attached directly to the wing base worked these wings. Direct muscles are still dominant in cockroaches and are important in some Orthoptera, Odonata, and Coleoptera. Most present-day insects, however, exhibit a flight mechanism that has undergone two fundamental modifications. First, they have developed a musculature which allows the wings to fold backward over the abdomen. Second, although the muscles altering wing inclination remain attached to the wings themselves, the muscles responsible for wing flapping are attached to the thoracic walls. Thus, the wings are flapped through indirect action by muscles changing the shape of the thoracic box to which they are hinged. Since at the same time wing inclination is synchronously changing, the overall wing flight pattern becomes much like that of a pair of small propellers directing an air stream downward and backward. In typical forward flight, each wing traces a figure 8 relative to the body at its base; many insects can hover or loop (Fig. 3-3) by changing the inclinations of this figure 8 relative to their body.

Apparently, a two-winged condition is more efficient in flight than the primitive four-winged one. Many forms of wing coupling may be found, linking the fore and hind wings so that they move together as a unit. Other insects have become functionally two-winged through the reduction, modification, or complete loss of one pair of wings or the other.

By necessity, as insects became smaller their wing movement rate increased. While a house fly may have a rate of about 200 beats per second, mosquitoes have a rate of up to 600 beats. Tiny ceratopogonid midges have been clocked at a wing vibration speed of over 1000 beats per second. How can this be possible? No known animal nerves are physically capable of transmitting stimuli fast enough to cause contraction and relaxation at these high speeds. The elastic nature of the insect thorax and the action of resonating flight muscles hold the key to this paradox. In many insects, especially certain Diptera and Coleoptera, the wings have two stable positions—completely elevated and completely depressed. As the wings move downward, normal thoracic elasticity resists this motion until a certain point is reached. At this "click point," three things happen simultaneously. First, the resistance vanishes and the wings click into a new position below the thorax. Second, their inclination automatically changes in readiness for the upstroke. Finally, the muscles which have been contracting are suddenly released. As they relax, the opposing muscles are suddenly stretched, which causes them to contract instantly. In this remarkable oscillating process, the insect has developed a system which does not require the synchronous nervous

**Figure 3-3** A green lacewing, *Chrysopa*, somersaults from a leaf. Multiflash photographs have shown that these delicate neuropterans frequently perform vertical take-offs, loops or half-loops, often landing directly behind or upon their launch site. Whether such flight patterns are accidental or controlled maneuvers remains unknown. (Drawing by J. W. Krispyn, based on Dalton, 1975.)

control for every contraction that is characteristic of dragonflies, locusts, and butterflies. Once initiated, this "improved model" can be operated at almost any speed, depending on thoracic elasticity, and can be modified by secondary controls as circumstance dictates.

For further information on insect flight, consult Nachtigall (1974a), Pringle (1974, 1975), Rainey (1976), and Wilson (1971). Dalton (1975) has produced some of the most remarkable photos of insects in natural flight ever taken; the highly readable essay on insect wings by H. E. Evans that introduces Dalton's book is also recommended.


## ORIENTATION

Having briefly viewed *how* insects move, let us turn our attention to *why* they move, first as individuals and then as populations. The subject of spatial adjustment is a critical one touching many facets of the life of an organism. A major part of an insect's behavior is in fact orientation to factors such as food, mate, prey, host, etc. Thus, it is unsurprising to find that the study of orientation and navigation is a dynamic part of modern biology, with a body of data which is growing rapidly. We can only be concerned here with some of its more general tenets. Useful reviews of this very broad topic include those of Fraenkel and Gunn (1961), Markl (1971), Jander (1963, 1975), and Birukow (1966).

**Spatial orientation** is the self-controlled maintenance or change of an organism's body position relative to environmental space. It occurs when certain stimuli in the environment elicit a responsive sequence of behaviors that results in an nonrandom pattern of locomotion, direction of body axis, or both. The fact that orientation is self-controlled in this way distinguishes it from passive transport. That it includes position maintenance means that orientation also may be taken to include postural adjustments such as response to gravity.

Common to all orientation capabilities is the requirement that one set of receptor organs able to detect information about spatial relations transmit this information to other effector systems capable of changing these relations. Thus, as with so many other classes of behavior patterns involving single organisms, orientation fairly begs for analysis of neural mechanisms. The student wishing access to the vast literature on this subject would find the book edited by Wehner (1972) helpful.

A confusing terminology has arisen through attempts to adequately describe all the varied types of orientation that exist. Traditionally, biologists have found it useful to view the subject according to the reaction mechanisms involved, a classification which results in categories such as kineses, taxes, and transverse orientations. But in another view, orientation basically involves the positioning of an organism in response to various stimulus fields. This includes behavioral responses to at least six major classes of stimuli: heat, magnetism, light, gravity, pressure, and chemicals. Thus, one may consider chemical orientation, gravity orientation, astronomical orientation, orientation to polarized light, or any of a host of other orientation subdivisions. (For discussion of a variety of sense organs used in perception of chemical, visual, and mechanical stimuli, refer to Chapters 5, 6, and 7.)

However, a third important observation is that orientation occupies an interface between behavior and ecology. For any organism, the environment contains both positive and negative factors—not only resources needed for sustaining life or their absence but stress sources such as the bright sunlight, which can be rapidly debilitating in the absence of compensating behaviors. A maximally fit organism will behave in a manner that consistently works to minimize its body distance from resources (food, shelter, etc.) and maximize its distance from sources of

**Table 3-1**  Ecological categories of spatial orientation in motile organisms[a]

| | |
|---|---|
| Organism stays in one position, maintaining it by compensating for disturbances. POSITIONAL ORIENTATION | 1. transverse orientations to light and gravity<br>2. temperature regulation by body position<br>3. compensations for displacement by air or water currents while swimming or flying |
| Organism moves with two major orientation phases: search and approach (or avoidance) | |
|    A. Resources or stresses in the environment are "objects," i.e., patchy in distribution | 1. organism exhibits simple search and approach or avoidance—OBJECT ORIENTATION<br>2. organism shows elaboration of approach phase of the orientation—TOPOGRAPHIC ORIENTATION<br>3. organism shows lengthening of search (straight run) phase, grading into dispersal and migration—GEOGRAPHIC ORIENTATION |
|    B. Resources or stresses are extensive in at least one dimension | 1. horizontal gradation—ZONAL ORIENTATION<br>2. layering or vertical gradation—STRATAL ORIENTATION |

[a] After Jander (1975).

stress. This point of view recently has led to a far more complete classification of orientation than past attempts (Table 3-1). By considering orientation from the viewpoint of its adaptive significance, this scheme should be extremely valuable as a focal point for future research.

In the pages that follow, we sample from each of these classification schemes to give the reader a brief overview of some of the more interesting aspects of insect orientation. Because of their historical importance, we begin with the traditional taxes and kineses before turning our attention to the wider perspective offered by other classification schemes.

## Kinesis

If woodlice are distributed over an area which has both humid and dry areas, they soon congregate in the humid areas. Can they directionally follow a humidity gradient? Do they "know" how to find damp areas? Studies of woodlice in the laboratory have shown that their response to humidity, despite its appearance, is actually nondirectional. The woodlice do not actively seek out humid areas, but are simply more active in the drier areas. When they encounter more humid areas in the course of their random running about, they become less active. Eventually, activity slows down more and more until some individuals stop altogether (Fraenkel and Gunn, 1961).

A response like that of the woodlice is termed a **kinesis**. This is perhaps the simplest type of locomotor response an animal can make to a stimulus—moving in a way that is related only to the *intensity* of that stimulus, disregarding any spatial properties which the stimulus might possess.

## Taxis

To escape predators, a male grayling butterfly will fly upward toward the sun; if blinded in one eye, he will "escape" in circles. Caterpillars move down the stems of their food plant when about to pupate in the ground. Sexually mature female crickets turn to face and approach the recorded song of a male cricket (see Fig. 7-9).

Movements such as these would seem to be among the most straightforward types of orientation to study, since the insect's track appears obviously related in direction to some physical or chemical polarization of the environment. In fact, such movements were one of the earliest types of orientation response to attract serious biological interest.

Jacques Loeb (1918) proposed a simple mechanistic theory based on these types of response to explain the causes of all animal movements.

For any bilaterally symmetrical animal, Loeb argued, a stimulus that registered unequally on the animal's two sides would simply cause the animal to turn until the stimulus was equalized. Loeb called such a directed movement a **tropism**, a term which has since come to refer primarily to movement in plants, being replaced by the term **taxis** when used in reference to an animal. Taxis now generally refers to any oriented heading of an animal, whether moving or stationary.

A number of specific taxes have been described based on (1) whether an animal moves toward or away from certain environmental characters, (2) the complexity of the organism's sensory appararatus, and (3) the manner in which the animal moves. Thus, for example, an insect that reacts to light is said to show a *phototaxis*. If, as with a maggot, it reacts by bending more vigorously on the lighted side of its body, an action which will move the creature into darker areas, the phototactic orientation is called *klinotaxis*. If instead it can be demonstrated that the movement is mediated by the organism's two eyes (so that removal of one eye causes it to move in circles in the light but not in the dark), the behavior is called *tropotaxis* or, more correctly, *phototropotaxis* (Fig. 3-4). Here, like the two reins of a horse, each receptor has a one-way turning action; pulled equally, the horse goes straight but if just one is pulled, the animal turns. In still other cases where orientation with form vision is clearly demonstrable (so that some degree of orientation to the light is still possible after unilateral blinding), the taxis is now termed *telotaxis* or, more correctly, *phototelotaxis*. In addition to all this, movements toward a stimulus are generally prefaced by the word positive (e.g., positive phototaxis), while those oriented away from the stimulus are termed negative (e.g., negative phototaxis).

At first, Loeb's theory had such appeal that taxes were considered to be forced movements over which the animal had little or no control. But gradually the accumulating data began to give biologists a new appreciation of the complexities of animal orientation, and Loeb's simple theory was discredited on several counts. For one, taxes are obviously as variable as the rest of behavior. A given taxis may depend on environment, context, and/or the organism's internal state (nutritional, sexual, developmental, etc.). For example, although some pupating caterpillars move downward, their first responses as emerging moths may be to climb upward as high as possible. In addition, a tactic response may change in type or sign at short notice. Thus, the blinded circling grayling butterfly will immediately follow in a straight line should a female grayling pass by. Nor are tactic responses genetically fixed. A cock-

**Figure 3-4** Phototelotaxis in a honey bee with left eye blackened. After blinding of one eye, the bee tends to move in circles toward the intact eye. After repeated trials it learns to compensate, however, and again moves directly toward the light. Only a sample of the twenty-some trials required are shown; the light is directly above the starting point in each case. (Drawing by J. W. Krispyn, based on work of Minnich cited in Fraenkel and Gunn, 1961.)

roach, for example, will begin to avoid its usual dark habitat if it is fed under light and given a mild electric shock whenever it approaches darkened areas.

Tactic responses are often far less simple than they appear, and tactic interactions are common. For this reason investigations of taxes must be controlled very carefully and experiments must be very painstakingly designed. On the one side, taxes grade into kineses, from which they differ in being *directed* responses of the insect relative to the stimulus source. In another direction, they overlap with such longer-range phenomena as migration, discussed later in this chapter; at times the two are difficult to distinguish. When one turns to the broad subject of postural control, it is evident that in other manifestations they grade into still another quite sophisticated set of responses—the whole subject of positional orientations in general and transverse orientations in particular—which may or may not be called a subcategory of tactic responses.

**Postural Control**

Locusts in flight maintain an even body keel partly through visual reactions to inclination of the horizon. Flies mechanically sense angular acceleration and angular motion by rapid oscillations of their gyroscope-like halteres, modified knoblike vestiges of their hind wings. Through continual compensatory reactions to a directed light source, a dragonfly banking its wings while flying a curve still keeps its head in a perfectly upright position.

All such examples involve positional orientation, that is, compensatory maneuvers for body stabilization against displacement by wind, water, etc. The most widespread forms of positional orientation in insects are the **transverse orientations**—those in which the body is positioned at a fixed angle relative to the stimulus.

Among the best known transverse orientations are the **dorsal** and **ventral light reactions**. In this type of light orientation, well illustrated by dragonflies, moths, and butterflies, both the long and transverse axes of the body are kept perpendicular to a directed source of light at all times. (Thus, these light reactions contrast with phototaxis, where orientation is parallel to the light rays.) For example, the dragonfly shows a dorsal light response mediated by the upper ommatidia of its eyes which assures that the upper part of its head remains turned toward the light. If blinded in one eye and illuminated equally from all sides, the insect will roll continuously toward its seeing side. These reactions are particularly common among both flying and swimming insects. For example, water beetle larvae normally swim to the surface for air; placed in an aquarium illuminated from below, they swim to the bottom and suffocate (Fig. 3-5).

Light reactions are particularly important in relation to gravity perception, where they act in tandem with other sources of sensitivity to gravity. Insects lack specialized gravity receptors which function like the inner ear in vertebrates or the statocysts found in various crustaceans. Instead, orientation to gravity is by use of relatively unspecialized sensory hairs (**proprioceptors**) usually clumped into plates in positions where they can measure the relationship between body parts differentially affected by gravity. For example, in ants, gravity receptors are located at points of body articulation (neck, antennal joints, thorax and petiole, petiole and gaster, and joints between thorax and coxae); stimulation of any one point alone is sufficient for gravity orientation, although the different joints are not equally reliable. Detailed information on proprioceptors in available in Mill (1976), Weis-Fogh (1971), and Wendler (1971).

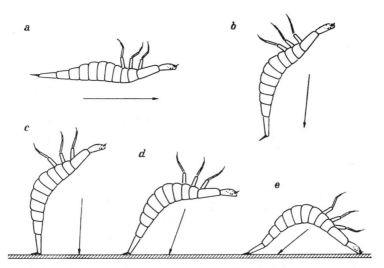

**Figure 3-5** Dorsal light response of *Acilius* beetle larvae. Diagrams show successive positions in response to light from below; the larvae swim to the bottom with their backs down, attempting to get air as though it were the surface. Unless illumination is reversed, the animals will die. Arrows indicate direction of swimming movements. (From H. Schöne, 1951.)

Visual input may often effectively substitute for gravitational forces since, for example, the horizon is always related in the same manner to the gravitational force for a flying insect. In fact, phototaxis and geotaxis, two of the best studied taxes, have several aspects in common. For most insects the sun is upward, and positive phototaxes and negative geotaxes are the norm. One of the most remarkable features of honey bee communication is the two-way transfer from the angle between sun and food source to the angle between vertical and direction of the straight part of the waggle dance, then back again (see Fig. 7-14). For the dancing honey bee, positive phototaxis is coupled with negative geotaxis; if the food is in the same direction as the sun, the straight run of the waggle dance is directed upward, opposite the direction of gravity. When information about gravity is experimentally altered, the dance performed by the bee reflects the changed input (Fig. 3-6).

Probably the single most striking aspect of postural control among insects is its information overload, that is, the manner in which it depends upon input from a great number of sources acting in concert to the point of redundancy. For example, an ant can correctly orient to gravity using any one of its five proprioceptive joint systems alone if the others are fixed in position with wax. In fact, experiments such as the

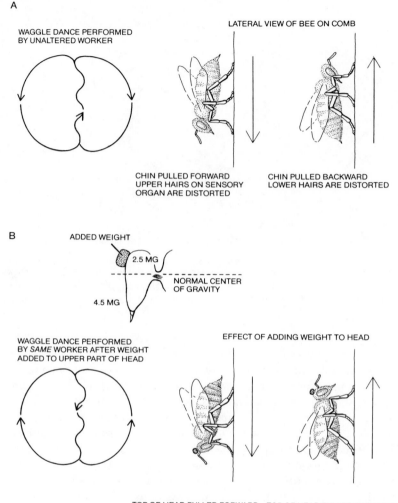

A

WAGGLE DANCE PERFORMED
BY UNALTERED WORKER

LATERAL VIEW OF BEE ON COMB

CHIN PULLED FORWARD
UPPER HAIRS ON SENSORY
ORGAN ARE DISTORTED

CHIN PULLED BACKWARD
LOWER HAIRS ARE DISTORTED

B

ADDED WEIGHT

2.5 MG

NORMAL CENTER
OF GRAVITY

4.5 MG

WAGGLE DANCE PERFORMED
BY *SAME* WORKER AFTER WEIGHT
ADDED TO UPPER PART OF HEAD

EFFECT OF ADDING WEIGHT TO HEAD

TOP OF HEAD PULLED FORWARD
LOWER HAIRS ARE DISTORTED
BEE SENSES ITSELF MOVING
*UPWARD*

TOP OF HEAD PULLED BACKWARD
UPPER HAIRS ARE DISTORTED
BEE SENSES ITSELF MOVING
*DOWNWARD*

**Figure 3-6** Gravity perception in the honey bee. Because of the way it is connected at the neck, the lower half of a honey bee's head weighs almost twice as much as the upper half. When the bee is dancing on the vertical surface of the honey comb, gravity causes this lower portion to swing downward, tilting against the sensitive proprioceptive hairs on the neck, and stimulating nerves at their base. Severing these nerves causes total disorientation. When a tiny weight glued to the top of the head alters its balance, the proprioceptive information received by the bee's brain results in a dance which is the reverse of normal. (From Alcock, 1975, based on Lindauer, 1971.)

86

weighting of the honey bee's head (Lindauer, 1971) may not produce the same results on every occasion, probably because other sensory areas are providing input.

## Orientation Responses to Radiant Energy

When an ant on its way back to the nest is placed in a dark box for a period of time, how will it orient when released? *Lasius niger* proceeded in the same course as before, relative to the sun; but the sun had shifted position in the sky, so the ant's orientation was incorrect in terms of its nest. Such an orientation so that locomotion occurs at a fixed angle relative to light rays is termed the **light-compass reaction**: it has been demonstrated in a wide variety of insects, including caterpillars, bees, and certain beetles and bugs. The polarization of light rays often serves as an orienting cue (see Wehner, 1976).

Have you ever noticed that various crawling and flying insects nearly always travel in a straight line across roads at right angles to their direction? This appears to occur irrespective of compass directions or other external stimuli. One hypothesized explanation of this so-called **transecting behavior** (which awaits rigorous experimental confirmation) is that the insects orient by balancing their reception of a symmetrical source of shortwave (infrared) radiant energy. Although roads are relative newcomers to the environment of insects, the adaptive significance of such behavior may be a survival advantage conferred in crossing large bare spots of earth or bodies of water in the shortest possible time with minimal energy expenditure and exposure to the elements or predators (Gunter, 1975).

Interestingly, evidence that some nocturnal insects are orientationally sensitive to the far infrared portion of the electromagnetic spectrum (10–1000 $\mu$m) has been documented by Callahan (1965, 1971); the ability may prove to be widespread as detectional systems are improved. For example, certain wasps parasitic on beetle larvae found in dead timber have also been shown to detect their hosts by means of infrared receptors on their antennae (Richerson and Borden, 1972).

## Magnetic Field Orientation

The ability to sense the earth's magnetic field has long been suspected for some insects. Some Australian termite species, for example, build large "magnetic" mounds oriented perfectly north–south (Fig. 3-7). However, existing evidence to date has been very tenuous and indirect. A pattern of systematic "residual errors" has been demonstrated in both

**Figure 3-7** Mound of the "magnetic termite," *Hodotermes meridionalis,* in a field in northern Australia. The long axes of all such nests are oriented north–south. (Courtesy of H. E. Evans.)

*Formica* orientation and honey bee waggle dances, for example (von Frisch, 1967; Lindauer, 1967). In the latter case, correction could be obtained by placing the hives in artificial magnetic fields that compensated for the earth's magnetic field. Some possible direct evidence for this ability has been found in the comb orientation of honey bees.

With wax secreted from specialized abdominal glands, the honeybee builds vertical combs that have hexagonal cells on both sides separated by a thin middle wall. The cells are tilted slightly upward at about 13° to the horizontal, an adaptation which serves to keep honey in. Modern beekeepers insert parallel artificial walls (frames) with a prestamped hexagonal pattern into the hive in order to more easily harvest the honey-filled combs. In so doing, the beekeeper determines the direction in which combs will be built. However, when a swarm takes possession of a new home such as a hollow tree, they will reconstruct, literally overnight, a complete series of combs having the regular parallel construction. How do the bees "decide" how their new combs shall be oriented?

When bees were transferred from a conventional hive into a plain empty cardboard cylinder with the entrance hole centered in the bottom, the bees produced new combs whose orientation exactly corresponded to that of the combs in the original parent colony. Since all directional landmarks had been removed in the round cylinders, it appeared that the bees were somehow able to orient to the earth's magnetic field. To test this hypothesis Lindauer and colleagues artificially disturbed and deflected the natural magnetic field with the aid of a powerful magnet placed outside the experimental nest cylinder. Invariably, the same bees that had previously reconstructed faithful new combs in experimental cylinders now built combs that differed from those in the previous nest by 40° exactly—the angle of artificial magnetic deflection (von Frisch, 1974). As yet unanswered is how the bees perceive the magnetic field (see Martin and Lindauer, 1974).

## Emitted Energy Orientation

Most orientation and navigation depend upon perception of environmental features which exist quite independent of the insect itself. Occasionally, however, an organism's own actions may form the basis for such navigational clues. Such **emitted energy orientation**, or reading of others or one's own energy patterns for clues as to the topography of the immediate environment, occurs in a wide diversity of animal species.

Groups of small whirligig beetles (*Gyrinus*) spin crazily about over the surface of most freshwater ponds. As long as they all keep moving, they seem to avoid contact; but if one stops, it may be bumped by another beetle. What keeps them all from colliding in one spectacular crash? Apparently each beetle, through impulses registered by receptors on a modified second antennal segment, is aware of the location and direction of the others by the vibrating waves they set up as they ripple over the surface (Tucker, 1969).

Gallery orientation by larvae of long-horned wood-boring beetles (Cerambycidae) is another example of emitted energy orientation. Larvae are apparently responsive to sound stimuli emanating from other larvae and change direction when exposed to artificially simulated gnawing sounds originating anteriorly or slightly to one side. Examination of gallery complexes reveals that the amount of turning and irregularity of burrows is directly proportional to the degree of infestation—with highly contorted burrows especially characteristic of heavy infestations (Saliba, 1972).

## DISPERSAL

Over and over, from certain areas in their homeland, swarms of up to 10 billion locusts crackle through the sky to land and devastate hundreds and thousands of acres of vegetation. Almost no green thing is left standing where they rattle to earth; even the bark of young trees is destroyed. The weight of a large swarm has been estimated at 15,000 tons, and its daily food consumption is equivalent to that of 1.5 million people. The migratory locust, feared since early Biblical times—what forces lie behind its seemingly diabolical and sinister behavior? Foraging army ants, *Eciton hamatum* (see Chapter 10), sometimes called the Kings of the Amazon, commonly move out of the nest in columns along branching trails to seize and carry back to the nest all small prey in their way. However, as new brood matures within the colony, instead of simply returning to the nest, workers reverse to lead a mass exodus that carries the whole colony away along one of the day's trails. During this migration the ants neither react to prey nor branch off.

Up to this point, we have been viewing insect spatial adjustment primarily as a phenomenon involving a single individual within its immediate environment. Now we step back and view insect movement on a larger scale, considering these behaviors in terms of the population or species.

### Dispersal Mechanisms

Gypsy moth, *Porthetria dispar*, females are unable to fly, and natural dispersal occurs primarily by young larvae being blown on their silken threads by wind. (A similar mechanism, **ballooning**, is employed by many newly hatched spiderlings.) Gypsy moth larvae vary in their dispersal propensity, and the behavior of first instar larvae depends on several variables such as the female parent's nutritional status, density of larvae, and food availability.

Young individuals quite commonly leave the area in which they were born. Usually such dispersal occurs either passively or under the juvenile's own volition, but in some cases dispersal may be encouraged by the indifferent or even hostile behavior of their parents or nest mates. As a result, through time all populations have a tendency to spread out spatially. Many different mechanisms may be involved, from relatively simple responses to gradients of certain environmental factors, to various active or passive dispersal mechanisms associated with the search for a mate or food.

The distinction between passive and active dispersal is made mainly for convenience; in reality a continuum exists between them. For example, a *Pemphigus* aphid, which lives on the roots of the sea aster growing in salt marshes, is for most of its life photonegative. However, first instar nymphs are photopositive, climbing up the sea asters until they set themselves adrift on the rising tide. Sea breezes send them scudding across open water to be deposited at low tide on another mud bank where they seek out and colonize new plants. As a result of their waterborne dispersal experiences they reverse their reaction to light and become photonegative (Kennedy, 1975).

Scelionid wasps ride upon the backs of female grasshoppers; ultimately, those tiny parasites will oviposit upon the grasshopper's eggs. Human bot flies will attach their eggs to the legs and body of mosquitoes, in this way transporting them to a human host for hatching and larval development. Some Trichoptera larvae undergo their development within gelatinous capsules upon the bodies of chironomid midges. Tiny wingless Mallophaga attach themselves to the bodies of the hippoboscid flies which parasitize their bird hosts, in this manner being carried from one host to another (Fig. 3-8). A great variety of mites ride upon beetles, ants, and other insects; the insects are probably not injured unless the numbers of mites become excessive. Such transport of flightless insects or other arthropods by others is known as **phoresy**. The subject has been reviewed by Clausen (1976).

Repeatedly in several groups of wasps a unique type of phoretic dispersal has independently evolved. All involve parasitic females that have become so highly modified for tunneling into the soil or food that they have permanently lost their wings. During copulation the genitalia in these species lock together, so that the males carry the smaller females about, suspended in this way, for considerable periods of time (Fig. 3-9). This phenomenon, termed **phoretic copulation**, allows adaptation for a burrowing life combined with effective dispersal of inseminated females into areas where new populations of hosts may be discovered.

## Migration

Behaviorally, insect flight may be divided into two broad categories. **Appetitive flight**, also sometimes termed "trivial flight," involves local movements of varying length and orientation concerned with food and mate finding, escape from potential enemies, location of suitable oviposition sites, territorial defense, and other such "vegetative" activities. In some cases, trivial flights may lead to some disperson, but often no effective displacement occurs at all, despite a good deal of flight activity.

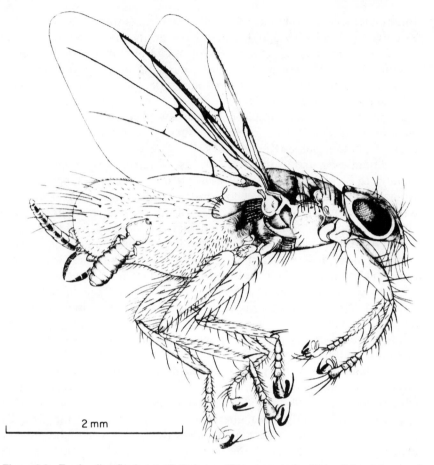

**Figure 3-8**  Feather lice firmly attach their mandibles to cuticular folds of the abdomen of hippoboscid flies, thereby obtaining phoretic transport to new bird hosts; 20 lice upon a single fly is not a rare number. (From Askew, 1971, redrawn from Rothschild and Clay 1952. With permission from *Parasitic Insects* by R. R. Askew, 1971, Heinemann Educational Books Ltd. and American Elsevier.)

**Migratory flight**, in contrast, involves a phase in adult life during which flight activity dominates over all other forms of behavior. In many insects, such activity is restricted to a short period, after which only appetitive flights occur; in some, the flight muscles may break down after migration so that no further flight is possible.

**Migration**, one of the most important forms of insect dispersal, has been succinctly defined by Kennedy (1975) as "adaptive traveling." It is an active mass movement functioning to displace populations. Whether

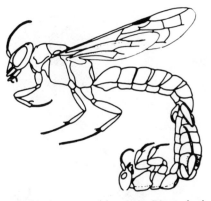

**Figure 3-9** Phoretic copulation in a parasitic wasp, *Dimorphothynnus haemorrhoidalis* (Tiphiidae). The male functions in both insemination and dispersal of the smaller, short-legged wingless female, often carrying her to sources of nectar or honeydew and in some cases actually feeding her by regurgitation. (From Evans, 1969.)

or not the males are in accompaniment, migration always involves the female sex; migrant females are generally sexually immature. Behaviorally, migration is characterized by persistent, enhanced locomotion in a straightened-out manner, and migrating individuals do not typically respond to stimuli for "vegetative functions" such as feeding, reproduction, etc. For example, the flight of certain scolytid bark beetles cannot be arrested by the host plant's odor until after they have been flying for many minutes. After long-distance flight the thresholds for vegetative activities are lowered, and further migration is inhibited.

Older studies on insect migration understandably concentrated on the long-distance flight and return of large and spectacular insects such as butterflies (Fig. 3-10). Brower and Huberth (1977) have produced an outstanding film dealing with the most spectacular of these, the monarch butterfly; Urquhart (1976) and Brower (1977) summarize recent developments in this fascinating saga (see also Fig. 8-9). However, it has become increasingly obvious that migration is a far more widespread phenomenon than previously suspected, including many small species whose movements, relatively speaking, are neither far nor spectacular. But in all of these, migration is a distinct behavioral and physiological syndrome closely intertwined with reproductive timing and strategy. Most migration takes place prior to egg development, and while the development of the flight system is maximized, that of the reproductive system is minimized, a phenomenon that results in migration occurring chiefly in young female adults. C. G. Johnson (1963, 1966, 1969) who has

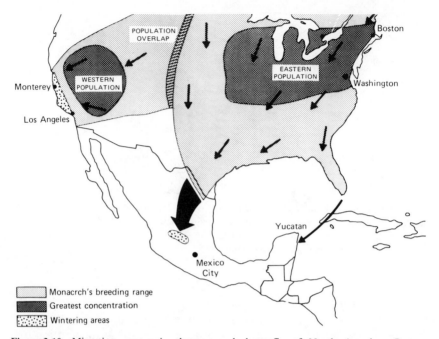

**Figure 3-10** Migration routes in the monarch butterfly of North America, *Danaus plexippus,* a strong flier and glider. Traveling southward each autumn, the butterflies congregate in great crowds, particularly in pine trees, only to return north in the spring, laying eggs along the way. New generations replace the old, and with the coming of fall, the butterflies again undertake their journey of perhaps 2000 or more miles. Discovery of their Mexican winter roost, a mountain enclave at an elevation of 9000 feet, by Urquhart climaxed a life-long study of this mysterious species. (Redrawn after Urquhart, 1976.)

done the most to crystallize migration theory, has termed this the **oogenesis flight syndrome**.

Adult life span has important implications for migratory behavior. First, there are many insects (including locusts, termites, aphids, and butterflies) that leave the breeding site, oviposit elsewhere, and die, all in a single season. A second type of migration occurs with short-lived adults that emigrate and return. For example, many dragonflies depart from breeding to feeding sites; after the eggs mature, females return to the vicinity of their original breeding site and oviposit. The third category includes long-lived adults that hibernate or aestivate away from the original breeding site, then return to it the following season. Several beetles, many noctuid moths, and the monarch butterfly fall into this category.

Why do insects migrate? Such a question is more properly two questions. First, what triggers the migratory "urge" and how is it maintained? Second, what ecological and/or evolutionary conditions might favor development of a migratory mode of life?

Two current hypotheses have attempted to explain the general causes of migrations. The first suggests that insects respond to the onset of adverse conditions by flying away. The second suggests that certain endocrine changes occur in correlation with particular environmental effects (crowding, food deficiency, short days, etc.). Although evidence presently favors the second hypothesis, both probably are accurate explanations for the situation in different species or in the same species under different conditions (Schneider, 1962).

The complexities of migratory behavior are well illustrated by the migratory locust. In some countries the migratory locust problem involves a single species; more often, several are involved. The best known is *Schistocerca gregaria*, which extends across north and central Africa to the Middle East, Arabia, and India. But, oddly, each such species occurs in two different races (phases) biologically and morphologically distinct. The sedentary or solitary phase, choosy about its food, is colored inconspicuously green, gray, or reddish. The swarm-forming, migratory phase, marked with more contrasting colors and usually having longer wings, devours any and all plants. Which phase is present depends upon how crowded they, and their mothers, have been. Previously uncrowded locusts live quietly, tend to repel one another and tend to remain upon clumps of vegetation. Hoppers that have been continuously crowded, however, are very active, strongly attract one another, and dodge clumps of vegetation, being arrested only briefly even by food plants. In its early stages, either phase can be formed from the other without intervening generations. For example, insects of the solitary phase may transform into migratory individuals if placed with swarm-forming companions. Migrations usually begin before the locusts even become adults. As the unwinged nymphs begin to march, the mass pattering of their feet is clearly audible. However, it is with the winged swarms that conspicuous migratory behavior reaches its peak.

At first sight, a migratory swarm appears to be a vast army flying single-mindedly toward a fixed goal. The illusion is strengthened by the fact that within the swarm, groups of locusts tend to be similarly oriented. However, photographic analysis has shown that in the swarm as a whole the locusts are randomly oriented toward one another. Such randomness, in combination with disruption from air turbulence, would be expected to lead to dispersal of the swarm, were it not for a striking phenomenon. All the locusts at the edges of the swarm orient toward the body of the swarm, a behavior probably mediated by visual and auditory cues (Camhi, 1971).

Inevitably, the locust swarm tends to be displaced downwind at a slow and variable speed, and the locusts are brought into those areas where there is a net excess of inflowing air over outflowing air across the boundaries. One such area is the Intertropical Convergence Zone between winds originating on either side of the equator. Borne along on the converging winds, the locusts accumulate (Fig. 3-11); swarms may continue to fly within these zones, but because of the irregularity of the winds they do not tend to become displaced. In the convergent areas, rising air currents are produced, which in turn lead to precipitation. In this way, as the zone moves back and forth across the equator once each year, the seasonal rains that promote vegetational greening and growth are caused. Thus the locusts, by their behavior, encounter the ecological requirements for successful reproduction—the moisture necessary for egg survival and fresh vegetation for the hatching larvae.

How can you tell if a flight is truly migratory? The definition of migration should offer an objective and experimental test: during the flight the insect should not be responsive to stimuli triggering vegetative behavior. The test has seldom been applied, but where it has, there does appear to be a reciprocal interaction between migratory flight and vegetative activities. Stimuli that evoke flight inhibit settling, and stimuli that evoke settling inhibit flight. One example where such migratory–vegetative interaction has been approached experimentally is that of the milkweed bug.

CASE STUDY: THE MILKWEED BUG, *Oncopeltus fasciatus*

The milkweed bug is a wide-ranging species occurring from Canada to Central America. It arrives in the northern reaches of its range between spring and early summer, settles on patches of milkweed, and mates. The eggs, laid close to the developing milkweed seed pod, are followed by five nymphal instars and, after a few weeks, by the adult. Throughout the summer the population increases rapidly, but with the shortening days of early autumn, numbers decline as maturing adults begin to leave. Flying south on the prevailing winds, these adults are able to avoid the oncoming winter, while those unable to complete their adult molt before the first severe frost are killed.

Are these bugs truly migrants? Dingle (1972) and associates have sought to answer this question through detailed field and laboratory investigations. Rearing studies soon confirmed the first criterion—a

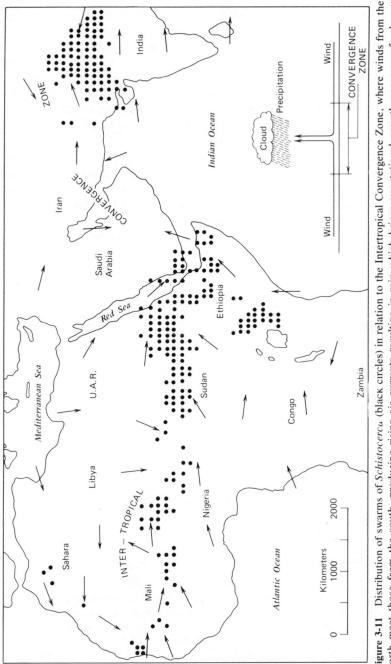

**Figure 3-11** Distribution of swarms of *Schistocerca* (black circles) in relation to the Intertropical Convergence Zone, where winds from the north meet those from the south, producing rising air currents resulting in rain which brings vegetational growth necessary for locust breeding. Arrows indicate the general wind direction. (After Rainey, 1969.)

separation of flight and vegetative activities. Flight activity peaks eight to 10 days after the adult molt, while oviposition begins only after 13 to 15 days (Fig. 3-12). Flights are also well separated from feeding. A few hours after the adult molts, females enter into a high rate of feeding which, until day 7, lacks periodicity. By day 8, however, a fully developed circadian rhythm is evident. Thereafter, the peak feeding activity for females occurs at the end of the day and proceeds simultaneously with mating. Interestingly, for males the feeding rate falls markedly after day 6 to persist at a very low rate for the remainder of life.

Temperature may have a very direct effect upon migration. Raising the temperature from 23° to 27°C, a warmth about optimal for population growth, Dingle found that a lower proportion of the population now exhibited tethered flights of 30 minutes or longer (the operational criterion for migration). This suggested that once the bugs reached a thermally favorable environment, they would tend to settle there.

Laboratory studies clearly indicated that peak flight preceded repro-

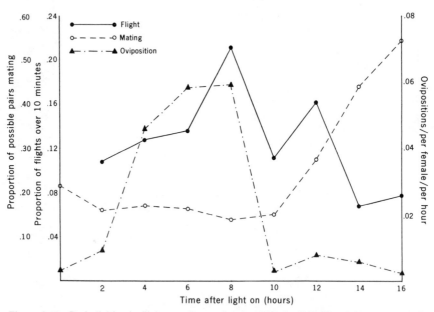

**Figure 3-12** Periodicities in flight, mating, and oviposition in the milkweed bug, *Oncopeltus fasciatus*, maintained under 16 hours of light and 8 hours of dark at 23°C. Flight and oviposition show nearly identical daily periodicities but do not overlap in time (temporal segregation). However, mating behavior is separated from flight not only temporally but also by a circadian periodicity peaking at the end of the daily light period. (From Dingle, 1972. Copyright 1972 by the American Association for the Advancement of Science.)

duction. Would prolonging the prereproductive period also prolong migration? By altering day length in the laboratory, Dingle knew that he could delay oviposition from 15 days to 45 days after the adult molt. Raising bugs under both regimes, he tested comparable groups of females for duration of tethered flight 25 days after adult molt. The early ovipositing females generally flew for only a few minutes or less. The delayed females, which showed no signs yet of reproductive development, performed like typical migrants. Under field conditions, Dingle reasoned, this phenomenon would have important consequences. One would be that in the autumn females would be capable of migrating for much longer periods, thus improving their chances for escaping the oncoming winter.

In what other ways might a short photoperiod affect migratory behavior? In tethered flights, Dingle compared sets of short-day and long-day bugs of both sexes repeatedly between eight and 30 days after adult molt. Invariably, a greater proportion of the short-day bugs flew for long periods (at least 30 minutes, usually 2 to 3 hours). Significantly, the results held true for both sexes. Evidently, in addition to its indirect effect via ovarian development, photoperiod was having a direct effect upon migration as well.

Environmental factors obviously had great importance as determinants of migratory behavior. What about hereditary influence? Under strong selection, Dingle found that he could increase the proportion of migrants of a population of milkweed bugs from 25% to over 60% in one generation. Clearly, migratory capability in *Oncopeltus* could be altered rapidly.

**Theoretical Considerations**

A fundamental concern for most insects is finding the optimal habitat in which to live and reproduce. For dispersing individuals, the problems are particularly complex and acute. At what point should an individual stop expending energy on the search and settle for whatever situation is available? When should selection favor the evolution of ways for individuals to return to specific sites after displacement from them? How should areas of fluctuating or disparate resources be best exploited? When is it worth defending one's own space against others? Such theoretical questions have been receiving greatly increased attention in recent years; good reviews include those of Dingle (1973, 1974) and Johnson (1969, 1974).

But in order to even attempt an answer, a fundamental question needs attention: What is the real nature and function of migration? Migration has long been viewed as a means of escape from one habitat to another, more suitable one. On first examination such a simple answer seems sufficient. Faced with a temporarily unfavorable period or untenable habitat, an insect species can adopt one of two evolutionary strategies. It may develop a migratory life style or it may go into *diapause*, a period of dormancy akin to hibernation. If the change in habitat is reversible, as with seasonal changes, diapause is favored. However, when habitat changes are irregular, migration has a clear advantage over diapause. But is migration really an act of massed fugitives, a desperate attempt to flee a hostile environment or to relieve population pressures? In recent years, a recrystallization of ideas drawing from the fields of physiology, behavior, and population ecology has suggested that migration is not just a means of escape from unfavorable environments but a positive act of dispersal over all available habitats. Under such a view, insect migrants are more accurately to be viewed as colonizers than as refugees, and migration as an evolved adaptation, not a reaction to current adversity.

The colonization aspect of migration has received a great deal of attention from ecologists. Abundant evidence confirms a relation between migration and habitat; migration occurs most often in those insects which occupy "temporary" habitats (Southwood, 1962). Often, these are habitats in early stages of ecological succession, but the impermanence of a habitat may result also from seasonal or irregular climatic changes or simply from the ephemeral nature of the habitat (such as flowers, fungi, carrion, etc.). In other cases, alterations in the insect's requirements at different life stages may be responsible. Acridid grasshoppers, for example, have different requirements for feeding and for oviposition. Another way of viewing the situation would be to say that migration enables a species to keep pace with changes in the location of its habitats. The tendency to exodus is, of course, abetted by responses to a variety of proximate stimuli, as well, such as photoperiod, crowding, alterations in the physiology of a food plant, etc.

From the view of insect migrants as colonizers, attention has logically turned to an examination of migration strategies. The basic theoretical question is this: In what ways might a migrant species be expected to evolutionarily modify its life history statistics to maximize its colonization success? First, the migratory situation obviously favors high reproductive productivity. A high rate of population increase (a high $r$, see Chapter 9) can be attained either by increasing fecundity or by reproducing earlier. Second, when it migrates an individual should theoretically have a high reproductive value, or degree of expected contribution to population growth. Young prereproductive adults, having survived the causes of juvenile mortality but still having their full reproductive life

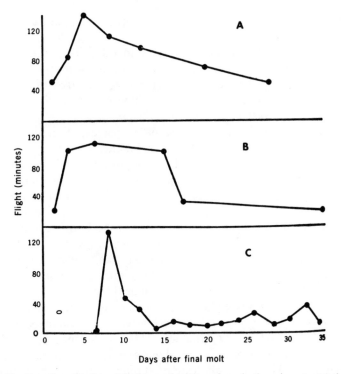

**Figure 3-13** Duration of tethered flight as a function of age in three insects: (a) the frit fly, *Oscinella frit,* (b) the fruit fly, *Drosophila funebris,* and (c) the milkweed bug, *Oncopeltus fasciatus.* The decline in flight occurs concurrently with an increase in reproduction (the oogenesis-flight syndrome). Similar effects have been demonstrated in a wide range of insects, including bugs, flies, mosquitoes, aphids, moths, grasshoppers, and beetles. (From Dingle, 1972. Copyright 1972 by the American Association for the Advancement of Science.)

ahead, would have a much higher reproductive value than juveniles or older adults. Thus, it is not surprising to find that dispersal in most insect species occurs just prior to reproduction and that flight activity correlates strongly with age (Fig. 3-13).

## THERMOREGULATION

The sphinx moth *Celerio lineata* stabilizes its thoracic temperature during flight over a range of ambient temperatures. The regulation of body temperature by basking has been described in a wide variety of insects, from butterflies and beetles to cicadas and arctic flies. Male tettigoniid grasshoppers elevate their thoracic temperature prior to singing.

In insects, as in other animals, body temperature is closely attuned to activity and energy supplies, affecting all aspects of life from the rate at which food can be located and harvested to the facility with which predators can be avoided. Most people think of insects as being purely "cold-blooded," passively reflecting the temperatures that surround them in their environment. However, many species, often highly mobile, utilize high-energy fuels and intense metabolic rates to produce body heat at rates sufficient to increase their body temperature beyond that of their environment. Others, primarily by behavioral means, control their body temperatures within a far narrower range than that of their surroundings.

Actually, **endothermy**—the ability to increase body temperature beyond that of the environment—has been long known in insects. However, until quite recently, such regulation was thought to be accomplished solely by behavioral means. An increasing number of studies now show that some insects use physiological means as well.

On the basis of body weight, most flying insects have higher rates of metabolism and hence of heat production than other animals. However, thermoregulation is a costly behavior. The maintenance of a high body temperature when an animal is not physically active requires much from energy supplies. The metabolic rate of stationary bumble bees maintaining their body heat when ambient temperatures are very low is very nearly identical to that in flight.

## Heat Production in Endothermic Insects

Small flies such as midges and fruit flies have rapid heat loss and little buildup of body heat during flight; their wingbeat frequency and flight speed varies nearly directly with ambient thermal conditions. Some of the larger insects, such as bumble bees and some moths, however, must warm their flight muscles to about 40°C before they can attain sufficient wingbeat frequency and lift to support themselves in free flight. Because of their rapid metabolism, while flying their body temperature is usually 5 to 10°C (and sometimes 20 to 30°C) above ambient temperatures.

In insects, essentially all endothermic increases of body temperature have the same causal mechanism—heat produced by the active flight muscles. Because these are the most metabolically active tissues known, endothermy in flight is largely an obligatory phenomenon. It has been estimated that the mechanical efficiency of the flight mechanism, in both insects and birds, is on the order of 10–20%. Thus, 80% or more of the energy expended during flight is degraded to heat.

High muscle temperature is, however, not just a consequence of muscle activity. In many situations, especially flight, it is also a prerequisite. Until the temperature of the muscles is sufficiently high, there is little overlap in the contractions of the antagonistic muscles, the wingbeat frequency is very low, and the insect remains grounded. Different relative wing sizes and power requirements determine the muscle temperature and wingbeat frequency necessary for a given insect to become airborne; this varies, for example, with body size, between different muscles used for different activities, and between different species. Large wings are one way around the problem, for they allow an insect to fly with a low wingbeat frequency. Thus, some butterflies are able to initiate flight without prior endothermic warm-up and to continue flight by gliding; this also reduces the energy expenditure of locomotion in such insects. Another response is to evolve dense pile "coats," which may halve the rate of convective heat loss from the insect's body. In this manner, for example, bumble bees conserve heat loss; the combination of high metabolic rate, relatively large body size, and good insulation helps them not only to passively elevate thoracic temperatures during free flight but also to maintain a sufficiently high thoracic temperature to be able to fly at very low ambient temperatures. Records show that queens flying at an air temperature of 3°C have a thoracic temperature of 36°C (Heinrich, 1974).

The requirement for a high thoracic temperature to *start* flight poses a real behavioral problem. When an insect comes to rest in the shade, its body temperature rapidly becomes practically equal to ambient temperature. Without some means of increasing muscular temperature, the insect could remain permanently grounded.

Warm-up, or **shivering**, one widespread mechanism for such preflight thoracic temperature increase, involves numerous patterns of flight muscle activation. Apparently, shivering has evolved independently on numerous occasions; it has been documented in a wide variety of unrelated insects from at least four orders. Among many Lepidoptera, the rates at which wings vibrate during such shivering have been shown to be directly correlated with muscle temperature (Fig. 3-14).

Shivering is well suited for variable rates of heat production, for it can occur at a wide range of activation frequencies. In addition, an insect may regulate its body temperature through intermittent activity—intermittent flight, intermittent shivering, or some combination of the two. Some insects are behaviorally better suited to make use of this option

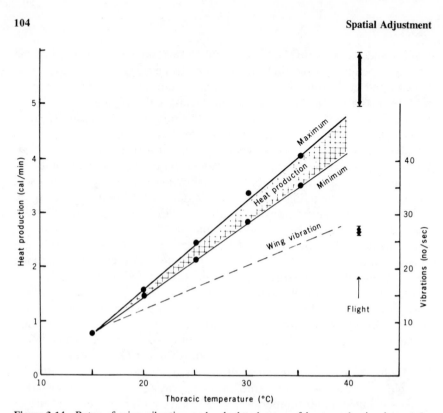

**Figure 3-14**   Rates of wing vibration and calculated rates of heat production by a 1.5-g sphinx moth, *Manduca sexta,* in relation to thoracic temperature. As thoracic temperature increases throughout the warm-up, so does the rate of wing vibration and, concomitantly, the rate of heat production. Finally, when the metabolic rate is near that observed in free flight (arrows), the insect begins to fly. (From Heinrich, 1974. Copyright 1974 by the American Association for the Advancement of Science.)

than others, however. For example, a hovering sphinx moth or dragonfly in continuous flight is in a less advantageous position in this regard than is a bee that lands on flowers at frequent intervals while foraging in the field.

Honey bees and bumble bees are, in fact, the most thoroughly studied examples of shiverers where intermittent muscle activity is possible. As their foraging activities show, intermittent activity may be very rapidly paced. Bumble bees, for example, may visit 20 or 30 different flowers per minute. In some cases, their thermoregulation is also adapted for ends other than flight. For example, individual founding queen bumble bees are able to behaviorally raise the temperature of their brood (Fig. 3-15). Thermoregulation in endothermic insects is discussed by Heinrich (1974) and Heinrich and Bartholomew (1972).

Ambient temperature:3–33°C

Thorax
35–38°C

Abdomen
31–36°C

Brood clump
24–34°C

Honeypot
temperature 3–33°C

**Figure 3-15** Brood incubation by the queen bumble bee, *Bombus vosnesenskii.* In characteristic brooding posture, the bee's legs are wrapped around the initial brood cluster and her abdomen closely appressed to it and extended lengthwise. Queens remain in this position for many hours or even days at a time, interrupting only to feed or shift position slightly. When contrasted to ambient temperatures, the temperatures of the bee, brood cluster, and adjacent honeypot reveal clearly the efficiency and effectiveness of incubation behavior. (From Heinrich, 1974. Copyright 1974 by the American Association for the Advancement of Science. Courtesy of B. Heinrich.)

## Regulation of Heat Gain

Experiments with spiders reveal that some will adjust their web position in response to incident illumination from various directions (Fig. 3-16). Certain *Pheidole* ants forage above ground all day during cool, cloudy periods but restrict their foraging to the period from late evening to early morning when the weather is hot and sunny.

Many arthropods that are exposed to insolation, such as some web-building spiders, various cryptic insects that rest by day on a matching background, and predators who use a lie-in-wait strategy, have at least a potential problem with heat gain. Not surprisingly, various behavioral

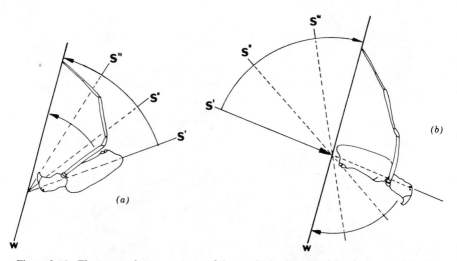

**Figure 3-16** Thermoregulatory postures of the tropical spider *Nephila clavipes* on its web (w) in relation to sun position. Whatever position the sun takes (or is redirected to with mirrors), the spider turns its abdomen to point directly toward the source of the light. (*a*) In response to light shining toward the plane of the web, the spider turns "upside down," keeping its ventral side up. (*b*) In response to illumination through the web, the spider orients "right side up," with dorsal side uppermost. With light redirected by mirrors, the spider can reposition from (*a*) to (*b*) in less than 20 seconds. (Modified from Robinson and Robinson, 1974.)

adaptations have evolved that involve postural adjustments to minimize the body surface area exposed to the heat source or that lead to avoidance of solar radiation totally during certain periods. Structural features also help. For example, the long legs of many ants and beetles living on sand in direct sunlight lift their bodies above the substrate, while light body pigmentation reduces heat input from above.

Most insects do not cool themselves as we do, with active evaporative mechanisms. Perhaps the availability of behavioral means of cooling plus the need for water retention have selected against such a mechanism. One exception is the tsetse fly, *Glossina morsitans*, of equatorial Africa, a blood-sucking fly able to feed while in sunshine on the hot hide of a mammal. While taking blood, the fly has a temporary abundance of liquid; at high ambient temperatures, the tsetse fly opens its spiracles, allowing the water to evaporate and body temperature to decline as much as 1.6°C.

Thermoregulation by colonies of social insects involves a variety of architectural and behavioral devices which maintain the nest temperature (and humidity) within carefully controlled tolerance ranges, irrespective of season or outside temperature (see Fig. 10-26). While such insects are individually "cold-blooded," or *poikilothermic*, socially they are nearly as "warm-blooded" (*homeothermic*) as birds and mammals. Honey bee workers maintain the hive interior at temperatures between 34.5 and 35.5°C (just below human body temperature) by fanning with their wings to promote air circulation and cooling by water evaporation. Winters are survived by clustering loosely at warmer temperatures and very tightly during extreme cold.

## SUMMARY

For nearly all insects, the ability to change position within the environment is essential. However, many aspects of insect spatial adjustment seem strange when extrapolated to man's scale. For example, because of their small size, insects' muscle power appears tremendous. Light weight relative to water surface tension allows some to propel themselves in various ways over the water surface. Small body size also causes rapid heating and cooling, of importance in both the physiology and behavior of thermoregulation.

Terrestrial locomotion, the ancestral condition, depends either upon thoracic legs, which can function only when the skeleton is relatively rigid, or upon fleshy abdominal prolegs. Aquatic locomotion, according to the nature of the insect's respiratory habit, may involve either surface propulsion or true underwater swimming; a great variety of modifications enhances efficiency of both methods.

Insects are the only invertebrates capable of flight. Insect wings, which evolved as four lateral thoracic expansions during the Devonian or Mississippian Periods, were used at first for gliding only. Direct musculature that allowed flapping arose and then underwent a twofold modification to allow backward wing folding. A two-winged condition and high wing speed represent further evolutionary advances.

Orientation to factors such as food, mate, prey, and host is a major component of an insect's behavioral repertoire. Spatial adjustment primarily involves a single individual within its immediate environment. On a larger scale—population or species—insect movement includes dispersal and/or migration.

Spatial orientation, the self-controlled maintenance or change of body position relative to environmental space, may involve any of at least six stimulus fields: heat, light, gravity, magnetism, pressure, and chemicals. Kineses, taxes, and transverse orientations provide one classification of spatial orientation, but a more comprehensive catalog is based on the adaptive significance.

Through time all populations have a tendency to spread out spatially. Dispersal mechanisms display a continuum of methods from passive to active. Insect flight includes two behavioral categories: appetitive ("trivial") and migratory. Migration always involves the female (usually sexually immature), whether or not males are in accompaniment. The general causes of migration appear to be certain endocrine changes which occur in correlation with environmental effects such as crowding, food deficiency, and changing day length.

During migratory flight, all other forms of behavior are suppressed. Defining when a flight is truly migratory is objectively done by testing whether the insect is responsive to stimuli triggering vegetative behavior. The test has seldom been applied, but an excellent illustration is provided by experimental studies of the milkweed bug.

Migration and diapause represent two evolutionary strategies for dealing with an unfavorable period or untenable habitat. Regular environmental changes tend to favor diapause, while irregular changes and/or "temporary" habitats tend to favor a migratory strategy. Migration has a number of evolutionary correlates, including high rate of population increase. Migratory individuals tend to have a high reproductive value, and flight activity correlates strongly with age. Migrants are better viewed as colonizers than as refugees.

Rather than being purely and passively "cold-blooded," insects practice thermoregulation through both physiological and behavioral means. Endothermy, the ability to increase body temperature beyond that of the environment, depends upon heat produced by the active flight muscles; it is largely an obligatory phenomenon. Cooling commonly depends upon strictly behavioral mechanisms; the tsetse fly is a striking exception in possessing a physiological mechanism based on active evaporative cooling.

## SELECTED REFERENCES

Alcock, J. 1975. *Animal Behavior. An Evolutionary Approach.* Sinauer Associates, Sunderland, Mass., 547 pp.

Askew, R. R. 1971. *Parasitic Insects.* American Elsevier, New York, 316 pp.

Birukow, G. 1966. Orientation behaviour in insects and factors which influence it. In *Insect Behaviour*, P. T. Haskell (ed.). Royal Entomol. Soc. London Symposium No. 3, pp. 1–12.

Brower, L. P. 1977. Monarch migration. *Nat. Hist.* **86**: 40–53.

Brower, L. P. and J. C. Huberth. 1977. *Strategy for Survival: Behavioral Ecology of the Monarch Butterfly*. Harper and Row Media, New York, 30-minute 16-mm color film.

Callahan, P. S. 1965. Intermediate and far infrared sensing of nocturnal insects. Part I. Evidence for a far infrared (FIR) electromagnetic theory of communication and sensing in moths and its relationship to the limiting biosphere of the corn earworm. *Ann. Entomol. Soc. Amer.* **58:** 727–45.

Callahan, P. S. 1971. Insects and the unsensed environment. Proc. Tall Timbers Conf. Ecol. Anim. Cont. Hab. Manag., Feb. 25–27, 1971, pp. 85–96.

Camhi, J. M. 1971. Flight orientation in locusts. *Sci. Amer.* **225:** 74–81 (August).

Chapman, R. F. 1969. *The Insects: Structure and Function*. American Elsevier, New York, 819 pp.

Clausen, C. P. 1976. Phoresy among entomophagous insects. *Annu. Rev. Entomol.* **21:** 343–68.

Dalton, S. 1975. *Borne on the Wind. The Extraordinary World of Insects in Flight*. E. P. Dutton, New York, 160 pp.

Dingle, H. 1972. Migration strategies of insects. *Science* **175:** 1327–34.

Dingle, H. 1973. Migration. In *Syllabus. Introductory Entomology*. V. J. Tipton (ed.). Brigham Univ. Press, Provo, Utah, pp. 377–84.

Dingle, H. 1974. The experimental analysis of migration and life history strategies of insects. In *The Experimental Analysis of Insect Behaviour*, L. Barton Browne (ed.). Springer-Verlag, New York, pp. 329–342.

Evans, H. E. 1969. Phoretic copulation in Hymenoptera. *Entomol. News* **80:** 113–24.

Fraenkel, G. S. and D. L. Gunn. 1961. *The Orientation of Animals: Kineses, Taxes and Compass Reactions*. Dover, New York, 376 pp.

Frisch, K. von. 1967. *The Dance Language and Orientation of Bees*. Belknap Press, Harvard Univ. Press, Cambridge, Mass., 566 pp.

Frisch, K. von. 1974. *Animal Architecture*. Hutchinson, London, 306 pp.

Gunter, G. 1975. Observational evidence that short wave radiation gives orientation to various insects moving across hard surface roads. *Amer. Natur.* **109:** 104–7.

Heinrich, B. and G. A. Bartholomew. 1972. Temperature control in flying moths. *Sci. Amer.* **226:** 70–77 (June).

Heinrich, B. 1974. Thermoregulation in endothermic insects. *Science* **185:** 747–56.

Hughes, G. M. and P. J. Mill. 1974. Locomotion: terrestrial. In *The Physiology of Insecta*, Vol. 3, 2nd ed., M. Rockstein (ed.). Academic Press, New York, pp. 335–79.

Jander, R. 1963. Insect orientation. *Annu. Rev. Entomol.* **8:** 95–114.

Jander, R. 1975. Ecological aspects of spatial orientation. *Annu. Rev. Ecol. Syst.* **6:** 171–88.

Johnson, C. G. 1963. The aerial migration of insects. *Sci. Amer.* **209:** 132–138 (December).

Johnson, C. G. 1966. A functional system of adaptive dispersal by flight. *Annu. Rev. Entomol.* **11:** 233–60.

Johnson, C. G. 1969. *Migration and Dispersal of Insects by Flight*. Methuen, London, 766 pp.

Johnson, C. G. 1974. Insect migration: aspects of its physiology. In *The Physiology of Insecta*, Vol. 3, 2nd ed., M. Rockstein (ed.). Academic Press, New York, pp. 280–334.

Kennedy, J. S. 1975. Insect Dispersal. In *Insects, Science, and Society*, D. Pimentel (ed.). Academic Press, New York, pp. 103–19.

Lindauer, M. 1967. Recent advances in bee communication and orientation. *Annu. Rev. Entomol.* **12**: 439–70.

Lindauer, M. 1971. *Communication Among the Bees.* Harvard Univ. Press, Cambridge, Mass., 161 pp.

Loeb, J. 1973. *Forced Movements, Tropisms and Animal Conduct.* Dover, New York, 209 pp. (Reprint of 1918 publication).

Markl, H. 1971. Proprioceptive gravity perception in Hymenoptera. In *Gravity and the Organism*, Gordon, S. A. and M. J. Cohen (eds.). Univ. of Chicago Press, Chicago, pp. 185–94.

Martin, H. and M. Lindauer. 1974. Orientierung im Erdmagnetfeld. *Fortschr. Zool.* **21**: 211–28.

Mill, P. J. (ed.). 1976. *Structure and Function of Proprioceptors in the Invertebrates.* Chapman and Hall, London, 686 pp.

Mittelstaedt, H. 1962. Control systems of orientation in insects. *Annu. Rev. Entomol.* **7**: 177–98.

Nachtigall, W. 1974a. *Insects in Flight,* McGraw-Hill, New York, 153 pp.

Nachtigall, W. 1974b. Locomotion: mechanics and hydrodynamics of swimming in aquatic insects. In *The Physiology of Insecta,* Vol. 3, 2nd ed., M. Rockstein (ed.). Academic Press, New York, pp. 381–432.

Pringle, J. W. S. 1974. Locomotion: flight. In *The Physiology of Insecta,* Vol. 3, 2nd ed., M. Rockstein (ed.). Academic Press, New York, pp. 433–76.

Pringle, J. W. S. 1975. *Insect Flight.* Oxford Biology Reader. Oxford Univ. Press, Oxford, 16 pp.

Rainey, R. C. 1969. Effects of atomospheric conditions on insect movement. *Quart. J. Roy. Meteorol. Soc.* **95**: 424–34.

Rainey, R. C. (ed.). 1976. *Insect Flight.* Royal Entomol. Soc. London Symposium No. 7, 298 pp.

Richerson, J. V., and J. H. Borden. 1972. Host finding behavior of *Coeloides brunneri. Can. Entomol.* **104**:1235–1250.

Robinson, M. H. and B. C. Robinson. 1974. Adaptive complexity: the thermoregulatory postures of the golden-web spider, *Nephila clavipes,* at low latitudes. *Amer. Midl. Natur.* **92**: 386–96.

Romoser, W. 1973. *The Science of Entomology.* MacMillan, New York, 449 pp.

Rothschild, M., Y. Schlein, K. Parker, C. Neville, and S. Sternberg. 1973. The flying leap of the flea. *Sci. Amer.* **229**: 92–100 (November).

Saliba, L. J. 1972. Gallery orientation in cerambycid larvae. *Entomologist* **105**: 300–304.

Schneider, F. 1962. Dispersal and migration. *Annu. Rev. Entomol.* **7**: 223–42.

Schöne, H. 1951. Die Lichtorientierung der Larven von *Acilius sulcatus* L. und *Dytiscus marginalis* L. *Z. Vergleich. Physiol.* **33**: 63–98.

Southwood, T. R. E. 1962. Migration of terrestrial arthropods in relation to habitat. *Biol. Rev.* **37**: 171–214.

Tucker, V. A. 1969. Wave making by whirligig beetles (Gyrinidae). *Science* **166**:897–99.

Urquhart, F. A. 1976. Found at last: the monarch's winter home. *Natl. Geogr.* **150**: 161–73.

Wehner, R. (ed.). 1972. *Information Processing in the Visual Systems of Arthropods.* Springer-Verlag, New York, 334 pp. (Includes several papers on orientation control systems and pattern recognition).

Wehner, R. 1976. Polarized-light navigation by insects. *Sci. Amer.* **235:** 106–115 (June).

Weis-Fogh, T. 1971. Flying insects and gravity. In *Gravity and the Organism,* Gordon, S. A. and M. J. Cohen (eds.). Univ. of Chicago Press, Chicago, pp. 177–84.

Wendler, G. 1971. Gravity orientation in insects: the role of different mechanoreceptors. In *Gravity and the Organism,* Gordon, S. A. and M. J. Cohen (eds.). Univ. of Chicago Press, Chicago, pp. 195–201.

Wigglesworth, V. B. 1964. *The Life of Insects.* The New American Library, New York, 383 pp.

Wilson, D. M. 1966. Insect walking. *Annu. Rev. Entomol.* **11:** 103–22.

Wilson, D. M. 1971. Stabilizing mechanisms in insect flight. In *Gravity and the Organism,* Gordon, S. A. and M. J. Cohen (eds.). Univ. of Chicago Press, Chicago, pp. 169–76.

# 4
# Feeding
# Behavior

The earliest insects apparently lived and fed chiefly in moist forest floor litter, a mode of life still continued by the bristletails and springtails today. However, when the first flowering plants appeared early in the Mesozoic Period, an almost limitless new food source became available, and insects were quick to take evolutionary advantage of it. A great blossoming in insect variety and numbers occurred, and today about half of all the Insecta dine upon plants (a condition termed **herbivory**, or **phytophagy**). Foliage feeding is most common, but one can also find sap feeders, root feeders, seed feeders, and, in the case of woody plants, bark feeders. A few groups induce gall development in the host plant, thus obtaining shelter in addition to food. To all these may be added the complex of insects which feed on the rotting debris below the plants and even the opportunists who partake of the sugary honeydew excreted by the sap feeders. Each of these species serves as a potential host for one or more parasitic insects which may in turn be attacked by their own specific enemies, *hyperparasites*. Thus, even a single plant species can form the base of a complex food web (Fig. 4-1).

A general classification of insect feeding habits is given in Table 4-1. Important exceptions occur in most taxa. In addition, such a classification is complicated by the fact that many holometabolous species feed at different trophic levels at different stages in their life cycle. For example, mosquito larvae feed upon plankton and suspended organic matter, but adults suck vertebrate and invertebrate blood (Fig. 4-2). However, only the females feed on blood, so additional complication arises from sexual differences in feeding habits. Furthermore, many insects that appear to feed on one type of food may upon closer scrutiny be found to be entirely dependent upon another. The fruit fly, *Drosophila*, is an example; larvae cannot be raised on a sterile culture medium. Although one associates them with decaying organic matter such as overripe bananas, they are actually eating microorganisms associated with the decaying fruit.

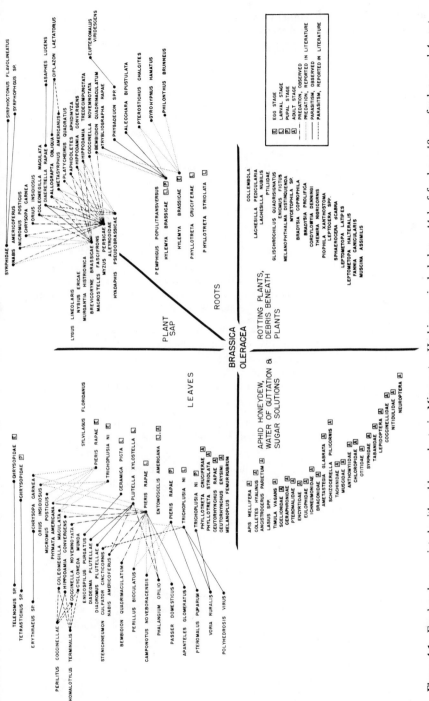

**Figure 4-1** Food web associated with cabbage plants in Minnesota. Herbivores included are 11 leaf feeders, 10 sap feeders, and 4 root feeders. Other trophic levels were less thoroughly studied, but included at least 21 species of detritivores, 79 honeydew (sugar) feeders, and 85 species of predatory, parasitic, or hyperparasitic carnivores. (From Weires and Chiang, 1973.)

113

**Table 4-1**    A classification of the feeding patterns of selected insect taxa on the basis of principal nutrition source[a]

| | |
|---|---|
| | **HERBIVORES** (plant feeders) |
| *Phytophagous* (green plants) | Orthoptera, most Lepidoptera, Homoptera, most Hemiptera, Thysanoptera, Phasmida, some Isoptera, some Coleoptera (Cerambycidae, Chrysomelidae, Curculionidae), Hymenoptera (Symphyta, Apoidea), some Diptera (e.g., Agromyzidae), some larval Trichoptera |
| *Mycetophagous* (fungi) | Some Diptera (e.g., Mycetophilidae), some Coleoptera (e.g., Ciidae, Platypodidae), some Hymenoptera (e.g., Siricidae and certain Formicidae), some Isoptera |
| | **CARNIVORES** (animal feeders) |
| *Predators* | Odonata, Mantodea, some Hemiptera (e.g., Phymatidae, Notonectidae), larval Neuroptera, some larval Trichoptera, some Mecoptera, some Diptera (adult Asilidae, Empididae), some Coleoptera (Adephaga, larval Lampyridae, Coccinellidae), some aculeate Hymenoptera (Sphecidae, Pompilidae, Vespidae) |
| *Ectoparasites* | Adults and sometimes immatures are parasitic; hosts usually vertebrates which are not killed. Include Anoplura, Mallophaga, adult Siphonaptera, some Dermaptera, some Hemiptera (e.g., Cimicidae), some adult Diptera (bloodsucking forms); a few Lepidoptera, a few Coleoptera |
| *Parasitoids* | Parasitic only in immature stages with adults free-living; hosts usually other individual insects which are killed by the parasite larva feeding either internally or externally. Include the "parasitic" and some aculeate Hymenoptera, including social parasites, some Neuroptera (e.g., Mantispidae), some Coleoptera (e.g., Meloidae, Sylopidae), some Diptera (Bombyliidae, Conopidae, Tachinidae) |
| | **DETRITIVORES** (feeders on dead and decaying organic material, e.g., dung, carrion, leaf litter, etc.) |
| | Collembola, Thysanura, Diplura, some Coleoptera (e.g., Staphylinidae, Scarabaeinae, Silphidae), some larval Diptera (e.g., Calliphoridae, most Phoridae), some Psocoptera, Blattodea, immature Plecoptera and Ephemeroptera, some Hemiptera (e.g., Corixidae), larval Siphonaptera |

[a] Derived from various sources, but especially Brues (1946), Askew (1971), and Price (1975).

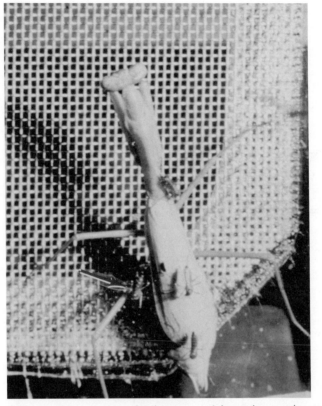

**Figure 4-2** Female *Aedes* mosquitoes feeding on an adult preying mantis; most feeding punctures are through intersegmental membranes, and a fully engorged individual is visible at the arrow. The extent of such invertebrate blood feeding by mosquitoes in nature remains to be documented, but these laboratory females were able to develop fertile eggs in the same manner as vertebrate blood-fed individuals. (Photograph by the authors.)

Most insects continue to feed throughout their entire lives, thus supplying the essential materials for body maintenance, growth, and reproduction as these needs arise. Sometimes, however, adaptation to special conditions has led to a partitioning of the feeding functions. For example, insects such as the silk moths, bot flies, and mayflies do not feed at all as adults. In such cases, the larva acts as a rapidly growing "feeding machine," storing huge quantities of reserves for adult life. Sometimes adult feeding serves solely to provide energy for locomotion; in the adult male blowfly, there is no growth, not even wound healing.

How choosy are insects about their diet? Most accept only a limited range of foods and usually prefer one or two (**oligophagy**). Often this is

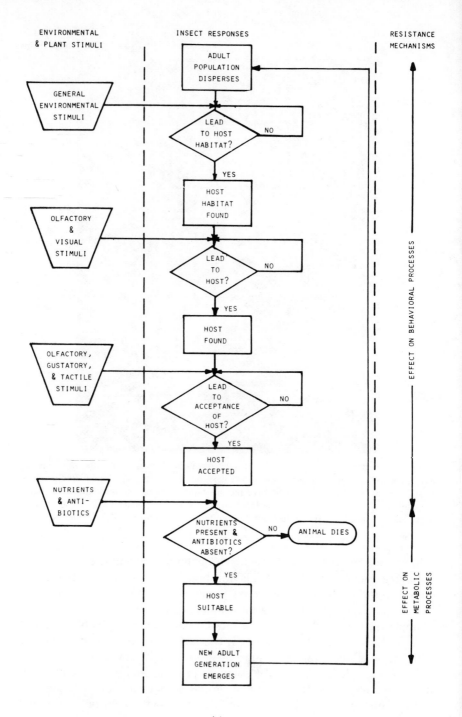

ENVIRONMENTAL & PLANT STIMULI

INSECT RESPONSES

RESISTANCE MECHANISMS

ADULT POPULATION DISPERSES

GENERAL ENVIRONMENTAL STIMULI

LEAD TO HOST HABITAT?    NO

YES

HOST HABITAT FOUND

OLFACTORY & VISUAL STIMULI

LEAD TO HOST?    NO

YES

HOST FOUND

OLFACTORY, GUSTATORY, & TACTILE STIMULI

LEAD TO ACCEPTANCE OF HOST?    NO

YES

HOST ACCEPTED

NUTRIENTS & ANTI-BIOTICS

NUTRIENTS PRESENT & ANTIBIOTICS ABSENT?    NO    ANIMAL DIES

YES

HOST SUITABLE

NEW ADULT GENERATION EMERGES

EFFECT ON BEHAVIORAL PROCESSES

EFFECT ON METABOLIC PROCESSES

*(a)*

116

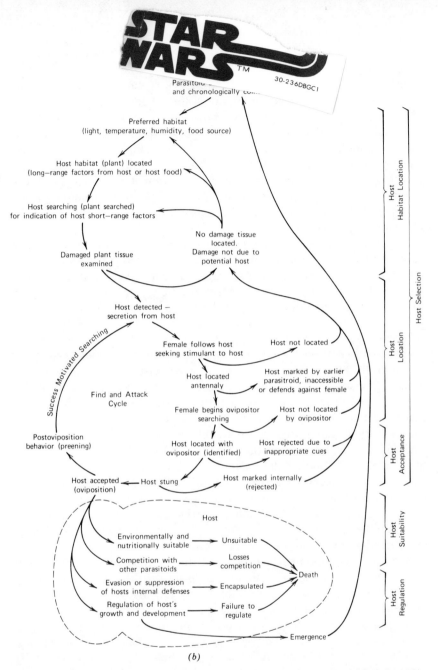

**Figure 4-3** Diagrammatic representation of the food selection process of (*a*) a phytophagous insect and (*b*) a parasitic insect. At each step, a decision involving sensorial receptors and integrative neural processes must be made. (A from Kogan, 1975; B courtesy of S. B. Vinson.)

reflected in the common name they are given. Tobacco hornworm moth larvae, for example, feed on various solanaceous plants but prefer tobacco or tomato. Some insects, however, accept a wide variety of foods, although they may still show decided preferences (**polyphagy**). Migratory locusts are a familiar and striking example. Robber flies and preying mantises are polyphagous predators, while cockroaches are polyphagous detritivores. And, as if for completeness, some insects exhibit strict specificity to one food, often a single host (**monophagy**). An example is the nocturnal halictid, *Sphecodogastra texana* (see Fig. 2-12), which gathers its pollen exclusively from the evening primrose. Others include certain leaf-mining caterpillars and gall insects which can develop successfully on only one species of host plant. Parasitic species in which both adult and immature stages are dependent on the host tend to also show a high degree of host specificity. A few monophagous specialists occur in almost every insect group.

Much of the support enjoyed by entomologists derives from the economic importance of insect feeding upon man's food plants, fiber sources, stored foods, waste products, farm animals, and even upon man himself. In turn, the propensity of other insects to feed upon man's pests has made them noticeable allies. However, no matter what the diet—and with the exception of coal there is hardly a source of organic carbon not used to some extent by some insects—the ability to feed successfully resolves into a remarkably constant chain of behaviors, each link of which facilitates the next (Fig. 4-3). Five major links may be recognized*: (1) food habitat location; (2) food finding; (3) food recognition; (4) food acceptance; (5) food suitability. While this chain can involve any or all of the senses, the chemical sensory systems predominate. The first two links require energy expenditure, sometimes a considerable amount. However, not surprisingly, most applied studies of insect feeding focus on the third and fourth steps in the chain.

Consider a common herbivorous pest, the potato beetle, *Leptinotarsa decemlineata*. Although the mother beetle actively chooses a site for egg laying, the larvae do not usually begin feeding near the place of hatching. Instead, they begin searching, guided mainly by vision, until they encounter a potato plant; at this point, they begin randomly biting the substrate over which they are walking. Close-range olfactory stimuli emanating from the plant are attractive to the larva. If the proper host plant stimuli are present, the larva clips off any hairs on the plant

---

* For some parasitic species, a sixth link—host regulation—may operate by means of which the host's physiological development is accelerated, retarded, or otherwise modified by the parasite's presence (Vinson, 1975).

surface and pierces the epidermis. Whether the plant is rejected or accepted depends upon taste stimuli. Finally, the nutritional value of the plant and the presence or absence of antibiotic factors determine whether the plant is adequate to sustain physiological growth and development processes (Hsaio, 1974).

The behavioral sequence just outlined is far from unique. For example, the prey-catching behavior sequence of solitary wasps is fundamentally similar. The bee wolf, *Philanthus triangulum*, hunts honey bees which it captures and paralyzes with its sting prior to transport to the nest, in a chain of behavioral reactions in which each distinct link is mediated by a different sensory mode. As Tinbergen (1972) has described the hunting behavior sequence:

"A hunting female of this species flies from flower to flower in search of a bee. In this phase she is entirely indifferent to the scent of bees; a concealed bee, or even a score of them put out of sight into an open tube so that the odor escaping from this is clearly discernible even for the human nose, fails to attract her attention. Any visual stimulus supplied by a moving object of approximately the right size, whether it be a small fly, a large bumble bee, or a honey bee, releases the first reaction. The wasp at once turns her head to the quarry and takes a position at about 10–15 cm to leeward of it, hovering in the air like a syrphid fly. Experiments with dummies show that from now on the wasp is very susceptible to bee scent. Dummies that do not have bee-odor are at once abandoned, but those dummies that have the right scent release the second reaction of the chain. This second reaction is a flash-like leap to seize the bee. The third reaction, the actual delivery of the sting, cannot be released by these simple dummies and is probably dependent on new stimuli, probably of a tactile nature."

It is apparent that the wasp's feeding behavior is dependent upon successive reactions to "bee look," "bee smell," and "bee feel." If the first test is satisfied, it leads to the second, the second to the third, and so on. While chemical cues are important, stimuli from other sensory modes are also brought into play and integrated into the sequence of feeding behaviors.

## HERBIVORES AND PLANTS: COEVOLUTIONARY STRATEGIES

In northeastern Australia in the early 1900s, adventive prickly pear cactus almost ruined the range lands until the successful introduction of a moth whose larvae feed exclusively upon the cactus. Throughout the world, pollen- and nectar-hungry bees, wasps, butterflies, flies, and beetles visit flower blossoms thus providing an essential service to many plants. Beneath the ground, some ants culture extensive "gardens" of unique fungi; other ant species habitually tend sedentary aphids, protecting them from predators while harvesting the sugar-rich honeydew the aphids excrete. Unexpected, often bizarre insect–plant relationships also exist (Fig. 4-4).

Obtaining and maintaining an energy pool is the major preoccupation of every individual in any species; to share this energy with organisms of other species is evolutionarily maladaptive unless one gains, or expects to gain, at least indirect benefit in return. Therefore, protecting one's own energy pool against other would-be consumers is a necessity while concurrently attempting to break through the defenses of other species, particularly those at a lower trophic level, to take advantage of a lucrative potential energy source. For each species, this double-edged sword results in a more or less continual "battle" of attack and defense or counterattack waged over the ultimate trophy of enough energy for successful reproduction. As in war, strange alliances also sometimes result. When two or more populations interact so closely that each acts as a strong selective force on the evolution of the other, reciprocal stepwise adjustments occur that result in coadaptation or, as it is more commonly called, **coevolution**.

Most Americans employ the term **symbiosis**, or living together, in the broad sense, embracing all close and protracted interactions between individuals of different species. Three principal types of ectosymbioses

**Figure 4-4**  A flightless, long-lived *Gymnophilus* weevil from mountaintop cloud forests in New Guinea, showing the fungal and algal growth covering the pronotum and elytra. Specialized depressions on the weevil's back appear to be associated with encouragement of the plant growth. Certain mites, nematodes, rotifers, psocopterans, and diatoms occur among the plants. (See Gressitt et al., 1963. Photograph by the authors.)

are commonly recognized: **commensalism**, in which one species benefits and the other is unaffected; **mutualism**, in which both species benefit; and **parasitism/predation**, in which one (the parasite/predator) benefits at the expense of the other (host). Intergrades between these occur, and it is not always possible to decide to which category a particular symbiosis belongs. Also, each category is subject to subdivision. For example, under parasitism, one can distinguish between parasitoids which consume the host's larvae and those which develop primarily upon the host's food store, and social parasites which steal labor rather than food. We consider only nonsocial associates here, reserving symbioses between fully social species for Chapter 10.

Evidence such as the dramatic insect control of cactus in Australia and Klamath weed in the northwestern United States demonstrates that insect herbivory at least has the potential to act as a powerful selective agent on plants (Peschken, 1972). Likewise, outbreaks of agricultural and forest pests emphasize the selective potential of insect feeding (Kulman, 1971). That the challenge of herbivores should be regarded as a *major* selective pressure in plant evolution was perhaps first explicitly emphasized in the mid-1960s by Ehrlich and Raven (1964, 1967). In the years since, this aspect of insect–plant coevolution has emerged as a major interdisciplinary focus, and the number of pertinent publications and symposia have accumulated at a geometric rate. However, it is important to remember that plants are under *multiple* evolutionary pressures, as are the insects. Broadly speaking, the effects of herbivores on plants are not unlike those of parasites on their hosts. In essence, then, this is but another example of the coevolutionary interaction referred to as parasitism/predation.

Plant species have been likened to islands defended by chemical, mechanical, and biological barriers penetration of which may allow an insect species to exploit a new "adaptive zone" (Janzen, 1968; Opler, 1974). Presumably, the metabolic costs of maintaining these barriers and the depredations of those species which have overcome and even exploited them are more than offset by the general degree of protection these defenses provide against other potential herbivores and pathogens. One's meat is another's poison. According to coevolutionary theory, a succession of defenses against herbivores is produced in a stepwise fashion by a particular plant species. At each step, some herbivores go extinct, while others circumvent the botanical innovations, forcing the plant to evolve further deterrents. Through time, the process may be regarded as a series of filters or revolving doors through which the plant and remaining successful herbivore taxa are alternately "released" for bouts of speciation and adaptive radiation. The behavioral flexibility of insects has undoubtedly been one of their most important assets for achieving coevolved relationships with plants.

## Biochemical Deterrents

It is clear that plant odor normally mediates the first steps in herbivory, determining whether the insect detects the plant as a potential host for sampling and whether it takes the first bite. As Dethier has said (1970b), "The first barrier to be overcome in the insect/plant relationship is a behavioral one. The insect must sense and discriminate before nutritional and toxic factors become operative." Removal or blocking of olfactory organs in a variety of insects, for example, leads to acceptance of nonhost plants otherwise treated as inedible.

Inedible should not be equated with nonnutritious, as is demonstrated by some experimental studies on sinigrin (Nault and Styer, 1972). This mustard oil glucoside, characteristic of the plant family Cruciferae, acts as a feeding stimulant for the turnip aphid and the cabbage aphid. When leaves of 10 nonhost plant species were treated with sinigrin, aphids readily fed upon all, and five turned out to be nutritionally adequate to support the growth and development of at least one generation of aphids.    Considerable experimental and observational evidence confirms that many plant chemicals are utilized by insects as distinctive cues to identify and discriminate among their host plants for feeding and breeding. These chemicals are of almost universal occurrence in plants and are called **secondary substances**—an array of diverse chemicals not known to have any function in plant growth or metabolism. These include alkaloids, terpenoids, essential oils, and quinones. Even though they may differ widely in other respects, the various food plants of a particular insect species, genus, or even family often share similar secondary substances. The concentration of these chemicals often differs in different parts of the plant. Intuitively, for example, one might predict that seed eaters would be likely to exert considerably stronger selection pressures than leaf feeders, since in the former case the vital reproductive units of the plants are being directly assaulted. For example, the natural insecticide pyrethrin is concentrated in flower heads.

Historically, secondary plant substances have been a subject of intense controversy among biologists. One view holds that these are either metabolic by-products (waste products) or metabolic precursors needed for some as yet unknown physiological functions. Another school contends that such metabolically expensive "curiosities," which are often produced in large quantities, have a *primary* significance to the plant as defensive substances, and to label them as "secondary" is in a sense an ironic injustice.* From the 1950s on, evidence has rapidly accumulated in favor of the latter interpretation.

Secondary plant substances have a variety of other interesting effects on the behavior of insects. Some parasites of phytophagous insects first find their host's habitat through attraction to secondary plant chemicals. For example, the primary parasite of the cabbage aphid initially responds to the odor of sinigrin, the same compound that serves as a feeding stimulant for its host (Read et al., 1970). Sometimes, plant chemicals also provide the herbivore with protection against its enemies. Monarch butterfly larvae sequester vertebrate heart poisons from the milkweed host plant, which chemicals later serve to confer protection from birds (Reichstein et al., 1968; see Fig. 8-9). In other cases, secondary compounds are used in sexual communication. For example, male butterflies of many species of Danainae and Ithomiinae congregate to feed at the dead shoots of plants containing dehydropyrrolizine alkaloids and then biochemically modify the ingested alkaloids to produce aphrodisiacs used during courtship (Pliske, 1975). Likewise, virgin female polyphemus moths will not begin to call unless stimulated by *trans*-2-hexenal which emanates from leaves of their host plant, oak; in other tree species, the activity of this chemical is apparently masked by other odors (Riddiford, 1967).

Insect response to plant chemicals falls into two categories: immediate and delayed reactions (Table 4-2). Immediate responses are basically behavioral, whereas delayed reactions are largely physiological and include developmental anomalies, toxicity effects, and hormonal changes.

It is of interest to consider whether any plants exist that are relatively free from insect attack. Two such cases are sometimes cited: the ginkgo tree and ferns, both survivors of relatively ancient plant groups. Indeed, it has been suggested that the reason the ginkgo still survives is that it has outlived all of its major enemies (Major, 1967). Generations of readers have used ginkgo leaves as bookmarks because of their reputed ability to ward off silverfish and booklice. Chemical analysis of ginko leaves has confirmed the validity of such practice by revealing the leaves to be highly acidic and inhibitory to insect feeding.

Although popularly thought to be free of insect enemies, many ferns are in fact regularly attacked by insects ranging from bark beetles to

---

* Some authorities prefer the term "plant natural products" for this reason, but usage of "secondary substance" seems to be more widely accepted and understood.

**Table 4-2**    Some ways in which plant-produced chemicals may exert effects on phytophagous insects.

| Chemical type | Class of effects | Response |
|---|---|---|
| **Biosemants:** Those affecting behavioral responses (immediate) resulting from detection | Repellents | Cause oriented movements away from source |
| | Attractants | Cause oriented movements toward the source |
| | Arrestants | Cause aggregation at source |
| | Stimulants | Elicit feeding, mating, or oviposition |
| | Deterrents and suppressants | Inhibit feeding, mating, or oviposition after attraction to source |
| **Antibiotics:** Those causing adverse physiological effects (delayed) resulting from ingestion | Phytohormones | Cause developmental anomalies such as failure to molt or metamorphose properly |
| | Growth regulators | Inhibit or accelerate growth and/or development |
| | Toxins | Cause sickness or sometimes death |
| | Nutrients | Quality may reduce fecundity and longevity |
| | Sterilants | Incapacitate reproductive system (none presently reported) |

lepidopteran larvae (Fig. 4-5). Nevertheless, the fact that common and widespread ferns such as the bracken fern apparently suffer little attack by herbivorous insects led to a study which revealed the presence of high concentrations of two major insect molting hormones, alpha-ecdysone and 20-hydroxyecdysone (Kaplanis et al., 1967). Such plant-produced ecdysones are now known to be of widespread occurrence, especially in the ferns (Polypodiaceae) and in two families of gymnos-

**Figure 4-5**  Christmas fern frond bunch knotted (corner insert) at the tip as a result of feeding by a pyralid moth larva (arrow). None of the vascular bundles of the uncoiled rhachis has actually been cut by the feeding larva. (Photograph by the authors.)

perms (the Taxaceae and Podocarpaceae) (Williams, 1970). These hormone mimics may function as potent deterrents and antifeeding agents; in some cases they are effective at dosages strikingly less than expected. Some insects have evolved enzymes capable of detoxifying such compounds. Polyphagous species appear more apt to produce mixed-function oxidases than do monophagous ones (Krieger et al., 1971).

Another major class of insect hormones, the juvenile hormones, have also been mimicked by certain plants. A classic example of phytohormonal effects is exemplified in the so-called "paper factor" story (Williams, 1970). The setting for this episode was the Harvard University laboratory of C. M. Williams in 1964, when K. Sláma came from Czechoslovakia to spend a year and brought along his favorite laboratory insect, a native European bug, *Pyrrhocoris apterus*. Very soon it became clear that cultures of *Pyrrhocoris* were not faring well in William's laboratory. Instead of metamorphosing into adults at the end of the fifth instar, they molted to an extra larval stage to form giant sixth instar nymphs, which ultimately died without becoming sexually mature.

Every precaution had been taken to eliminate all sources of hormone contamination from the culture containers, yet the problem seemed to be clearly of a hormonal nature. Baffled, Sláma and Williams undertook a systematic analysis of every item in the bug's culture environment, searching for evidence of suspected juvenile hormone. By a process of elimination the culprit was at last found—the paper toweling used to line the cage floors. Substitution of the filter paper that Sláma had always used in Prague completely eliminated the problem. Further detective work traced the origin of the hormone to paper pulp derived from the Balsam fir (*Abies balsamea*), a principal pulp tree indigenous to North America. This bit of serendipity was the first indication of the existence of juvenile hormonal materials occurring naturally in plants. As was true for ecdysone, juvenile hormone analogs have since turned up in a diverse array of plants.

## Mechanical Deterrents

Foliage of the Neotropical passion flower vines is eaten by larvae of flashy heliconiine butterflies. However, when vines of *Passiflora adenopoda* were exposed to *Heliconius* attack in the laboratory, no feeding damage occurred, and the larvae that had been placed on the plants were found dead and desiccated by the following day, although other *Passiflora* species similarly exposed sustained heavy feeding damage. The cause of the immunity of *P. adenopoda* was traced to the cloak of hooked trichomes (hairs) which cover the plant's surface. Scanning electron micrographs verified the defensive function of these trichomes which made numerous puncture wounds in the larval integument, immobilizing them and causing them to starve to death (Gilbert, 1971; Rathcke and Poole, 1975).

An arsenal of spines, thorns, pubescences, and tough cuticles characterizes many plants from cactus to acacias. Such structural features of plant surfaces have long been assumed to confer a certain measure of resistance to herbivore attack, and their prevalence seems to bear testimony to the intensity of herbivore pressure, though some may have arisen in response to other selective pressures. For example, the waxy texture of succulent leaves undoubtedly helps to prevent desiccation. Some types of trichomes are glandular and secrete various chemicals, including many secondary plant substances. In some cases these secretions appear to have a simple function of mechanical entanglement (Fig. 4-6). Levin (1973) further discusses the role of trichomes in plant defense.

In different strains of crop plants ovipositional behavior of some female insect pests has been shown to vary according to density of pubescence. Hence, plant breeders are well aware of the importance of these kinds of physical defenses in plant resistance. For example, gravid female cereal leaf beetles lay significantly more eggs on smooth-leaved strains of wheat than on densely pubescent forms; furthermore, larval

**Figure 4-6**  Leaves of an Australian sundew plant, which possess large, stalked glands with a sticky secretion that have trapped a small moth (arrow). Interestingly, an American *Drosera* is attacked by caterpillars of a plume moth, *Trichoptius parvulus*, which feeds with impunity on the leaves, including the glands and their secretions, and even consumes dead insects trapped by the plant (see Eisner and Sheppard, 1965). (Photograph by the authors.)

survival is only 20% on densely pubescent plants compared to 92% on glabrous strains (Kogan, 1975). Other types of physical barriers to herbivore attack may be more subtle. Various morphological traits of the bean family (Leguminosae)—pods with a flaky surface that scale off eggs, gummy sap, pods that explode when penetrated, scattering the seeds, dense, hard seed coats, etc.—all lower the success of bruchid (pea and bean) weevils that attempt to oviposit in or on the seed pods (Janzen, 1969a).

## Mutualism as a Biological Defense

One of the stranger types of defense employed by plants against herbivores involves other organisms as a front guard or standing army. These mercenaries are paid, in turn, through benefits they receive from the plant, and a mutualistic relationship results. Some of the best illustrations may be found among the ants, many of which have become specifically adapted to live upon certain plants (Hocking, 1975). For example, in the Tropics, ants obligatorily restricted to living in association with particular plants have independently evolved at least 13 times.

The mutualism between these plant ants and ant plants is highly coevolved and costs each participant a substantial energy outlay, but each also derives appropriate benefits. In a very real sense, plant ants may be regarded as analogous to the secondary compounds found in the foliage of most plants, the primary defense of their plant against herbivores.

## CASE STUDY: MUTUALISM OF *Pseudomyrmex* ANTS AND ACACIAS

In disturbed areas in the lowlands of Mexico and Central America, a common shrubby tree is the bull's horn acacia, *Acacia cornigera*, so named for the pair of swollen, hornlike thorns that occur at the base of most of its leaves. A rapidly growing woody plant that cannot tolerate shading, it quickly springs forth as sucker growth from old root stocks in pastures, along roadsides and in natural disturbance sites such as river banks and arroyos. Close examination of the plants reveals that they crawl with small *Pseudomyrmex* ants, which pour forth from holes in the thorns whenever the acacia is touched (Fig. 4-7).

**Figure 4-7** Nest compartment of *Pseudomyrmex* ants in swollen thorn of *Acacia*, opened at the base to show brood and workers inside. The entrance hole near the tip is clearly visible, as are extrafloral nectaries appearing as swellings on the adjacent leaf petioles. (Courtesy of R. E. Silberglied.)

A hundred years ago in Nicaragua, the naturalist Thomas Belt discovered a similar relationship between an ant and acacia and noted that the ends of each leaf segment were modified into peculiar little oval structures (Fig. 4-8). These "Beltian bodies" seldom survived long, for the ants soon cut them off and either ate them or fed them to their young. Subsequent studies have shown that they have an unusually high food value for foliar tissue, on the order of yeast in quantity and quality of nutrients.

Following Belt's discovery there was a great deal of armchair speculation on the exact nature of the relationship between ant acacias and acacia ants. That the ants depend upon the acacias was clear, for most

**Figure 4-8**  Portion of an *Acacia* showing a heavy production of Beltian bodies on the tips of the leaflets which are being harvested by *Pseudomyrmex* workers. (Courtesy of D. H. Janzen.)

acacia ant species had been recorded only from living ant acacias. But do the acacias benefit from the presence of the ants? Two viewpoints persisted: one stated that the ants were merely exploiting the acacia, while the other regarded the relationship as a true symbiosis. This question became the focus of intensive eastern Mexican field studies by Janzen (1966, 1967). One of Janzen's early observations was that these acacias were commonly left unbrowsed in pastures with cows and other domestic animals. From a local farmer he borrowed a pet native deer that was thought not to have had any previous experience with *A. cornigera*. When he offered it foliage that had been cleaned of all ants, the deer ate the foliage readily, including some of the thorns. After several days of feeding the deer unoccupied foliage, Janzen offered a branch complete with some very agitated ant workers. As the deer began to eat the foliage, ants ran onto her face and stung; immediately she stopped feeding and withdrew to clean them off. Results of similar simple experiments with cows and burros (who may have had previous experience with *A. cornigera*) did not differ materially but suggested additionally that these herbivores may learn to recognize the alarm odor of the ants and learn to avoid contact with the plant.

The survival of the bull's horn acacia is dependent upon rapid growth so as to remain unshaded. Thus, Janzen began to examine the relative growth and development of occupied versus unoccupied shoots of *A. cornigera* by treating selected shoots with an insecticide.

A number of striking differences emerged between the experimentally defaunated shoots and those with intact *Pseudomyrmex* colonies. First, the frequency and extent of phytophagous insect damage was greatly increased in shoots from which ants were removed, which resulted in a great lowering of the growth rate of the shoots. In contrast, ant-occupied shoots remained virtually free of phytophagous insects, for the latter were quickly attacked and removed from the shoot by worker ants. Janzen found that at least 40 species of insects fed on unoccupied shoots of *A. cornigera*, whereas only eight species were not efficiently deterred from feeding on occupied acacia shoots, and even the cumulative sum of their feeding was not serious. Phytophagous insect activity increased greatly during the rainy season; differential growth rates of occupied versus unoccupied shoots were even more striking, with unoccupied shoots showing almost no growth while the occupied shoots grew vigorously. With a lowered growth rate, the unoccupied acacias quickly became shaded by surrounding vegetation which further contributed to stunting of growth. A high mortality of unoccupied shoots also occurred.

Data from experimental plots also confirmed another observation that Janzen had made, namely, that occupied shoots are almost invariably free of living vines although the acacia plant appeared ideal for the support of vine growth. Whenever foreign vegetation contacted the acacia plant, the ants would maul it and chew off the growing tips. In this way they prevented vines and lateral branches of neighboring plants from growing into the canopy of occupied *A. cornigera*. Unoccupied acacias, in contrast, accumulated heavy masses of vines during the rainy season. Vegetation-mauling activity of worker ants also resulted in a bare circle on the ground about the base of the shoot. Janzen felt that this behavior was of significance in protecting the shoot from fires as well as lowering the incidence of phytophagous insects reaching the shoot. Even when the shoot was killed, suckers from protected root-stock in the bare basal circle quickly sprouted and were immediately colonized by the mature ant colony, giving the new shoot an immediate competitive advantage in the postfire succession.

Based on these data and observation of occasional naturally unoccupied acacias, Janzen concluded that a shoot of *A. cornigera* must be occupied by a colony of *P. ferruginea* for a substantial part of its life to produce seeds and become a part of the reproductive population. While morphologically *P. ferruginea* is a more or less representative pseudo-myrmecine ant, behaviorally it possesses several unique characteristics associated with its interaction with swollen-thorn acacias (Table 4-3). Because of the interdependence of the ant and acacia for normal population development, the interaction between them can be properly called one of obligatory mutualism.

Recently, Janzen (1975b) has discovered a "parasite" of the mutualism, *Pseudomyrmex nigropilosa*, which harvests the resources of the swollen-thorn acacias but does not exhibit the behaviors that serve to protect the acacia, neither attacking foreign objects nor cleaning its foliage of debris. Its colony structure has adjusted to a parasitic way of life by producing fewer workers and by producing reproductives sooner than does a colony of the long-lived obligate acacia ants.

Many other ant–plant relationships have been recorded (Bequaert, 1922; Wheeler, 1942). Each will require the same sort of detailed investigation in order to understand the dynamics and evolutionary adaptations on the part of the participants before they can be truly termed mutualisms. One of these is the interaction between *Cecropia* plants and certain species of *Azteca* ants throughout Central America (Fig. 4-9). Like the acacia, *Cecropia* produces glycogen-rich food bodies (trichilia or Mullerian bodies) which are harvested by the ants.

There are, of course, many forms of mutualism between ants and other organisms. One of the best known is that between ants and a diverse assortment of sap-sucking Homoptera (Way, 1963). These in-

**Table 4-3**     Coevolved traits of *Pseudomyrmex* ants having obligate relationships with acacia plants compared to those of species not involved in obligatory mutualistic relationships with plants[a]

| A. General features of *Pseudomyrmex* of importance to the interaction | B. Specialized features of obligate acacia ants (coevolved traits) |
|---|---|
| 1. Fast and agile runners, not aggressive | 1. Very fast and agile runners, aggressive |
| 2. Good vision | 2. Same as A 2 |
| 3. Independent foragers | 3. Same as A 3 |
| 4. Smooth sting, barbed sting sheath not inserted | 4. Smooth sting, barbed sting sheath often inserted |
| 5. Lick substrate, form buccal pellet | 5. Same as A 5 |
| 6. Prey items retrieved entire | 6. Same as A 6 |
| 7. Ignore living vegetation | 7. Maul living vegetation contacting the swollen-thorn acacia |
| 8. Workers without morphological castes | 8. Same as A 8 |
| 9. Arboreal colony | 9. Same as A 9 |
| 10. Highly mobile colony | 10. Same as A 10 |
| 11. Larvae resistant to mortality by starvation | 11. Same as A 11 |
| 12. One queen per colony | 12. Sometimes more than one queen per colony |
| 13. Colonies small | 13. Colonies large |
| 14. Diurnal activity outside nest | 14. 24-Hour activity outside nest |
| 15. Few workers per unit plant surface | 15. Many workers active on small plant surface area |
| 16. Discontinuous food sources and unpredictable new nest site | 16. Continuous food source and predictable new nest sites |
| 17. Founding queens forage far for food | 17. Founding queens forage short distances for food |
| 18. Not dependent on another species | 18. Dependent on another species group |

[a] From Janzen (1966).

sects excrete a sugary substance, called honeydew, which is avidly collected by ants as well as many other insects. Certain aphids, in particular, by virtue of their gregarious behavior produce prodigious amounts of this semiprocessed phloem sap. The concentrated food bonanza is exploited in a highly systematic fashion by certain ants that has been likened to cattle farming by some early naturalists. In return for the honeydew, which is produced on demand in response to stroking

**Figure 4-9** A section of the stem of a *Cecropia* plant showing the *Azteca* ants and their queen residing in the hollow stem internodes. Weaknesses in the internode wall permit connections between chambers. Ants provide protection from herbivores, which "compensates" the plant for the energy expended in the maintenance of the ant colony. The same species of *Cecropia* occurs on the island of Puerto Rico; here, the trees are unoccupied by *Azteca* or any other ant and lack trichilia. (See Janzen, 1969a; courtesy of R. L. Jeanne.)

by the ant, the aphids receive protection from their predators and other services (see Fig. 5-12).

That the aphids may themselves be involved in a mutualistic relationship with the host plant has recently been suggested by Owen and Wiegert (1976), who propose that consumers such as aphids often have the overall effect of *increasing* plant fitness and that plants accordingly regulate and even "solicit" a spectrum of consumers. Owen and Wiegert regard nitrogen supply to be one of the main factors limiting a plant's productivity and reproductive success. The large quantity of honeydew excreted by aphids stimulates free nitrogen fixation beneath the plant which speeds up the cycling of this essential nutrient, increasing productivity. Such a system, if supported by experimental evidence, will require modification of the idea that plants are fighting an evolutionary battle with consumers and that their adaptations are primarily defensive in nature.

### Mutualism for Pollination

Mutualism between plants and insects for herbivore defense is rather exceptional. The most common insect–plant mutualisms have evolved in the context of pollination. Insect-pollinated plants may be grouped into several insect–flower syndromes (Table 4-4). Each has characteristic flower morphology and a correspondingly adapted major group of insect

**Table 4-4** Insect–flower syndromes. Shape, color and scent of flowers are adaptations for attracting or excluding various classes of pollinators[a]

| Pollination classes | Types of insects | Anthesis | Predominant colors | Odor | Flower shape | Flower depth | Nectar (honey) guides | Nectar | Other |
|---|---|---|---|---|---|---|---|---|---|
| 1. Cantharophily | Beetles | Day and night | Variable, usually dull | Strong, fruity or aminoid | Actinomorphic | Flat to bowl shaped (rarely closed) | None | None or open | Pollen or food bodies. Flower parts in large numbers |
| 2. Sapromyophily | Carrion and dung flies | Day and night | Purple-brown or greenish | Strong, often of decaying protein | Usually actinomorphic | None or deep if of trap type | None | Open or none (sometimes pseudonectaries) | Often no food provided. Transparent windows or other features contribute to temporary traps. Mobile appendages to flowers or "tails" to petals often present |

| | | | | | | | | |
|---|---|---|---|---|---|---|---|---|
| 3. Myophily | Syrphid and bombyliid flies | Day and night | Variable | Variable | Usually actinomorphic | None to moderate | None | None or present | Bombyliids have tongues to 50 mm; syrphids can chew pollen |
| 4. Melittophily (several subclasses) | Bees | Day and night or diurnal | Variable except pure red | Present; usually sweet | Actinomorphic to zygomorphic; morphology of flowers can be recognized | None to moderate; often a broad tube (gullet) | Present | None or present (usual); open to concealed | Larger bees can open "closed" flowers. The most variable of the syndromes |
| 5. Sphingophily | Hawkmoths | Usually nocturnal or crepuscular | Usually white or pale (or green) | Strong; usually sweet | Usually actinomorphic; held horozontal or pendent | Deep narrow tube to corolla (or spur) | Usually none | Present; concealed: ample | Rim of corolla may be reflexed and dissected, often no alighting platform. Anthers versatile |
| 6. Phalaenophily | Small moths | As in 4 | As in 4 | May be less strong than in 4 | As in 4 | Less deep than in 4 | None | Present; concealed | Less extreme modifications than in 4 |
| 7. Psychophily | Butterflies | Day and night or diurnal | Variable (pink being very common) | Moderately strong; sweet | Usually actinomorphic; usually held upright | Deep; narrow tube to corolla (or spur) | Often present | Present; concealed: ample | Usually flat wide margin to corolla tube; not dissected |

pollinators. Authorities are in general agreement that the first insects to become anthophilous were the Coleoptera, with the habit appearing later in the Lepidoptera and Diptera and reaching its culmination in the Hymenoptera, especially the bees (Baker and Hurd, 1968).

A fig is more than just a fruit; it actually represents a complete influorescence, like a deep cup with flowers placed on the inside. (The tiny hard grains we notice in the dried fig are actually seeds of the individual fruits.) Certain tiny wasps belonging to the family Agaonidae (Chalcidoidea) (Fig. 4-10) are found only in association with figs and have evolved a perfectly timed, intricate relationship so highly developed that if one partner should die out, the other would inevitably follow—a beautiful case of mutualism in which the fig obtains pollination services in return for providing a nursery and food source for the production of new generations of agaonid wasps.

The life cycle of the Smyrna fig may be taken as representative of this complex symbiosis although in some ways it is simpler than in the

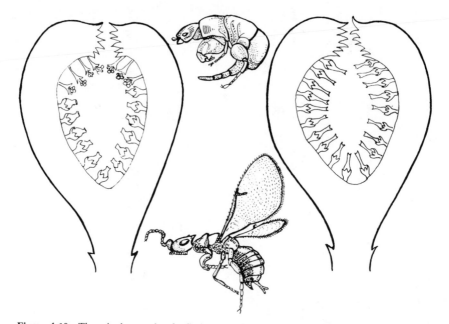

**Figure 4-10** The wingless male of a fig insect, *Blastophaga psenes,* above, and the winged female, below. On the left, a diagram of the inedible "goat fig" in longitudinal section is shown. The short-necked female flowers, where the wasp galls are formed, line the bottom of the cavity while the male flowers are clustered around the entrance to the cavity. On the right, a section through the edible fig, which contains only long-necked female flowers. (From Wigglesworth, 1964.)

majority of other fig species. In the spring, a wild fig tree produces inedible figs called "goat figs" or "profichi," each of which contain numbers of both male and female flowers. However, the two flower sexes are not synchronized in development; female flowers develop first and are receptive to pollination and oviposition at the time of female penetration. Through an intricate pore, females enter the fig and lay eggs inside the modified short-styled female flowers. Flowers into which eggs are deposited develop into tiny galls, each with a developing wasp larva feeding inside.

In the summer, as the profichi reach maturity, so do the wasps within. Wingless males chew holes in the sides of their galls, through which they

**Table 4-5**   Coadaptations between figs *(Ficus)* and fig wasps (Agaonidae)[a]

| *Ficus* | Fig wasps |
|---|---|
| 1. Intricate ostiola, which permit only specialized pollinators to enter and exclude predators such as ants | 1. Extreme morphological adaptations in both sexes (see Fig. 4-10): |
| | Flattened heads and bodies |
| 2. Female flowers adapted for development of symbiont larvae in "galls" | Mandibular appendages |
| | Curved spines on second antenna segment and teeth on the fore and hind tibia |
| 3. Asynchronous development of the two flower sexes—female flowers ripen at time wasps enter fig and oviposit, but male flowers in same inflorescence ripen with the emergence of the young adult wasps | Development of specialized corbiculae (pollen baskets) for pollen transport in several genera |
| 4. Ripening of fig delayed until after wasp emergence, thereby minimizing pollinator mortality while at the same time allowing other animals (birds, bats) to serve as dispersal agents | Correlation of female ovipositor length with style length of host flowers |
| | Telescoping abdomen of males for copulating with preemergent females inside their development galls |
| 5. Year-around fig production so figs in all stages are available among a population of plants, although those of a particular tree tend to be synchronized in their development | 2. Sex ratios strongly biased in favor of females and strong sexual dimorphism (see Fig. 4-10) |

[a] Adapted from Ramirez (1970).

escape. Finding females still imprisoned in adjoining galls they chew into these and quickly mate with their sisters. To leave the fig where they were born, the fertile female wasps must pass through a ring of male flowers surrounding the fig's entrance, becoming well dusted with pollen as they pass. The female wasps search for new figs in which to oviposit, but encounter an evolutionary surprise—the tree is now producing not profichi, but edible figs instead. These "fichi" have large numbers of female flowers with long styles. Ovipositing in them is all but impossible, but in them repeated vain attempts the females successfully pollinate all the flowers in the fig. Leaving, each female vainly continues to attempt egg deposition until a whole succession of edible figs have been pollinated.

Finally, in early fall, the fig tree begins to produce still another type of fig, small, inedible, and found only on its upper branches. These "mamme" or mother figs contain only short-styled female flowers in which the female wasps are finally able to oviposit. Young male and female wasp larvae hibernate within, to begin the profichi–fichi–mamme cycle anew the following spring. The fig–agaonid relationship is a closely coevolved one (Table 4-5). Those interested in additional information on this unusual symbosis may consult the papers of Baker (1961), Ramirez (1970), Galil and Eisikowitch (1968), and Galil et al. (1973).

## FOOD LOCATION

Procuring and recognizing food are items of overwhelming importance on the life agenda of all insects, whether they live by capturing and/or parasitizing prey, chewing or sucking upon plants, or scavenging. Let us examine a few strategies by which this problem has been approached.

### Active Search

Blood-sucking Diptera can be attracted with $CO_2$. Phytophagous aphids come to light of longer wavelengths, especially yellows and greens. Dragonfly larvae often stalk, then snap at, nearby moving objects within certain size limits; hungry preying mantises behave in a similar fashion.

What of those species in which the mother oviposits directly on or in the plant (e.g., gall midges) or animal host (e.g., parasitic wasps)? Location of food would appear to be no crisis for these species, for the young are literally surrounded by food. But in reality the burden of food location has simply been shifted to another life stage.

For by far the majority of insects food location involves some manner of active search such as we saw for *Philanthus*. It is usually mediated by several sensory modes; visual cues often operate over a distance, with chemical cues called into play over shorter ranges. Movement is probably the most common attribute which predatory insects exploit in initial recognition of potential prey. The most widely shared characteristic of prey animals, movement is particularly useful to relatively polyphagous predators.

Because active search for food requires energy expenditure and time involvement, one would expect selection to favor behaviors that increase its efficiency. In organisms such as bumble bees, which rely on nectar for their energy, foraging behaviors can be directly analyzed in terms of caloric costs and benefits (Heinrich, 1973, 1975; Heinrich and Raven, 1972). The amount of food which can be obtained from any one flower is rather limited, and for maximal efficiency one might expect bees to be able to exploit new food resources as they become available and to differentiate between more and less rewarding flowers in bloom at any one time.

Bumble bees generally collect pollen from several plants simultaneously, and as Heinrich (1976) explains it, "They are playing a game analogous to market investment: they do not know beforehand which is the best commodity (flower) and their best strategy is to invest primarily in the flower that appears to be the most remunerative while simultaneously investing some energy in several "minor" species." By experimentally enriching certain "minor" flowers with sugar syrup, Heinrich has been able to demonstrate that an individual forager will immediately switch from its previously most preferred flower and adopt the fortified flower for as long as the sugar syrup is added.

The widely held view of nectar as "just sugar water" has been revised (Baker and Baker, 1973). Part of the adaptive significance of nectar lies in the considerable quantities of various amino acids needed nutritionally by adult insects who are apparently unable to synthesize them (Table 4-6).

When food is mobile, it is advantageous to catch more food in less time, to catch larger food items in the same time required to take smaller ones, and to avoid chasing food or hosts likely to escape or prove unsatisfactory for some other reason. The variety of means by which such ends have been approached is nearly limitless. Where food items are stationary, selection has favored increased speed and efficiency of location (see Fig. 2-10).

## Cleptoparasitism

When an insect lays an egg in the nest of another species and leaves it for the host to rear, usually at the expense of the host's own young, we

**Table 4-6**    Protein-richness of a unit volume of nectar from various flowers
grouped by major pollinator insect. On a histidine scale ranging from 1 to 10,
~~~it represents a doubling of amino acid concentration; the average score
~~owers examined was 5.05. Flowers which have evolved a dependence
~~~inators with few or no other protein sources have concurrently tended
~~ develop a nectar rich in amino acids.[a]

| Number of flower species in sample | Major pollinator insect type | Mean histidine scale score of nectar |
|---|---|---|
| 208 | Short-tongued bees such as honey bees | 4.59 |
| 96 | Long-tongued bees such as bumble bees | 5.64 |
| 86 | Both bees and butterflies | 5.49 |
| 40 | Butterflies | 6.40 |
| 14 | Nocturnal flower-visiting "settling" moths | 5.75 |
| 25 | Hawk moths | 4.86 |
| 68 | Generalized flies such as hover flies | 4.35 |
| 9 | Specialized flies such as carrion flies and dung flies | 9.00 |
| 7 | Beetles | 7.86 |
| 6 | Wasps | 5.58 |

[a] Adapted from Baker and Baker (1975).

recognize this as a special case of brood parasitism called **cleptoparasitism**. Predators of a sort, cleptoparasites attack the host's young not primarily as food but rather to make the host's food store more fully available to their own offspring.*

Among the most abundant parasites of solitary wasps, for example, are a great number of small flies that look superficially like tiny house flies (see Fig. 1-7). Each of the several genera involved has its own particular method of attack, but in all of them the eggs apparently hatch just before they are laid. The wasp egg or small larva is usually attacked and promptly destroyed, but the maggots develop primarily upon the prey in the cell, reducing the cell contents to a putrid mass within a few days. These "food thieves" are not at all host specific and will attack solitary wasp nests of several families indiscriminantly if they are in the proper ecological location.

Not all cleptoparasites act this much like generalized predators. In

* The terms cleptoparasitism and social parasitism are occasionally used interchangeably, but the two are not usually synonymous. The latter uses its host primarily as a work force rather than as a direct source of food. For a discussion of social parasitism, see Chapter 10.

many cases cleptoparasitism probably began with simple prey stealing, such as the "brigandage" reported often among solitary wasps. Many cleptoparasites are closely related to their hosts and appear to have shared a recent common ancestry. Such is the spider wasp, *Ceropales,* which enters the nest burrow of close relatives after the rightful owner has prepared and stocked the hole and laid an egg. At this point, some species substitute their own egg after eating the host egg themselves; others leave the host egg for their own young to eat. The parasite egg may often be cleverly concealed (Fig. 4-11).

Often, the cleptoparasite has become "armored" in various ways, apparently as an adaptation for deflecting the bites and stings of its host. For example, the exoskeleton of cuckoo wasps (Chrysididae) is very hard and coarsely punctured; when disturbed or threatened, these wasps curl into a small tight ball which is nearly impenetrable and difficult to grasp.

**Figure 4-11** The cleptoparasitic Australian spider wasp *Ceropales ligea,* left, hides its eggs in the book lungs of the spider prey of its host, *Elaphrosyron socius,* right. Oviposition occurs very rapidly while the spider is temporarily dropped as the wasp searches for her nest burrow. (Drawing by J. W. Krispyn, based on a film by H. E. Evans.)

## Laying in Wait

Not all arthropods actively seek out their food. The bola spider twirls a sticky ball of silk from a silken thread. Moths approach the swinging ball, drawn by an impregnated attractant (Eberhard, 1977), and become stuck to it, whereupon the spider retrieves its catch and consumes it. Certain assassin bugs have forelegs covered with secretory hairs exuding small droplets of a highly viscous substance which looks like dew; when the bug spots small, fast-moving prey such as fruit flies, its forelegs are raised and held parallel to the ground. The flies, attracted to the "dew," are entrapped and quickly consumed. The Javan bug, *Ptilocerus*, has a tuft of bright-red hairs on its body, marking the spot where a gland opens beneath the abdomen; secretions from this gland are very attractive to ants. However, after partaking of the secretion, an ant collapses, apparently from narcotic action, whereupon the bug pierces the ant through the neck and promptly sucks it dry.

Attraction of prey may take other forms. Certain Malaysian preying mantids greatly resemble flowers; unsuspecting prey attracted visually to these "blossoms" become a quick meal. Similarly, female *Photuris* fireflies use deceit to sexually lure unrelated males, which are then eaten (see Fig. 6-3). Such behaviors, in which the predator is a proverbial wolf in sheep's clothing, are termed **aggressive mimicry**. (Despite the superficial similarity in terminology, such behavior is quite different from the classical types of mimicry treated in Chapter 8.)

So far we have mentioned examples in which prey is actually attracted to the predator. But while spectacular, such behavior is actually a relatively uncommon form of the lie-in-wait strategy. More common and extremely effective when employed by an agile predator is **ambush**, often combined with camouflage. Grand masters of this are undoubtedly the spiders, whose evolutionary repertoire includes not only web weaving but quite a wide range of other ambush methods (Fig. 4-12).

Web building is a very effective tactic the advantages of which are great enough that this behavior has evolved independently not only in two distantly related spider groups but in a number of unrelated arthropod taxa and among several marine gastropods. In addition, many predatory insects such as tiger beetles, ant lions, and worm lions construct devices such as pits or snares to increase the probability of locating and capturing prey (Fig. 4-13). Considerable time and energy are often invested in the building and maintenance of such devices (Tophoff, 1977). However, they can result in a good return in food which otherwise would be so widely dispersed as to be uneconomical to wait for and perhaps difficult to catch as well. Like other lie-in-wait tactics,

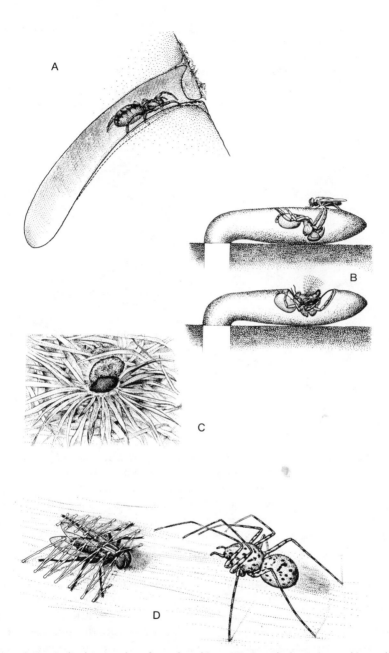

**Figure 4-12** Ambush strategies of certain spiders. (A) A typical trapdoor spider waits in its burrow for the approach of a prey. (B) *Atypus* captures a fly in its silken tube. (C) *Aganippe* constructs a burrow with a trapdoor surrounded by radiating signal lines. (D) *Scytodes* sprays a sticky substance from its mouthparts at passing prey, thus gluing them to the substrate; here it is about to devour a trapped fly. (From Alcock, 1975.)

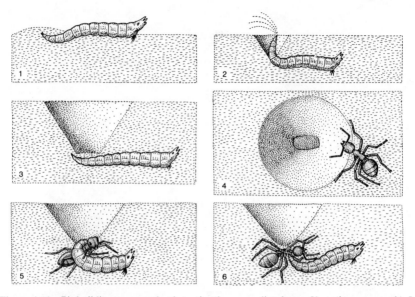

**Figure 4-13**  Pit building as an ambush tactic: the worm-lion larva, *Vermileo comstocki*. In frames (1) and (2), the fly larva is making a pit, in which it lies in wait (3). As an ant approaches the trap, (4), it slides into the pit where it is seized and paralyzed (5). Hauled under the surface of the sand, the ant's body fluids are sucked out by the worm lion (6). (From Alcock, 1975.)

these have several advantages, not only reducing searching energy expenses, but also allowing other concurrent activities such as environmental monitoring for mates and/or territorial intruders.

Obviously, a lie-in-wait strategy requires that the food itself be mobile. Usually this implies a diet upon animal prey. However, when wind or water currents exist, small plants and microorganisms may become an important diet item for lie-in-wait food gatherers. For example, Trichoptera larvae of the family Hydropsychidae use nets spun with their silk glands to capture drifting food particles in streams (Fig. 4-14). These feeding nets sometimes have an exceedingly fine mesh, allowing the caddisfly larvae to graze upon fine particulate organic matter, phytoplankton, and bacteria strained out of the moving water. The evolution of hydropsychid larvae reveals a tendency toward more complicated larval feeding structures and smaller capture-net mesh size. In some species, plant detritus may comprise more than half of the diet of younger instar individuals (Wallace, 1975).

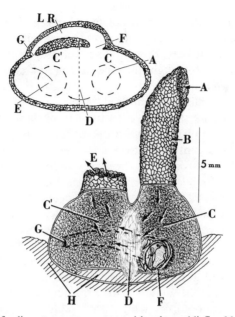

**Figure 4-14** Larval feeding structure constructed by the caddisfly, *Macronema transversum*. Below is a lateral view of the feeding structure with sagittal view of the interior of the chamber: A, Entrance hole (facing upstream) for inflowing water; B, raised sand and silk entrance tube; C, anterior portion of chamber, or feeding chamber; C′ the posterior or downstream side of the chamber; D, capture net spun across the inside of chamber; E, exit hole for outflowing water; F, anterior opening of larval retreat with larva in place; G, exit hole from larval retreat (for feces and water passing over gills); and H, sandstone substrate. Dashed lines between F and G represent the approximate location of the larval retreat chamber built adjacent to the larger chamber (C and C′). Arrows represent the main path of water. Above is a cross-sectional view; LR, larval retreat. *Macronema transversum* larvae possess densely pilose brushes on the fore tibiae and labrum used in grazing on food particles trapped in the net. (From Wallace and Sherberger, 1975.)

## Mutualistic Feeding Arrangements

Attine ants of the New World have entered into a unique mutualistic partnership with various species of fungi that do not occur outside of the ants' nests (Weber, 1966, 1972). The fungus serves as their primary and probably sole source of food, for these ants are unable to feed directly upon cellulose. If they are denied access to their fungus garden, a rapid deterioration of the fungus ensues, but under their care it flourishes in the chambers of their underground nests. The more primitive attine genera utilize a combination of plant debris and/or insect feces as substrate for their fungus. The leaf cutters go in files, often along well-defined trails, to cut leaves, flowers, and stems to be transported to their nests. A leaf fragment brought into the nest is first cleaned and scraped. Then it is chewed into a small pulpy mash by the workers who apparently add a salivary secretion. When finally inserted into the fungus garden, the mash is covered by a transplant of several tufts of fungal mycelium and one or more drops of fecal "fertilizer."

A widespread assumption that the ants weed out competing fungi and bacteria from their gardens has been discredited, and no biochemical support has been found for the hypothesis that the salivary secretion and/or fecal "fertilizer" possess fungistatic and bacteriostatic factors. However, the ants' secretions may be necessary for satisfactory growth, for the fungus appears to lack the full proteolytic enzyme complement necessary to metabolize polypeptide nitrogen (the predominant form in leaves). The ants' fecal material contains significant quantities of all 21 natural amino acids, a nutrient supplement of considerable importance, as well as providing the fungus with a proteolytic enzyme that breaks down polypeptides to forms which are usable by the fungus (Martin, 1970). Many aspects of the ants' behavior and morphology have been shaped by the fungus they culture. These include the leaf-cutting behavior, the fungus-culturing behavior, and the adult mouthparts, infrabuccal pocket (a blind pouch off the pharynx), and the pecten or comb used in grooming. Nest-founding queens carry a small mass of fungi in the infrabuccal pocket to innoculate the garden in the newly established nest.

Mutualistic feeding arrangements such as this are relatively common among insects. For example, many have a symbiotic relationship with microorganisms. Sometimes this is an internal symbiosis, with the microsymbiont living within several organs. In other cases the microsymbionts live outside the insect's body, though they may be temporarily stored in special organs of ectodermal origin for purposes of dissemination. Understandably, most endosymbionts of insects are microscopic. Their discovery dates back to the nineteenth century, but initially their existence as independent living beings was difficult to comprehend, and they were variously described as sporozoa, yeasts, metabolic products, and yolk spheres.

There is a distinct relationship between the occurrence of endosymbiosis and the type of nourishment of the insect host (Koch, 1967). The first, and the most numerous, category of endosymbiotic hosts includes insects that suck sap, which is rich in carbohydrates but poor in protein—scale insects, leaf lice, and aleurodid and psyllid "flies."

Among the bugs (Hemiptera), predaceous forms possess no symbionts, but those, such as stink bugs (Pentatomidae), which have switched over to plant juice for nourishment have symbionts.

Second, all insects that suck vertebrate blood for the whole of their lives have symbionts; this includes the bedbugs and their relatives. Vertebrate blood is deficient in certain vitamins which the symbionts synthesize. However, mosquitoes, which suck blood only as adults and have a bacteria-rich nourishment at their disposal during earlier developmental stages, have no symbionts. Symbionts also occur among that small group of insects whose whole life is spent on keratin-containing food, the Mallophaga, or feather lice.

A third group of insects having symbionts are the wood feeders. Although generalization is more difficult among the wood-eating groups than the others, these basically include not only termites but especially that subset which feeds not on the wood itself but on cellulose-rich substrates: the ambrosia cultivators, some bark beetles, some wood wasps, and others. Finally, there is a small omnivorous group of symbiont bearers: the cockroaches, the most primitive termites, and a few representatives of the ants.

One of the most interesting aspects of endosymbionts is the varied arrangements that host organisms have developed for housing them; these are species specific in both origin and development. Within these, the symbionts must be rigidly retained, but they also must be passed on to offspring. Depending upon the location of these guests within their host, this may be accomplished in at least three different ways: oral uptake of symbionts by young brood, smearing of eggs with symbionts, or infection of eggs or embryos before laying.

A special type of ectosymbiotic relationship occurs among some of the wood-inhabiting insects—the ambrosia beetles, wood wasps of the families Siricidae and Xiphydriidae, and some bark-feeding bark beetles. Sometimes, the fungus is eaten together with wood particles; in the most impressive cases the insects feed upon their fungus alone (Hartzell, 1967).

Ambrosia beetles generally have an extremely wide host range but tend to occur as secondary insects on diseased trees or felled logs. Each species is symbiotically associated with one or more specific fungi indispensible for the development of its brood. Identification of fungal symbionts is a difficult matter, for they do not produce fruiting bodies; however, it appears fairly certain that the true ambrosia fungi are highly specialized forms which cannot grow in the host plant in the absence of the symbiotic insect. In return, the fungus provides a rich and available food containing all important vitamins for the insect and its larvae.

How is fungus transmitted from an old host plant to a new one? Early workers thought spores were simply carried in the beetle's gut or upon its integument. It now appears, however, that certain specialized organs of variable location and structure but usually confined to one sex are involved (Batra and Batra, 1966). They serve to protect the fungi from desiccation, provide secretions necessary for germination and arthrospore formation, and ensure the mechanical dissemination of the fungus on the tunnel walls.

Wood wasps also attack weakened trees or freshly cut logs. However, their symbiosis with wood-inhabiting fungi is entirely different from ambrosia cultivation. During their whole adult stage, wood wasps are on the wing. Only the larvae bore in the wood, but larval development may last two or three years. Female wasps possess intersegmental pouches at the base of the ventral parts of the sting; during oviposition, symbionts are pushed out with the egg. Little is known of the meaning of this symbiosis for the wood wasp. Apparently, young larvae of some species can live on a pure culture of the fungus, and hyphae are often digested by the larvae along with the wood. Whether larvae are able to use the components of the cell wall directly as food is still an open question, however (Morgan, 1968).

## Nest Symbionts

The nests of social insects provide living quarters for a diversity of creatures besides those building them. Like our own homes despite our efforts to keep them clean, insect nests are besieged by everything from cockroaches to flies. In addition to these uninvited guests, some social insects purposely keep guests of other species in their nests, much as we keep various pets.

Depending on their hosts' identity, such symbionts or "symphiles" are referred to as termitophiles, mellittophiles, etc. Nest guests are particularly frequent in the ants, where all manner of organisms have specialized as myrmecophiles, literally, "ant lovers." Many bristletails (Thysanura), for example, live with ants. Feasting on debris, food scraps, and even ant corpses within their common dwelling, these bristletails reward the ants with a "tidy house" in turn for being inadvertently fed by their ant hosts. When army ants emigrate, bristletails may be seen running in the column (see Fig. 10-21) or riding along upon ant larvae and booty.

Representatives of the Coleoptera are also common in the nests of social insects. Sometimes, acting as scavengers they appear to be ignored by the nest owners. In other cases, the ants not only tolerate the interlopers but feed, groom, and rear their guests. Even when their intruders eat the host ants' young, they are treated cordially. For example, *Formica sanguinea* ants avidly seek out beetle larvae of *Lomechusa strumosa*, even though they usually destroy the whole population of the ant hill. The puzzle was expressed well by W. M. Wheeler (1923, p. 221) when he stated:

"Were we to behave in an analogous manner we should live in a truly Alice-in-Wonderland society. We should delight in keeping porcupines, alligators, lobsters, etc., in our homes, insist on their sitting down to the table with us and feed them so solicitously with spoon victuals that our children would either perish of neglect or grow up as hopeless rhachitics."

How have the myrmecophiles gained such acceptance? Let us examine one well-studied example of the thousands of myrmecophilous organisms—a staphylinid, or rove beetle.

### CASE STUDY: *Atemeles*, A ROVE BEETLE GUEST OF ANTS

Among the various Staphylinidae a number of different myrmecophilous relationships have evolved. Some only live along ant food-gathering trails or at ant garbage dumps outside the nest. Others live within the nest's outer chambers, but a select number have penetrated all the way inside the brood chambers. Of this last group, one well-known example is *Atemeles pubicollis*, a European rove beetle which spends its larval stage within the nest of the mound-making wood ants, *Formica polyctena*. In examining their behavior in some detail, Hölldobler (1971) has unraveled much of the communicative behavior between host and symbiont.

Watching the behavior of ants encountering *Atemeles* larvae, it seemed likely to Hölldobler that some signal was being given, for brood-tending ants respond to *Atemeles* by grooming them intensely. Was some chemical substance being transferred? Experiments with radioactive tracers indicated it was. Perhaps this chemical was acting as an attractant. Hölldobler variably coated large numbers of beetle larvae with shellac, then placed them at the nest entrance. As long as at least one body segment was left unpainted, the ants carried larvae into the nest and adopted them. When the entire larva was covered, however, the ants either ignored it or deposited it in the garbage dump. In an acetone bath, Hölldobler washed some beetle larvae; ants dumped most of these "deodorized" larvae as well. Soaking filter paper with the acetone extraction, Hölldobler left paper dummies at the nest entrance;

like real larvae, many were carried into the nest. Paper dummies soaked with fresh acetone, however, were either ignored or carried to the dump.

Within the brood chambers, adult ants scurry about feeding their young. Imitating the behavior of ant larvae, *Atemeles* larvae touched by an adult ant's antennae or mouthparts rear up and attempt to contact the ant's head. Successful contact causes the ant to regurgitate a droplet of food (Fig. 4-15). How does the beetle's success rate compare with that of the ant larvae? Hölldobler gave the ants food labeled with radioactive sodium phosphate. In a mixed brood of beetle and ant larvae, the beetles obviously obtained a more than proportionate share of food. Apparently, they were more intense as beggars than were the ant larvae. In addition, the predacious beetle larvae ate small ant larvae. How could the ant colony survive such intense competition and predation? Observation quickly provided a simple answer. The beetle larvae were not only

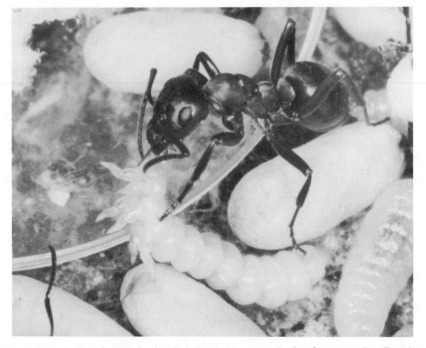

**Figure 4-15** Larva of *Atemeles* beetle being fed by regurgitation from a worker *Formica* ant. Contact between ant and beetle larva is enhanced by a chemical attractant secreted by a row of paired glands along the sides of the beetle abdomen which causes worker ants to groom their guests intensively. Such attention causes the beetle larva to rear up and attempt to make mouth-to-mouth contact with the ant. (See Hölldobler, 1971; courtesy of B. Hölldobler.)

predacious but cannibalistic. Unable to distinguish their fellow larvae from ant larvae, they cut down their own population ruthlessly, whereas ant larvae did not. Thus, the brood chambers soon contained clusters of ant larvae but only scattered, lone beetle larvae.

*Atemeles* beetles, it turns out, have not one but two ant homes—a summer woodland domicile with *Formica* and a winter grassland one with *Myrmica*. After the beetle larvae have pupated and eclosed in their original home, they beg for one final, ample food supply. Drumming rapidly with their antennae upon an ant to attract its attention, they touch the ant's mouthparts with their own maxillae and forelegs; since this mechanical signal is quite similar to that used by ants among themselves, the ant responds by regurgitating food. Now the adult staphylinids begin to migrate out of the nest. Guided primarily by light and odor, the beetles move into open grasslands and find *Myrmica* nests. When a staphylinid encounters one, it wanders around until it encounters an ant worker. Going through a brief ritual, the beetle moves in, carried by its host right into the brood chamber. Although the beetles are now adult, they are still sexually immature. Within these latter nests, they continue to be fed until sexual maturity the following spring, at which time they return to *Formica* nests for mating and laying of eggs.

In order to penetrate *Myrmica* nests, *Atemeles* must adopt a new set of skills and a second language. Hölldobler again suspected chemical cues, for as the two species first encounter one another, the beetle antennates the ant lightly, then raises its abdomen toward its host. In response, the ant licks the tip of the abdomen, seeming to grow calmer in the process, then moves on to the side of the abdomen. Finally, the beetle lowers its abdomen. The ant then grasps bristles around the beetle's sides and carries its tightly curled guest inside. True to prediction, Hölldobler found two types of glands and secretions (Fig. 4-16). At the tip of the abdomen, "appeasement glands" produce a partially proteinaceous secretion that apparently suppresses aggressive behavior in the ant. Along the sides, a series of "adoption glands" produces a chemical necessary if the ants are to welcome the beetle in. Apparently, this odor mimics the odor of members of the ant species.

Appeasement glands turn out to be farily widespread among the better-integrated social symphiles. They are of a number of novel forms (including glandular hairs) and exist in different locations, so apparently they have been independently derived. Appeasement devices are just one of a syndrome of changes in morphology and behavior that social symbionts belonging to various insect groups have independently undergone (Table 4-7). Some of these are worth special mention.

First, a great number of ectosymbionts look very much like their

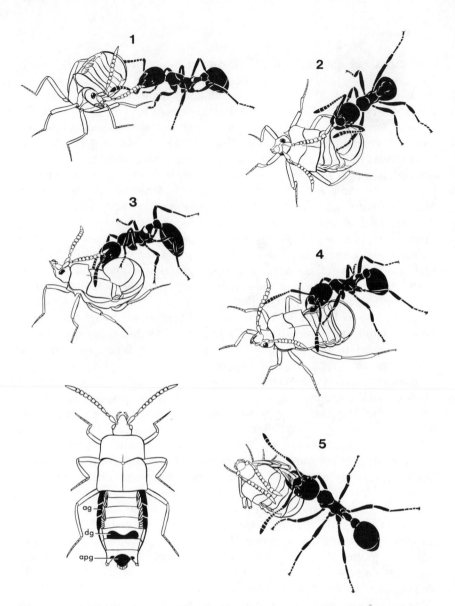

**Figure 4-16** Symphyly in the staphylinid beetle *Atemeles pubicollis* illustrating the adoption ritual used by adults to gain entry to nests of the ant genus *Myrmica*. Success depends on the interplay of both chemical and tactile cues. Secretions originate from glands along the sides and tip of the beetle's abdomen, bottom left (ag = adoption gland, dg = defensive gland, apg = appeasement gland). On encountering a potential host and worker, the beetle presents the tip of its abdomen and taps the ant lightly with its antennae (1). The ant responds by licking the gland the secretion of which apparently serves to suppress the ant's normally aggressive behavior toward intruders (2). After this the ant proceeds to lick the sides of the beetle's abdomen where it obtains the "adoption"

**Table 4-7**    Some morphological and behavioral characteristics of ectosymbiotic arthropods living in ant and termite nests[a]

*Morphological adaptations*

Lighter body coloration
Exocrine glands or glandular hairs secreting attractive substances
Mimicry of host's body shape
Physogastry
Limuloid or tear-drop body form

*Integrative behaviors*

Following trails of hosts
Regurgitation
Recognition and tactile interaction
Alarm
Phoresy of immature stages on hosts

[a] All characteristics are not possessed by a particular symbiont. Representatives of at least 17 orders including 120 families have become ectosymbionts of social insects. Based on Wilson (1971) and Kistner (1969).

hosts. Many commensal staphylinids, in particular, strikingly resemble ants, with a slender body form, antlike "petiole," and even body sculpturing and color (Fig. 4-17); this antlike appearance is found almost nowhere else in this large family of beetles. Early investigators, working from museum specimens, believed that the myrmecoid body form of the symphiles was tactile mimicry to deceive host ants, and that color was visual mimicry to deceive birds and other predators upon marching ant columns. However, in ants chemical identification is paramount, and an ant whose surface odor is disturbed even slightly will be immediately attacked by her sisters, though her morphology has not changed. Therefore, it seems unlikely that such mimicry is tactile and directed toward ants. A more likely hypothesis is that both the color and morphological mimicry have been selected by predators watching ant columns for edible morsels. This idea is supported by the fact that such mimicry tends to be absent among termitophiles, which spend their entire lives underground.

secretion (3, 4) necessary for releasing brood-carrying behavior in the ant. The tightly curled beetle is then picked up by the ant and carried into the nest brood chamber (5) where it has access to the ant's brood. (Drawing by Turid Hölldobler, courtesy of B. Hölldobler.)

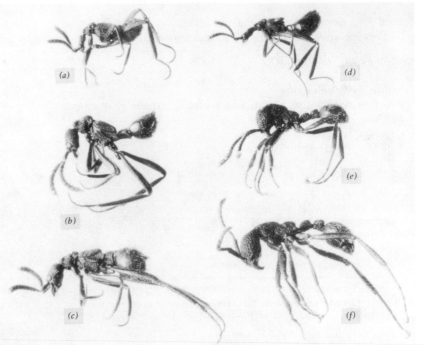

**Figure 4-17** (*a–d*) Some antlike staphylinid beetles associated with the army ant, *Neiva-myrmex sumichrasti*. (*e–f*) *N. sumichrasti* workers. Punctation of the head and thorax of the beetles as well as their coloration closely resembles that of their host. (From Akre and Rettenmeyer, 1966; courtesy of C. W. Rettenmeyer.)

A second morphological adaptation characteristic of many unrelated social symbionts is **physogastry**—the condition wherein the abdomen is greatly enlarged, particularly in its membranous parts. Physogastry appears to be one of the major ways in which termitophiles, particularly among Coleoptera and Diptera, have approached mimicry of their hosts. Such swelling often makes them superficially resemble the bodies of termites. However, direct study of termites and termitophiles poses problems not encountered with their hymenopterous counterparts. Termitophiles are extremely delicate, and it is usually necessary to crack open the termite nest to reach them, then make one's observations under quite artificial laboratory conditions. Thus, whether the role of physogastry is primarily visual or chemical is hard to answer. The variety of observations which exist appear to indicate that some physogastric genera are licked, but others may not be. In at least some cases, there appears to be a mutual exchange of exudates between termites and termitophiles, and histological studies of some termitophiles demonstrate the existence of various abdominal glands and pores. Whether such exudates may be beneficial to the termites is quite unknown. That the termitophiles depend on their hosts is more certain. Most physogastric species have such rudimentary mouthparts that they could no longer feed themselves and such a heavy dependence on the controlled temperature and humidity of the termite nest that they do not live long upon removal from it (Kistner, 1969; Wilson, 1971).

A limuloid (i.e., tear drop-shaped) body has also evolved convergently many times in both termitophiles and myrmecophiles. Generally, it is assumed that this body form has arisen for defense. The entire dorsum is smoothly streamlined, which makes it difficult for an ant to grasp. Such a shape suggests that these guests are subject to attack by either their hosts or other elements in the ant society. However, actual field observations of limuloid symphiles are even scarcer than those of physogastric species.

Interestingly, specialized symphiles are almost unknown in nests of the social bees and wasps. The few mites, beetles, and flies that live as scavengers and brood commensals are quite generalized in form and behavior by comparison with symbionts of ants and termites. Most are probably either attacked by their hosts or treated indifferently. Yet social Hymenoptera and Isoptera share the same fundamental communication forms, namely, trophallaxis, grooming, progressive care of larvae, nestmate recognition, alarm, and recruitment. No really great differences exist in colony size, nor necessarily in length or stability of colony life. Why then do social wasps and bees have such a paucity in number and variety of obligate commensals? Probably the most plausible explanation centers around nest structure and location, coupled with feeding habits. Ants and termites live in relatively open systems rich in refuse. Many chambers and galleries go unguarded from time to time, and the nest interiors are generally made of material not too different from the immediate environment—soil and rotting vegetable matter. Furthermore, young are reared clustered in groups. Social wasps and bees, in contrast, "run a tight ship," constructing compact, tightly sealed nests typically in arboreal locations. Thick envelopes of carton or wax pose formidable obstacles; nest entrances, often narrow and tightly guarded, may also be lined with sticky substances and/or repellants. Furthermore, refuse inside is usually sparse. Pollen and nectar are highly concentrated food sources, which do not produce much waste. Even when detritus occurs, workers of many species simply heave most of it out the nest entrance. Finally, young are reared individually in specially constructed cells which makes it more difficult for a symbiont to conceal itself.

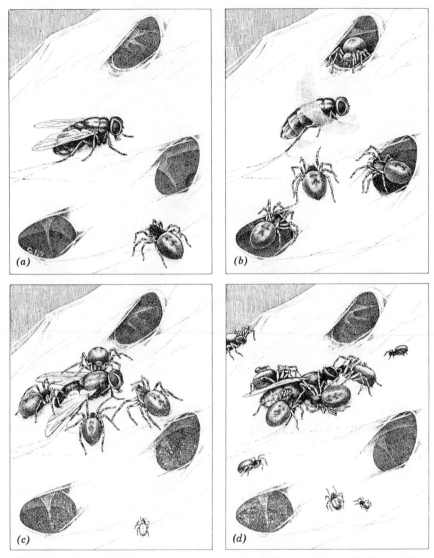

**Figure 4-18** Stages in the cooperative capture of a fly by several socially living spiders, *Mallos gregalis*. A fly landing on the web (*a*) causes nearby spiders to reorient toward it. Buzzing vibrations from the entangled fly elicit response from several individuals (*b*), who advance on the prey in quick short jumps (*c*). Their bites provoke frenzied buzzing from the fly, which further stimulates other spiders, and feeding begins (*d*), with younger spiderlings joining in the communal feast. (From ''Social Spiders'' by J. W. Burgess. Copyright © 1976 by Scientific American, Inc. All rights reserved.)

156

## SOCIAL FEEDING BEHAVIORS

A number of spiders live in cooperative groups numbering dozens of individuals. Some species construct a huge communal sheet web covering several square yards; when a large prey happens into the web, the spiders attack and subdue it together (Fig. 4-18), then dine upon it communally. Gregarious feeding is also characteristic of some herbivores, such as tentworm caterpillars and various sawflies (Fig. 4-19). Courtship feeding, another type of cooperative social feeding between conspecific individuals (see Chapter 9), is also common. It is among the truly social insects, however, that cooperative feeding behaviors are exhibited in their widest scope and variety. In fact, certain social feeding behaviors are more or less unique to the social insects; let us examine some of them.

### Trophallaxis

When a social wasp larva is fed by an adult, the larva almost always secretes a droplet of salivary fluid, which the adult imbibes. In many social insects, chemical communication signals are spread through the colony by an exchange of liquids between nest mates; in fact, the reciprocal exchange of liquid foods between colony members, or **trophallaxis**, is one of the most striking and easily visible characteristics of social life. This transfer occurs both among adults and between adults and larvae (Wilson, 1971).

Trophallaxis appears to occur generally throughout the eusocial wasps. Among bees and ants, on the other hand, its occurrence is highly variable, apparently determined both by phylogenetic position and ecological constraints. In termites, trophallaxis has multiple functions. Lower termites share both "stomodeal food" from the salivary glands and crop and "proctodeal food" from the hindgut. The former is the principal nutrient source for the royal pair and nymphs; the latter, a milky material quite different from feces, contains symbiotic flagellates. These break down the cellulose which the termites are otherwise unable to use as a source of nutrition. With each molting these hindgut symbionts are lost, and the nymphs must acquire new ones. Among the higher termites, the habit of proctodeal trophallaxis has been lost along with dependence on symbiotic flagellates for cellulose digestion. However, the nymphs have become entirely dependent on stomodeal exchanges and no longer even possess functional mandibles.

Since Wheeler (1918) first coined the term trophallaxis for this unilateral or bilateral liquid food exchange, there has been strong dispute

over the signal value of these substances. Some have even suggested that trophallaxis be used synonymously with communication since it provides a mechanism for maintaining a colony-specific odor (see Chapter 5). However, this exchange, though seemingly simple, actually may serve not only several different social functions but nutritional ones as well. The larvae of some vespid wasps, for example, take over certain metabolic functions for the colony through trophallaxis. Receiving proteinaceous food from the adults, they transform it into carbohydrates which are stored and, if required, later returned to the adults. Since the adults are apparently unable to synthesize these essential nutrients, such behavior is critical to colony survival during periods of food shortage, such as rainy periods (Ishay and Ikan, 1969).

Both prolonged, direct observations and radioactive tracer studies have shown that in the majority of social insects trophallaxis allows a material to be distributed throughout a colony with striking rapidity (Fig. 4-20). Trophallaxis is truly an open system—each individual shares with an unlimited number of nestmates. Although trophallaxis requires both a giver and a receiver, which individual performs which role largely depends upon the state of their crop contents at the moment, and roles can easily be switched. For example, a set of *Formica* ant workers showing predominantly begging behavior can be shifted to predominantly donor behavior simply by feeding them to satiation. Because of this rapid exchange, each worker shares virtually the same diet and is kept informed of the nutritional status of the colony as a whole. Therefore, when an individual worker reacts to feeding stimuli in terms of its own hunger or satiation, in the majority of cases it is also acting in a way that is appropriate for the colony as well.

What are the releasers for trophallactic behavior? In the honey bee, they appear to be the combination of tactile and olfactory cues provided by a honey bee head with antennae intact. A freshly severed head will elicit either begging or offering; if its antennae are removed, the head is less favored until insertion of imitation wire antennae of the proper length and diameter restores its effectiveness. In addition, both ants and

---

**Figure 4-19** Gregarious "escalator style" feeding by larvae of the Brazilian sawfly, *Themos olfersii,* whose host plant possesses extremely tough leaves. Aligning themselves in two convergent rows, the young larvae use their enormous heads and jaws to gain access to the thick edges of the leaves (*a*). As the larvae feed, the leading individuals are pushed forward until they collide (*b*). Upon collision, the leading larvae retreat to the ends of the rows (*c*) where they cannot readily feed, while others are moved forward for a turn at eating. This process continues for many hours. (From Dias, 1975; courtesy of B. Dias. Copyright 1975 by Editora Vozes Ltda, Petropolis, Rio de Janeiro.)

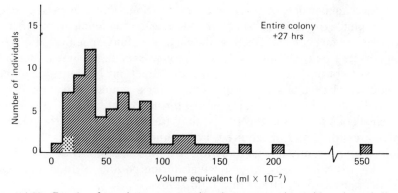

**Figure 4-20** *Formica fusca* is an ant species that engages in rapid oral trophallactic exchanges. Within a day after the feeding of a single worker with a small amount of honey mixed with radioactive iodide, evidence of the radioactive food was present in varying amounts in every colony member, including the 2 queens (stippled part). Over time, the frequency distribution of individual shares of the radioactive food became progressively more normalized. Thus, the combined crops of the adult population might be called a "communal stomach." (Modified from Wilson and Eisner, 1957.)

bees appear to favor larger individuals, with the result that queens, males, and larger workers tend to receive more than they give.

In social wasps, the release of trophallactic behavior is more structured and complex, for dominance relations place severe constraints on food exchange, slowing the rate of food distribution and increasing the variance in crop contents among workers. For example, an inert severed head does not release trophallaxis in *Vespula*; rather, the pair must engage in a very definite pattern of continuous reciprocal antennal signaling (Fig. 4-21).Dominant workers receive more food than they give. The mother queen always receives and seldom ever gives, and virgin queens dominate over their worker sisters. Males, possessing a quite different antennal form, are quite inept at begging and must rely primarily upon surreptitious sips of the regurgitated liquids being passed between others or upon stimulating larvae to give forth salivary secretions.

### Cannibalism and Trophic Eggs

Some *Reticulitermes* termite workers eat their apparently healthy nestmates when grooming is carried too far. If the cuticle of a leg is broken, for example, the leg is eaten, and then the whole termite is consumed. Winged *Coptotermes* reproductives which are unable to leave on a

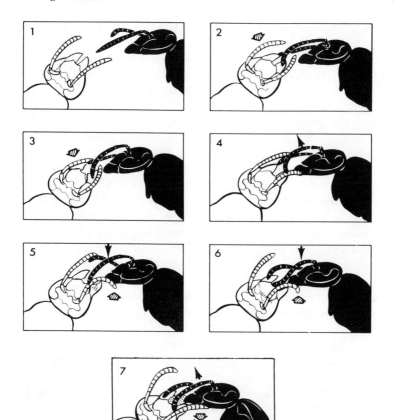

**Figure 4-21** Two workers of the yellowjacket wasp *Vespula (Paravespula) germanica* initiating trophallactic exchange. The solicitor (black) approaches the donor (white) and places the tips of her flexible antennae on the donor's mouthparts (1, 2). In response, the donor closes her antennae onto those of the solicitor (3). In turn, the solicitor begins to gently stroke her antennae up and down over the donor's lower mouthparts (4-6). This interaction must continue before the donor will begin to regurgitate so that the solicitor is able to feed (7). Arrows indicate the directions of antennal movement. (Reprinted by permission of the publishers from *The Insect Societies* by Edward O. Wilson, Cambridge, Mass.: The Belknap Press of Harvard University Press, Copyright © 1971 by the President and Fellows of Harvard College. Based on Montagner, 1966.)

normal nuptial flight are finally killed and eaten by workers. Alien conspecific workers that enter a termite nest are generally disabled, then consumed. In fact, all termites studied to date eat their own dead and injured on at least some occasions, a degree of cannibalistic behavior far more intense than in any other social insect group. Among the Isoptera, cannibalism probably functions as a protein-conserving device. The diets of termites under natural conditions are undoubtedly low in protein. Colonies of *Zootermopsis angusticollis*, for example, become intensely cannibalistic when reared on a laboratory diet of pure cellulose, but adding sufficient casein to the diet reduces the cannibalism to almost zero.

Among the social Hymenoptera, cannibalism of adults is rare in some groups, unknown in most. However, a related phenomenon is common—the eating of immature stages. In ant colonies, for example, injured eggs, larvae, and pupae are quickly eaten. When colonies are starved, workers begin attacking healthy brood as well. Hunger and the degree of such **brood cannibalism** are so precisely related as to suggest that the colony's store of immature stages functions normally as a last-ditch emergency food supply to keep the queen and workers alive.

A related phenomenon, widespread in the social Hymenoptera, is egg cannibalism, **oophagy**. In the more primitive social groups, the dominant queen eats the eggs laid by subordinates, thus ensuring that her own progeny will predominate. This exploitive character has been transformed into something quite different among the higher social Hymenoptera, however. In these groups it has become an important form of food exchange between cooperating members of the same colony. Sometimes, oophagy functions like brood cannibalism; once colonies are well under way, if workers are starved the eggs are the first brood stage to be eaten. In other cases, however, eggs laid by workers are usually eaten by the queen and/or larvae (or sometimes by other workers) shortly after they are laid. Such **trophic eggs** have been reported in a wide diversity of ants and are apparently laid in enormously varying frequencies among the various ant species. As a rule, however, the more frequent the exchange of trophic eggs, the less frequent the liquid exchange through trophallaxis. Wilson (1971) provides additional information on the various social feeding behaviors.


## FOOD RECOGNITION AND ACCEPTANCE

On what specific basis does an insect recognize that an item is food? The polyphagous chalcid wasp, *Trichogramma evanescens*, parasitizes the eggs of more than 180 insect species. For the female chalcid, oviposition requires only that the object be firm enough to walk upon, protrude from the surface, fall within certain size limits, and have no dimension greater than four times any other. Odor, color, and surface texture are irrelevant, but one important distinction is made. *Trichogramma* will not oviposit in eggs that have already been parasitized unless the egg is washed to remove the smell the previous female has left behind (Salt, 1937; also see Fig. 9-19). *Rhagoletis* fruit flies similarly mark their oviposition sites on apples.

The potato beetle larva's reaction to close relatives of the potato is governed by plant chemistry, particularly the presence or absence of a variety of alkaloids. In the potato, that alkaloid corresponding to deterrents in other solanaceous species is, for the *Leptinotarsa* larva, neutral in effect (Table 4-8). However, if its palps and antennae, which carry the olfactory receptors, are removed, the larva will eat substances positively harmful to it. Apparently, the attractive flavoglucosides which maintain continued feeding are widespread or universal in the Solanaceae (Hsiao, 1974).

Since the 1940s a great deal of research has been directed toward discovering the nature of the exact factors which underly insect food preferences. In the process, two schools of thought arose. Some scientists pointed out the paramount importance of secondary substances in host plant selection; because these compounds appeared to be acting as classical behavioral releasers, they were called sign, or

**Table 4-8**    The effects of various alkaloids from Solanaceae as deterrents to feeding by the Colorado potato beetle, *Leptinotarsa decemlineata*[a]

| Chemical | Plant origin | Degree of inhibition[b] |
|---|---|---|
| Solanine | *Solanum tuberosum* | — |
| Chaconine | *S. tuberosum, S. chacoense* | — |
| Demissine | *S. demissum, S. jamesii* | +++ |
| Leptine | *S. chacoense* | +++ |
| Soladucine | *S. dulcamara* | ++ |
| Solacauline | *S. acuale, S. caulescens* | ++ |
| Solamargine | *S. aviculare, S. sodomeum* | — |
| Solanigrine | *S. nigrum* | ++ |
| Solasonine | *S. sodomeum, S. carolinense* | — |
| Tomatine | *Lycopersicon esculentum* | +++ |
| Capsaicin | *S. capsicum* | ++ |
| Nicotine | *Nicotiana tabacum, N. rustica* | Toxic |
| Nicandrenone | *Nicandra physalodes* | +++ |
| Atropine | *Atropa belladanna* | + |
| Scopolamine | *Datura* spp. | ++ |

[a] From Hsiao, (1974).
[b] +++, Strong; ++, moderate; +, slight; —, no deterrent effects.

"token," stimuli (see Chapter 2), and food specificity in insects was thought to be based solely on their presence or absence. Roughly equivalent to "flavors," such tokens were defined as having no nutritional value per se. Other scientists, however, disagreed with this interpretation. Token stimuli might be important in the initial discrimination of different plant species, they argued, but factors more closely related to nutrition must mediate finer decisions such as the discrimination of leaves of different ages. Thus, food preference was a two-step, or "dual discrimination," process. When an insect showed food preferences, it relied upon stimuli which, even if not identical with the nutritionally important substances in the plant, were more closely related to nutrition than to "flavor" (Dethier, 1966).

Parts of both of these two theories now seem correct. As with the potato beetle, nutritionally nonessential (or even toxic) compounds may exercise an important influence as feeding stimulants. A wide variety have been reported, but the most nearly universal is certainly sucrose, which is an effective stimulus for most insects. Monophagous insects appear to respond to a stimulus species specific to their hosts; for phytophagous species, this may even be a secondary substance originally evolved to repel herbivores. Often, monophagous feeders have become physiologically incapable of surviving away from their usual host. Polyphagous insects, on the other hand, show such a broad responsiveness that they may respond directly to nutrients without the intervention of a mediating chemical stimulus; any apparent choice they exhibit is likely to be based on selective rejection due to repellent substances. One useful experimental technique for sorting out the relative importance of the different plant chemical constituents is the "feeding preference test" in which the insect is given a choice of different food combinations; the amount of feeding is used as the bioassay (Fig. 4-22).

Despite voluminous records of insect–plant and parasite–host relationships, a deeper understanding of the evolution of insect diet selection has been difficult to gain. For example, why are some insects monophagous and others polyphagous? A feasible explanation is found in the **congruency hypothesis** advanced by Dethier (1970b). As Dethier points out, both are continually evolving against a background of multiple pressures—the plant by synthesizing different chemicals, the insect by developing different sensory and central decision-making capabilities. Specific changes occur in both organisms as a result of random mutations. Congruency, or appropriateness between plant as food source and insect as feeder, occurs whenever these two independently mutating systems interact in such a way that formerly nonattracting chemicals

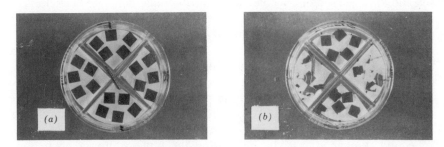

**Figure 4-22** Experimental design for determining insect food preferences. (*a*) Fresh 1-cm² leaf pieces of four different plant species are placed in a Petri dish. An insect, in this case the fall armyworm, is introduced into the center and allowed to feed for 24 hours. (*b*) Relative feeding damage, readily determined visually, may be quantified if desired by comparing the dry weights of leaf squares before and after feeding. With modification the same technique can also be used to bioassay the relative importance of various plant products by incorporating chemicals to be tested in standard agar discs made with powdered leaves. (Courtesy of W. G. Burnett.)

now stimulate feeding. Sometimes, the change may be in the insect, resulting in the addition, subtraction, or substitution of capabilities. For example, insect neural changes may cause formerly neutral chemicals to become attractive or repellents to be no longer detected. At other times, various plant chemicals may arise by mutation.

Like any model, the congruency hypothesis presents an oversimplified view. For one thing, the model assumes reproductive isolation of the new mutant, so that once a new insect–plant association is initiated the insect's survival is related to food availability rather than to competitive pressures from the normal allele. Second, it assumes that mutations represent quantum jumps rather than intermediate states. For phytophagous insects, little information exists upon which to test this, although a number of documented cases of sudden irreversible shifts in insect feeding habits suggest that the assumption is not unreasonable. Nevertheless, the congruency hypothesis provides explanations for a number of previously puzzling phenomena, such as why some plants such as ferns have feeding deterrents even though they evolved long before phytophagous insects did. The theory also explains why shifts in diet may occur in any direction, including from polyphagy to monophagy, and why some apparently suitable plants are not eaten.

Two related points are appropriate to this discussion. First, the host plant or animal is not only a source of food; often it equally may be a place to live. Thus, for a new feeding habit to be established, the host need not necessarily be better nutritionally. For many insects, plant selection is selection of a whole community—a microclimate, a shelter, a set of predators and diseases. Nor is the plant a passive partner to the association; rather, it is a respondent capable of developing protective mechanisms which interfere with the utilization of its tissues as insect food (see Table 4-2). With simultaneous development of plant resistance mechanisms on the one hand and insect tolerance and preference on the other, the picture at any given moment must be one of dynamic equilibrium, and feeding must be viewed as a compromise between nutritional and ecological optima.

Second, though it may be belaboring the obvious, for a plant and insect to interact they must be available to each other. This has been graphically illustrated by introduced pest species throughout the world, enthusiastically munching upon certain plants not formerly available to them. Nor does availability require that it be the pest that is introduced; again, a striking example is the Colorado potato beetle. This native American insect fed upon weedy scrub until the potato was introduced into its habitat; moving onto potatoes, it came to prefer these over native hosts.

## REGULATION OF FEEDING

Preying mantises will consume a relatively constant number of house flies each day when flies are continually available. Blowflies will maintain a relatively constant daily food intake at a constant sugar concentration; if the concentration is decreased, daily intake increases, and vice versa. In fact, most insects will feed to a point and then stop. What causes a feeding insect to finally stop eating? What determines the timing—how long and how often—of feeding?

One of the most thoroughly studied cases of feeding regulation involves *Phormia regina*, the blowfly, investigated by Dethier (1962, 1976) and his associates. The adult blowfly needs only water, carbohydrates, and oxygen for maintenance, receiving all other necessary materials during its larval stage; adult feeding occurs only to provide locomotive energy. Thus, the blowfly offers an ideally simple system for studying two essential aspects of feeding behavior: the nature of the "on/off" mechanism and the nature of quality control. The initiation of blowfly feeding involves several steps. First, taste receptors on the tarsi are stimulated as the fly steps upon potential food. These lead to the extension of the proboscis, which thus contacts the food solution. Combined mechano- and chemoreceptive hairs on the inside of the labellum experience the consistency and taste of the food. The food quality,

along with peri        ntral adaptive processes, determines the
sucking reactio        of feeding behavior involves a homeostatic
mechanism (         at, in the presence of excess food, a constant
amount is         day. One of the consequences of feeding,
namely, g         ovides negative feedback which reduces the
probabilit         eeding. As the gut is filled, stretch receptors are

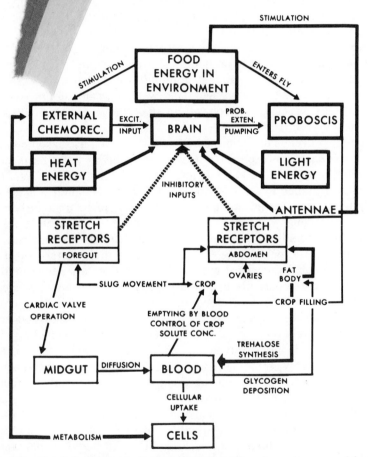

**Figure 4-23**  Metabolic homeostasis in the blowfly, *Phormia regina*: a model system
illustrating the interactions of external and internal stimuli and resultant feedback which
regulate the physiological aspects of feeding. The actual cessation of food intake is
probably mediated through negative feedback from internal receptors. The internal sensing
upon which this depends may be based on any of quite a number of variables, such as food
bulk and/or gut capacity, level of sensory stimulation, and length of time spent at
continuous feeding. (From Stoffolano, 1974.)

activated whose firing inhibits brain input from the external chemoreceptors that elicit feeding. These stretch receptors are situated in the
foregut, for filling the midgut and hindgut by means of an enema has no
effect on feeding inhibition. At this stage in feeding regulation, nonnutri-
tive or metabolically useless foods are not distinguished from nutritious
ones. A blowfly will, for example, take up the useless sugar fructose and
regulate the amount until it dies.

The question of the timing of insect feeding is more complex. In
general terms, the length of the intermeal period is usually related to the
quantity and nature of the previous meal and to the amount of energy
expended in the interim. For most insects, feeding occurs at relatively
short intervals of minutes or hours. However, some insects, such as
certain filter feeders, eat almost continuously. Others, such as some
parasites, may feed only at wide intervals of many hours, days, or
weeks. Many species have phases in their life cycle (e.g., diapause)
when they do not feed at all.

Even long-term changes in feeding behavior, such as seasonality, may
rely on the same basic physiological mechanisms as short-term feeding
regulation, however. A good example is provided by the face fly, *Musca
autumnalis*, studied by Stoffolano (1968). During the summer, face flies,
which feed upon cattle and lay their eggs in fresh, undisturbed cow
manure, go through several generations. Adult males, which apparently
feed on nectar, show the greatest feeding response to glucose. Like
many other flies, female face flies exhibit a cyclical pattern of protein
ingestion, which increases dramatically when eggs are being matured;
face flies cannot apparently live on protein alone, however. Female face
flies are responsive to both blood and glucose, as long as their ovaries
are undeveloped. However, as their ovaries begin to swell and act on
abdominal stretch receptors, female feeding begins to decline. Fully
gravid females do not feed at all. With responsiveness to blood and
glucose effectively removed, the lower level of responsiveness to
manure they have maintained all along now asserts itself. In the fall,
under the influence of short days and low temperature, the ovaries of the
last generation of flies of the season fail to develop. Now face flies of
both sexes ignore cattle and feed only on nectar. On this diet and under
changed hormonal influences, a face fly's fat body swells and its
abdomen distends until abdominal stretch receptors apparently respond
with impulses to the central nervous system, nullifying the sensory input
from tarsal receptors. Feeding stops. But gradually through the winter,
fat is utilized; the fat bodies shrink and the crop gradually empties. By
spring, diapause is over. Flies then feed on nectar until cattle are put to
pasture.

Among plant-eating insects, feeding periodicity is also common. One important evolutionary reason undoubtedly is that a plant is not a homogeneous chemical entity. Rather, it is a heterogeneous, ever-changing microchemical environment. Important plant constituents such as carbohydrates, fats, proteins, minerals, alkaloids and essential oils (the last two contributing the majority of odors and tastes) vary with a number of factors, such as time of day and season, plant growth stage and tissue, climatic and soil conditions, etc. For example, consider the seasonal changes that occur in oak leaves. New leaves contain relatively more protein, water, and sucrose and less tannin than older leaves. As tannin increases, protein availability is reduced until its decline becomes a limiting factor for most herbivores. Not surprisingly, most lepidopteran larvae attacking oaks feed early in the season on new foliage; by so doing, they make the best use of available protein and avoid most of the toxic tannins which slow growth rates and reduce fecundity (Feeny, 1970). Plants also show shorter-term cycles, the most familiar example of which is the diurnal rhythm of nectar production and flower opening in many plants. Honey bees, apparently able to remember not only the location where food is available but the time of day at which this occurs, continue to gather only during the previously learned hour.

Environmental variables are also important in the establishment and maintenance of feeding periodicity, and it would be a mistake to underestimate their effect. For example, most ants are more active at some times of day (or temperature and light ranges) than at others. Many ant species in temperate climates forage throughout the day, but in hotter lands some regularly stop feeding for a midday break. Many others forage only at night; 54 out of 58 species of ants living in the Sahara are nocturnal. The effect of such foraging rhythms, of course, is to confine the outside activity of the ants to times when temperature and humidity are least harmful and/or to when their food is most easily obtained. It is sometimes difficult to determine the degree to which an insect's foraging rhythm depends upon internal biological clocks, however. For example, leaf-cutting ants, *Atta cephalotes,* normally collect leaves all day, beginning early in the morning. Shading their nest entrance between 5:30 A.M. and 6:00 A.M. will delay the time of appearance of the first workers, but no amount of light before 5:30 A.M. will make foraging start. Some internal clock is apparently responsible for bringing the workers to the nest entrance to inspect for light. At the same time, many studies have shown that when ants are fed regularly at the same time for three to five days, they learn to search for food at that time on succeeding days. This implies that ants in the wild would learn to forage when food was most plentiful. Thus generalizations must be made quite cautiously. Probably the daily foraging activity of most insect species is the complex result of minute-to-minute conditions, an internal rhythm itself partly dependent on past conditions, and, in the case of social species, the individual past experience of the colony members.

## SUMMARY

Insects may be phytophagous, mycetophagous, predators, ectoparasites, parasitoids, or detritivores. About half of all insects feed upon plants, foliage feeding being the most common habit. Most accept only a limited range of foods and usually prefer one or two.

Successful feeding depends on a remarkably constant chain of behaviors, including food habitat location, food finding, food recognition, food acceptance, food suitability judgment, and sometimes host regulation. Chemical sensory systems predominate throughout the chain.

Herbivory is the best-studied habit. Coevolutionary theory provides a background for understanding the relationship between herbivores and plants. Symbiotic interactions may include commensalism, mutualism, and parasitism/predation. The first steps in herbivory are commonly mediated by odor, but inedible does not necessarily mean nonnutritious. The primary function of plant secondary substances involved has been subject of intense controversy. Sometimes these substances may mimic major insect hormones.

Mechanical feeding deterrents, including spines, thorns, etc., characterize many plants. Mutualism such as that exhibited by ants and acacias is another defense, but mutualism is more common in the context of pollination.

Food location strategies include active search, cleptoparasitism, laying in wait, and a variety of mutualistic arrangements including nest symbionts. Cooperative feeding behaviors reach their widest scope among the truly social insects. Trophallaxis, the reciprocal exchange of liquid foods among colony members, is one of the most striking characteristics of social life and serves both nutritonal and communicative needs. Cannibalism and brood cannibalism may serve as a protein-conserving device and/or emergency food supply.

Much research has been directed toward discovering the exact nature of the factors underlying insect food recognition and acceptance. Nutrition does not appear to be the prime director. The congruency hypothesis provides an evolutionary explanation of insect diet selection. Regulation of feeding involves homeostatic mechanisms; this phenomenon is best studied in Diptera. Feeding periodicity is also common in a wide variety of insects.

Those interested in further reading on the various aspects of the insect–plant relationship should consult the edited compilations of de Wilde and Schoonhoven (1969), Van Emden (1972), Gilbert and Raven (1975), Wallace and Mansell (1976) and Labeyrie (1977). Chapters 2–4 and 20 in Price's (1975) *Insect Ecology* also provide overall perspectives, as do the papers by Kogan (1975, 1977), Staedler (1977), and Swain (1977). Earlier contributions include those of Fraenkel (1959), Thorsteinson (1960), and Schoonhoven (1968). General works on insect feeding are those of Brues (1946), Dethier (1966), and Wigglesworth (1964). For summaries of studies on the regulation of feeding consult Gelperin (1971, 1972) and Barton Browne (1975).

## SELECTED REFERENCES

Akre, R. D. and C. W. Rettenmeyer. 1966. Behavior of Staphylinidae associated with army ants (Formicidae: Ecitonini). *J. Kans. Entomol. Soc.* **39**: 745–82.

Alcock, J. 1975. *Animal Behavior. An Evolutionary Approach.* Sinauer Assoc., Sunderland, Mass., 547 pp.

Askew, R. R. 1971. *Parasitic Insects.* American Elsevier, New York, 316 pp.

Baker, H. G. and I. Baker. 1973. Amino-acids in nectar and their evolutionary significance. *Nature* **241**: 543–45.

Baker, H. G. and I. Baker. 1975. Studies of nectar-constitution and pollinator-plant coevolution. In *Coevolution of Animals and Plants,* Gilbert, L. E. and P. H. Raven (eds.). University of Texas Press, Austin, pp. 100–140.

Baker, H. G. and P. D. Hurd. 1968. Intrafloral ecology. *Annu. Rev. Entomol.* **13**: 385–409.

Baker, H. G. 1961. *Ficus* and *Blastophaga. Ecology* **15**: 375–79.

Barton Browne, L. 1975. Regulatory mechanisms in insect feeding. *Adv. Insect Physiol.* **11**: 1–116.

Batra, S. W. T. and L. Batra, 1966. The fungus gardens of insects. *Sci. Amer.* **217**: 112–20. (November).

Bequaert, J. 1922. Ants in their diverse relations to the plant world. *Bull. Amer. Mus. Nat. Hist.* **45**: 333–621.

Brues, C. T. 1946. *Insect Dietary. An Account of the Food Habits of Insects.* Harvard Univ. Press, Cambridge, Mass., 466 pp.

Burgess, J. W. 1976. Social spiders. *Sci. Amer.* **234**: 101–107 (March).

Chapman, R. F. 1974. Feeding in leaf-eating insects. In *Oxford Biology Readers,* J. J. Head, (ed.). Oxford Univ. Press, 16 pp.

Dethier, V. G. 1962. *To Know a Fly.* Holden-Day, San Francisco, Calif., 119 pp.

Dethier, V. G. 1966. Feeding behavior. Royal Entomol. Soc. London Symposium 3, pp. 46–58.

Dethier, V. G. 1970a. Some general considerations of insects' responses to the chemicals in food plants. In *Control of Insect Behavior by Natural Products*, D. L. Wood, R. M. Silverstein, and M. Nakajima (eds.). Academic Press, New York, pp. 21–28.

Dethier, V. G. 1970b. Chemical interactions between plants and insects. In *Chemical Ecology*, E. Sondheimer and J. B. Simeone, (eds.). Academic Press, New York, pp. 83–102.

Dethier, V. G. 1976. *The Hungry Fly*. Belknap Press of the Harvard Univ. Press, Cambridge, Mass., 512 pp.

Dias, B. F. 1975. Comportamento pre-social de sinfitas do Brazil Central. I. *Themos olfersi* (Klug) (Hym., Argidae). *Studia Entomol.* **18:** 401–32.

Eberhard, W. G. 1977. Aggressive chemical mimicry by a bolas spider. *Science* 198:1173–75.

Ehrlich, P. R. and P. H. Raven. 1964. Butterflies and plants: a study in coevolution. *Evolution* **18:** 586–608.

Ehrlich, P. R. and P. H. Raven. 1967. Butterflies and plants. *Sci. Amer.* **216:** 104–13 (June).

Eisner, T. and J. Sheppard. 1965. Caterpillar feeding on a sundew plant. *Science* **150:** 1608–9.

Feeny, P. P. 1970. Seasonal changes in oak leaf tannins and nutrients as a cause of spring feeding by winter moth caterpillars. *Ecology* **51:** 565–81.

Fraenkel, G. 1959. The raison d'etre of secondary plant substances. *Science* **129:** 1466–70.

Fraenkel, G. 1969. Evaluation of our thoughts on secondary plant substances. *Entomol. Exp. Appl.* **12:** 473–86.

Galil, J. and D. Eisikowitch. 1969. On the pollination ecology of *Ficus sycomorus* in East Africa. *Ecology* **49:** 259–69.

Galil, J., W. Ramirez B., and D. Eisikowitch. 1973. Pollination of *Ficus costaricana* and *F. hemsleyana* by *Blastophaga esterae* and *B. tonduzi* in Costa Rica (Hymenoptera: Chalcidoidea, Agaonidae). *Tijdschr. Entomol.* **116**(11): 175–183.

Gelperin, A. 1971. Regulation of feeding. *Annu. Rev. Entomol.* **16:** 365–78.

Gelperin, A. 1972. Neural control systems underlying insect feeding behavior. *Amer. Zool.* **12:** 489–96.

Gilbert, L. E. 1971. Butterfly-plant coevolution: has *Passiflora adenopoda* won the selectional race with heliconiine butterflies? *Science* **172:** 585–6.

Gilbert, L. E. and P. H. Raven, (eds.). 1975. *Coevolution of Animals and Plants*. Univ. of Texas Press, Austin, 246 pp.

Gressitt, J. C., J. Sedlacek, and J. J. H. Szent-Ivany. 1965. Fauna and flora on the backs of large Papuan moss-forest weevils. *Science* **150:** 1833–1835.

Hartzell, A. 1967. Insect ectosymbiosis. In *Symbiosis*, Vol. 2, Henry, S. M. (ed.). Academic Press, New York, pp. 107–40.

Heinrich, B. 1971. The effect of leaf geometry on the feeding behaviour of the caterpillar of *Manduca sexta* (Sphingidae). *Anim. Behav.* **19:** 119–24.

Heinrich, B. 1973. The energetics of the bumblebee. *Sci. Amer.* **288:** 96–102 (April).

Heinrich, B. 1975. Energetics of pollination. *Annu. Rev. Ecol. Syst.* **6:** 139–170.

Heinrich, B. 1976. Bumblebee foraging and the economics of sociality. *Amer. Sci.* **64:** 384–395.

Heinrich, B. and P. H. Raven. 1972. Energetics and pollination ecology. *Science* **176:** 597–602.

Hocking, B. 1975. Ant-plant mutualism: evolution and energy. In *Coevolution of Animals and Plants*, Gilbert, L. E. and P. H. Raven (eds.). Univ. of Texas Press, Austin, pp. 78–90.

Hölldobler, B. 1971. Communication between ants and their guests. *Sci. Amer.* **224:** 86–93 (March).

Hsiao, T. H. 1974. Chemical influence on feeding behavior of *Leptinotarsa* beetles. In *Experimental Analysis of Insect Behaviour*, L. Barton Browne, (ed.). Springer-Verlag, New York, pp. 237–246.

Ishay, J. and R. Ikan. 1969. Gluconeogenesis in the Oriental hornet *Vespa orientalis* F. *Ecology* **49:** 169–71.

Janzen, D. H. 1966. Coevolution of mutualism between ants and acacias in Central America. *Evolution* **20:** 249–275.

Janzen, D. H. 1967. Interaction of the bull's horn acacia (*Acacia cornigera* L.) with an ant inhabitant (*Pseudomyrmex ferruginea* F. Smith) in Eastern Mexico. *Univ. Kans. Sci. Bull.* **67:** 315–558.

Janzen, D. H. 1968. Host plants as islands in evolutionary and contemporary time. *Amer. Nat.* **102:** 592–595.

Janzen, D. H. 1969a. Seed-eaters versus seed size, number, toxicity and dispersal. *Evolution* **23:** 1–27.

Janzen, D. H. 1969b. Allelopathy by myrmecophytes: the ant *Azteca* as an allelopathic agent of *Cecropia*. *Ecology* **50:** 147–153.

Janzen, D. H. 1975a. *Ecology of Plants in the Tropics. Studies in Biology*. Crane Russak & Co., New York 80 pp.

Janzen, D. H. 1975b. *Pseudomyrmex nigropilosa:* a parasite of a mutualism. *Science* **188:** 936–937.

Kaplanis, J. N., M. J. Thompson, W. E. Robbins, and B. M. Bryce. 1967. Insect hormones: Alpha ecdysone and 20-hydroxy ecdysone in bracken fern. *Science* **157:** 1436–1438.

Kistner, D. H. 1969. The biology of termitophiles. In *Biology of Termites*, Vol. 1, K. Krishna and F. M. Weesner (eds.). Academic Press, New York, pp. 525–57.

Koch, A. 1967. Insects and their endosymbionts. In *Symbiosis*, Vol. II, Henry, S. M. (ed.). Academic Press, New York, pp. 1–106.

Kogan, M. 1975. Plant resistance in pest management. In *Introduction to Insect Pest Management*, Metcalf, R. L. and W. H. Luckman, (eds.). Wiley, New York, pp. 103–146.

Kogan, M. 1977. The role of chemical factors in insect/plant relationships. Proc. XV Internat. Congr. Entomol., Washington, D.C., pp. 211–227.

Krieger, R. I., P. P. Feeny, and C. F. Wilkinson. 1971. Detoxification enzymes in the guts of caterpillars: an evolutionary answer to plant defenses? *Science* **172:** 579–580.

Kulman, H. M. 1971. Effects of insect defoliation on tree growth and mortality of trees. *Annu. Rev. Entomol.* **16:** 286–324.

Labeyrie, V. (ed.). 1977. *Comportement des Insectes et Milieu Trophique*. Coll. Internat. Centre National de la Recherche Scientifique No. 265, Paris, France, 493 pp.

Levin, D. A. 1973. The role of trichomes in plant defense. *Quart. Rev. Biol.* **48:** 3–15.

Major, R. T. 1967. The Ginkgo, the most ancient living tree. *Science* **157:** 1270–1273.

Martin, M. M. 1970. The biochemical basis of the fungus-attine ant symbiosis. *Science* **169:** 16–20.

Meeuse, B. J. D. 1961. *The Story of Pollination*. Roland Press, New York, 243 pp.

Montagner, H. 1966. Le mécanisme et les consequences des comportements trophallactiques chez les guêpes du genre *Vespa*. Thèses, Faculté des Sciences de l'Université de Nancy, France, 143 pp.

Morgan, F. D. 1968. Bionomics of Siricidae. *Annu. Rev. Entomol.* **13**:239–56.

Nault, L. R. and W. E. Styer. 1972. Effects of sinigrin on host selection by aphids. *Entomol. Exp. Appl.* **15**: 423–437.

Opler, P. A. 1974. Oaks as evolutionary islands for leaf-mining insects. *Amer. Sci.* **62**: 67–73.

Owen, D. F. and R. G. Wiegert. 1976. Do consumers maximize plant fitness? *Oikos* **27**: 488–492.

Peschken, D. P. 1972. *Chrysolina quadrigemina* (Coleoptera: Chrysomelidae) introduced from California to British Columbia against the weed *Hypericum perforatum*: comparison of behaviour, physiology, and colour in association with postcolonization adaptation. *Can. Entomol.* **104**: 1689–1698.

Pliske, T. E. 1975. Courtship behavior of the monarch butterfly, *Danaus plexippus* L. *Ann. Entomol. Soc. Amer.* **68**: 143–151.

Price, P. W. 1975. *Insect Ecology*. Wiley, New York, 514 pp. (especially Chapters 2–4 and 20).

Ramirez, B. W. 1969. Fig wasps: mechanism of pollen transfer. *Science* **163**: 580–581.

Ramirez, B. W. 1970. Host specificity of fig wasps (Agaonidae). *Evolution* **24**: 680–91.

Rathcke, B. J. and R. W. Poole. 1975. Coevolutionary race continues: butterfly larval adaptation to plant trichomes. *Science* **187**: 175–176.

Read, D. P., P. P. Feeney, and R. B. Root. 1970. Habitat selection by the aphid parasite *Diaeretiella rapae* (Hymenoptera: Braconidae) and hyperparasite *Charips brassicae* (Hymenoptera: Cynipidae). *Can. Entomol.* **102**: 1567–1578.

Reichstein, T., J. von Euw, J. A. Parsons, and M. Rothschild. 1968. Heart poisons in the monarch butterfly. *Science* **161**: 861–866.

Riddiford, L. M. 1967. Trans-2-hexenal: mating stimulant for *Polyphemus* moths. *Science* **158**: 139–140.

Robinson, M. H., H. Mirick, and O. Turner. 1969. The predatory behavior of some araneid spiders and the origin of immobilization wrapping. *Psyche* **76**: 487–504.

Salt, G. 1937. The sense used by *Trichogramma* to distinguish between parasitized and unparasitized hosts. *Proc. Roy. Soc. London Ser. B*, **122**:57–75.

Schoonhoven, L. M. 1969. Gustation and foodplant selection in some lepidopterous larvae. *Entomol. Exp. Appl.* **12**: 555–564.

Schoonhoven, L. M. 1968. Chemosensory bases of host plant selection. *Annu. Rev. Entomol.* **13**: 115–136.

Staedler, E. 1977. Sensory aspects of insect plant interactions. Proc. XV Internat. Congr. Entomol., Washington, D.C., pp. 228–248.

Stoffolano, J. G., Jr. 1968. The effect of diapause and age on the tarsal acceptance threshold of the fly, *Musca autumnalis*. *J. Insect Physiol.* **14**: 1205–1214.

Stoffolano, J. G., Jr. 1974. Control of feeding and drinking in diapausing insects. In *Experimental Analysis of Insect Behaviour*, L. Barton Browne (ed.). Springer-Verlag, New York, pp. 32–47.

Swain, T. 1977. The effect of plant secondary products on insect plant co-evolution. Proc. XV Internat. Congr. Entomol., Washington, D. C., pp. 249–56.

Thorsteinson, A. J. 1960. Host selection in phytophagous insects. *Annu. Rev. Entomol.* **5:** 193–218.

Tinbergen, N. 1972. *The Animal in its World. Explorations of an Ethologist, 1932–1972,* Vol. I. *Field Studies.* Harvard Univ. Press, Cambridge, Mass., 343 pp. (especially pp. 128–145).

Tophoff, H. 1977. The pit and the antlion. *Nat. Hist.* **86:** 64–71.

Van Emden, H. F. (ed.). 1972. *Insect/Plant Relationships.* Royal Entomol. Soc. London Symposium 6, 215 pp.

Vinson, S. B. 1975. Biochemical coevolution between parasitoids and their hosts. In *Evolutionary Strategies of Parasitic Insects and Mites,* Price, P. W. (ed.). Plenum Press, New York, pp. 14–48.

Vinson, S. B. 1976. Host selection by insect parasitoids. *Annu. Rev. Entomol.* **21:** 109–33.

Wallace, J. B. 1975. The larval retreat and food of *Arctopsyche*; with phylogenetic notes on feeding adaptations in hydropsychid larvae (Trichoptera). *Ann. Entomol. Soc. Amer.* **68:** 167–173.

Wallace, J. B. and F. F. Sherberger. 1975. The larval dwelling and feeding structure of *Macronema transversum* (Walker) (Trichoptera: hydropsychidae). *Anim. Behav.* **23:** 592–596.

Wallace, J. W. and R. L. Mansell (eds.). 1976. *Biochemical Interaction between Plants and Insects.* Plenum Press, New York, 426 pp.

Way, M. J. 1963. Mutualism between ants and honeydew-producing Homoptera. *Annu. Rev. Entomol.* **8:** 307–344.

Weber, N. A. 1966. Fungus-growing ants. *Science* **153:** 587–604.

Weber, N. A. 1972. *Gardening Ants: the Attines. Mem. Amer. Phil. Soc.* No. 92. Amer. Phil. Soc., Philadelphia, 146 pp.

Weires, R. W. and H. C. Chiang. 1973. Integrated control prospects of major cabbage insect pests in Minnesota—based on the faunistic, host varietal, and trophic relationships. Univ. Minn. Agr. Exp. Sta. Tech. Bull. 291, 42 pp.

Wheeler, W. M. 1918. A study of some ant larvae with a consideration of the origin and meaning of social habits among insects. *Proc. Amer. Phil. Soc.* **57:** 293–343.

Wheeler, W. M. 1923. *Social Life Among the Insects.* Harcourt, Brace and Co., New York, 375 pp.

Wheeler, W. M. 1942. Studies of Neotropical ant-plants and their ants. *Bull. Mus. Comp. Zool. Harv.* **90:** 1–262.

Wigglesworth, V. B. 1964. *The Life of Insects.* The New American Library, New York, 383 pp. (especially Chapter 4).

Wilde, J. de and L. M. Schoonhoven (eds.). 1969. *Insect and Host Plant.* North-Holland, Amsterdam, 337 pp. Also published in *Entomol. Exp. Appl.* **12:** 471–810.

Williams, C. M. 1970. Hormonal interactions between plants and insects. In *Chemical Ecology,* E. Sondheimer and J. B. Simeone (eds.). Academic Press, New York, pp. 103–132.

Wilson, E. O. 1971. *The Insect Societies.* Belknap Press of Harvard Univ. Press, Cambridge, Mass., 548 pp.

Wilson, E. O. and T. Eisner. 1957. Quantitative studies of liquid food transmission in ants. *Insectes Sociaux* **4:** 157–166.

# 5

# Chemical
# Communication

A flashing firefly flits through the evening shadows. In a tree, a caterpillar stiffens and sways back and forth. Behind a stone, a cricket chirps, while nearby ants scurry along in precise single file. What is the common thread in all these diverse actions? In each case, the insect is communicating, that is, sharing information coded in the form of signals. More precisely, each is producing a signal, measurable in the language of physics or chemistry, that alters the response patterns of another individual.*

The firefly's flash and the cricket's chirp may have the same function, to gain a mate. The caterpillar in essence sends the predator world the message that it is not food. The ants share their message that travel along this particular trail is apt to be rewarding. But while the firefly, cricket, and ants are communicating with their own kind, the caterpillar obviously is not. Thus, insect signals may be of two broad, somewhat overlapping categories; even when the method of signaling is the same, the adaptive result of these two types of communication may be quite different.

Communication is vital, first and foremost, in establishing an insect's credentials within its species and in recognizing the credentials of other insects as being like or unlike itself. One important function of intraspecific communication systems is species and sex recognition; others include identification of populations or individuals, alarm, and social coordination. Not only is it a waste of both time and energy to attempt matings likely to end in reproductive failures, but the strongest competitor for available resources is often a member of one's own species, so these must be recognized and either avoided, driven away, or cooperated with.

Once the species has been recognized, subtler distinctions may be

---

* In its very broadest application, *any* sensory signal that produces a behavioral change in an animal is communication. The message need not even emanate from another animal; it could equally come from a plant or an inanimate object.

made. Ants and honey bees, for example, are able to recognize the smell of the members of their own colonies. Although individual recognition is clearly less well developed among insects than among, for example, higher vertebrates, it does exist in some cases, often accompanied by a system of linear dominance or "peck order" not unlike the classic case in chickens which gave it that name. For example, in paper wasps with multiple foundress queens, one dominant female eventually becomes the sole egg layer, in a peck order based upon ovarian development and related behaviors and maintained by individual recognition (see Chapter 9).

But although the language associated with recognition and/or courtship is an essentially private one, many other types of communication transpire between widely different members of the animal kingdom. For example, the noise made by a death's head moth when picked up has caused more than one human to drop the specimen hurriedly! Interspecific communication such as this functions in a broad spectrum of attack- and escape-related responses, including alarm, threat, intimidation, and crypticity. Of course, in some cases the same message may be conveyed both between and within species, while in other cases the same signal may send a very different message concurrently within and between species.

It is customary to speak of signals as being sexual, aggressive, alarm, etc. However, problems arise with this. For one, it is often difficult to distinguish the responses that such categories of signals will actually elicit, for the environment, or "context," of a signal may alter its message. For example, the same chemical in harvester ants may elicit alarm under one set of circumstances but elicit approach in another situation. Otte's (1974) classification of animal communication systems (Table 5-1) is based on recognition that a given signal may have various context-dependent meanings. A second problem is that the respondent's first detectable behavioral response to many signals is to change positions; as it moves, it enters new stimulus situations. In analyzing the insect's ultimate behavior, it becomes difficult to separate the influence of these new situations from that of the initial signals.

Of all the various schemes that have been developed to classify communication, however, perhaps the most widely accepted and utilized has been the one based upon the receptor involved. In terms of the transmission and reception channels, communication may be *visual, acoustical, chemical, tactile* (mechanical), or *electrical*. It is around this classification that our discussion of insect communication is organized in the present and the two succeeding chapters. (Electrical communication is not known among the insects, and we treat mechanical and acoustical

**Table 5-1**    A classification of basic types of communicative signals based on the type of information conveyed. A given signal may contain several types of information. The italicized words relate to the critical information involved[a].

---

   I.   Deictic information (signals or parts of signals that draw attention to an individual or object)
        *Look!*
  II.   Identification or indexical information (individual or class information, e.g., individual, sex, species, colony, age, relationship, etc.)
        I am *X*
        I belong to *species X*
 III.   Spatial information (location in space, direction, distance)
        to the *right*
        *90° to the right of the sun*
  IV.   Response level information (information on the state of the emitter, on levels of probability of aggression, submission, sexual receptivity, etc., or on quantity or quality of exploitable resources such as food, nest sites, or shelter)
        The food source is *rich*
        A predator is *nearby*
        I am *very* angry
   V.   Temporal information (information on whether some event will occur, is occurring, or has already occurred)
        I will *soon* be ready
  VI.   Event information (information on events taking place, such as the absence, presence, or need for food; presence of shelter, predators; or information on behavior that will take place)
        I *need food*
        A *predator is nearby*

---

[a] From Otte (1974).

modes together in Chapter 7.) It must be remembered, however, that an insect may be receiving information simultaneously in a number of sensory modes, and in many cases the total message and its specificity may depend upon receipt of all these different channels together.

Chemical communication is the most thoroughly studied. One result has been the development of an extensive chemical terminology which has tended to artificially sunder these signals from other types. In actuality, the behavioral effect of a signal is similar no matter what the communicative mode. Some scientists have suggested that it might help if a single term such as biosemants were used to designate communication signals in general. Specific classes of signals might then be recognized—phonosemants, chemosemants, etc. The study of all animal communication then would become **zoosemiotics** (Sebeok, 1968).

## MECHANISMS OF CHEMICAL COMMUNICATION

It is a hazy summer morning just before dawn, and a man follows a footpath to a clearing in the middle of an apple orchard. Reaching up, he removes the tray that hangs from a wooden post and smiles with satisfaction: the sticky tray top is covered with ¾-inch-long rust-and-yellow moths. As he starts back toward his pickup truck, the man pauses to watch a flying moth. It zigzags, comes closer, then hovers over the tray, extending its genital claspers. As its wingtips brush the cardboard and are caught by the gummy surface, its abdomen curves upward, throbbing convulsively as the moth tries to mate with a polyethylene disc in the center of the tray. What has happened here? The scene was a carefully chosen test site, the man a field entomologist; the moths were male red-banded leaf rollers (*Argyrotaenia velutinana*). The plastic disc was dipped in a compound that smelled like a secretion produced by a virgin female red-banded leaf roller. Male moths were lured by the smell to the trap where, instead of mating, these destructive apple orchard pests were captured (Roelofs, 1975).

That odors secreted by some insects might function as stimulants inciting those of the opposite sex to mate may seem obvious, especially since some of the odors are apparent to humans. In fact, the German zoologist von Siebold proposed such an idea about 140 years ago. But during the 1800s the phenomenon of chemical communication was largely ignored or disbelieved. Even about 1890, when the French naturalist Fabre verified that a female great peacock moth caged in the dark could still attract large numbers of male moths, he was reluctant to attribute it to chemical communication. As did most others, Fabre easily accepted the idea that insects could detect the odors of other insects. But he could not believe that these odors, which the human nose was unable even to detect, could operate over such long distances; it simply strained credibility to think that a single female moth could inject into the vast atmosphere enough of a substance to be perceived by males kilometers away. This would be like tinting an entire lake with a single drop of dye. No, thought Fabre, it was more likely that for long-range communication the female moth emitted, not an odor, but some sort of olfactory radiation, something like the radio which was then in its infancy.

But though the idea strains credibility, the female peacock moth can and does emit a substance that can be detected by a potential mate a million times her own body length away. And she is not an isolated example (see McIndoo, 1917). As increasingly intensive entomological attention has focused upon this subject, it has become increasingly evident that the role of odors in communication is an extremely important one. In fact, it is now usually considered that, in the evolution of animal communications as a whole, chemical signals may have been among the earliest to arise, perhaps even before the formation of multicellular organisms. Any primitive communication between protozoan cells was almost certainly chemical, and some scientists have even postulated a lineal evolutionary relationship between these chemical releasers and the hormones through which cells in the metazoan body communicate. Whatever its relation to endocrine evolution, communication through chemistry has been found to be a persistent theme in almost all animal taxa (Wilson, 1970).

To date, the largest number of communication chemicals in the animal world has been found among the members of the largest class of animals, the insects. It is here that they have been studied longest and here that science knows most about their modes of action. But before examining them in detail, we should underscore two important characteristics of these chemical messengers. First, they occur far more widely than anyone would have suspected even a few years ago. Hundreds of such substances have been found in both sexes and all life stages of many species, not just of insects but spanning the zoological universe (Wilson, 1975). Second, they communicate matters vital to survival, such as the presence of danger, the identification of friends and foes, the call to arms or emigration, the availability of food, the urge to reproduce. They are chemical hotlines, not polite conversational perfumes.

Where do insect communication chemicals originate? How are they received? Are the insects smelling their chemical messages or are they tasting them? At the turn of the century, Mayer (1900) undertook some simple experiments to establish the mechanism by which male promethea (*Hyalophora promethea*) moths are attracted to females. First, when he placed five female moths in a glass jar topped with mosquito netting, Mayer noted that male promethea moths over 100 feet away were immediately attracted. Next, he inverted the jar and packed sand around its mouth so that air could not escape; now males were no longer attracted. So perhaps smell was involved; was sight also involved? Mayer wrapped female moths loosely in cotton to make them invisible; males not only came but grasped the cotton in their abdominal claspers in typical copulatory attempts. On another set of moths, Mayer replaced female wings with male wings; males still came and mated without hesitation. Having ruled out the role of sight in male attraction, he performed one more experiment to establish that only chemical cues were involved. He set up a small wooden box containing females, so that air blown into the box came out through a small chimney. True to prediction, the males were attracted not to the box but to the chimney top, even when the vicinity contained fumes of carbon disulfide and diethyl sulfide.

Where were the moth's odors being emitted? Mayer severed female moth bodies and placed the pieces in various locations. Males released five feet away flew to the abdomens and ignored the remainder. How were the odors being received? Males with their abdomens cut off still responded; so did males whose spiracles were covered with glue. However, males whose antennae were gummed up did not seek females. In fact, they showed no excitement even when held within 1 inch of virgin females.


## ODOR PRODUCTION AND CHEMORECEPTION

Perhaps the most outstanding characteristic about insect odor sources has been their awesome variety and complexity, especially among the social species (Fig. 5-1). Chemicals come not only from the abdomen but often from head or thorax. A variety of **exocrine glands**, clusters of secretory cells whose products are discharged to the outside of the body, send forth single liquids or blends, releasing them as streams, droplets, thin films, aerosols, or gases. Emission rates and concentrations may be controlled through adjustable nozzles, retracting applicators, evaporation pads, or other equally elaborate devices.

Insect receivers are no less complex and sophisticated. By convention, two different types of chemical receiver systems are recognized, taste and smell, analogous to gustation and olfaction in vertebrate animals. Strictly speaking, insect taste stimuli are waterborne compounds with a limited range of qualities, whereas odor stimuli vary widely and may be either airborne or waterborne (though usually they are gaseous). An additional difference between these two chemical detection systems lies in their thresholds. Insect taste cells require that stimulating molecules be in much higher concentration than do odor receptors to elicit a reaction. As all these facts taken together might indicate, in a situation analogous to that in most vertebrates, the chemosensory system of insects seems to rely far more heavily upon smell than on taste.

Insect chemoreceptor cells are surrounded by characteristic cuticular specializations which comprise the most obvious external parts of the chemosensory organ. Called **sensilla**, these occur in at least four chemically sensitive forms: bristles or hairs, pegs, plates, and pits (Fig. 5-2). Taste receptors from the labellum of the blowfly (see Chapter 4) are

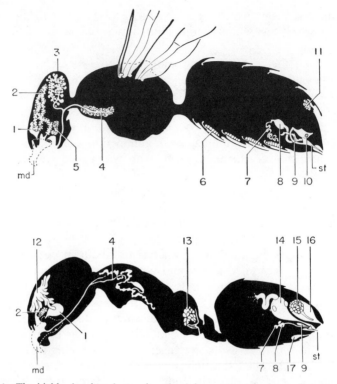

**Figure 5-1** The highly developed exocrine glandular system of a honey bee worker (top) and a typical worker ant belonging to the genus *Iridomyrmex* (bottom), are adapted to produce chemicals for communication. Some glands, such as the mandibular gland, have storage reservoirs and release their chemicals in bursts as needed; others secrete chemicals more or less continuously. (1) Mandibular gland; (2) hypopharyngeal (maxillary) gland; (3) head labial gland; (4) thorax labial gland; (5) postgenal gland; (6) wax glands; (7) poison gland; (8) vesicle of poison gland; (9) Dufour's gland; (10) Koschevnikov's gland; (11) Nassanoff's gland; (12) postpharyngeal gland; (13) metapleural gland; (14) hindgut (glandular nature uncertain); (15) anal gland; (16) reservoir of anal gland; (17) Pavan's gland; st = sting; md = mandible. (From Wilson, 1965. Copyright 1965 by the American Association for the Advancement of Science.)

among the best studied physiologically. A review of insect chemoreception is provided by Hodgson (1974).

Because insect antennae are by far the most common organ of olfactory reception, a brief glance at one well-studied pair (Fig. 5-3) is of interest. The peripheral part of any sensory cell reacts to an adequate stimulus with a temporary change in the electric charge of its membrane, in a measurable response termed the **receptor potential**. Studies of whole

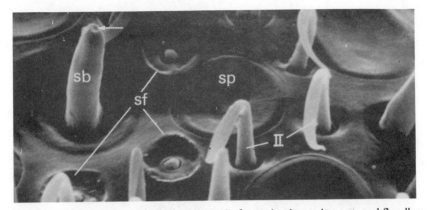

**Figure 5-2** Portion of the surface of segment 10 of a worker honey bee antennal flagellum showing four types of sensilla: sf = sensilla companiformia, sp = sensilla placodea, sb = sensilla basiconica, II = sensilla trichodea. A single antenna may have in excess of 8000 of the sensilla trichodea alone. These and other sensilla types (not shown) respond to a wide range of chemical and mechanical stimuli present in the bee's environment. Similar receptor structures also may occur on the mouthparts, ovipositor, cerci, and other structures among various insect groups. (Courtesy of A. Dietz; from Dietz and Humphreys, 1971, 3900× magnification.)

antennal responses frequently utilize the **electroantennogram** (EAG), which gives a measurement of the summed receptor potentials of a number of olfactory receptors responding to a stimulus.

By means of microelectrodes and odor-laden air blown through a glass tube, one can test the responses of various parts of an insect's antennae to various concentrations of a chemical, its isomers, and its homologs. Signals from the antenna are amplified and displayed on an oscilloscope. For example, 50% of the male silkworm moth's antennal odor-receptor cells are tuned to respond to a single substance, the sex attractant released by a receptive female. The antennae of the female do not respond to this scent at all.

Microelectrode probes reveal also that the only receptor cells that respond to the attractant and its isomers are those which are connected to the specialized long hairs of the antennae. Near the threshold level, experiments with tritium-labeled synthetic attractant present convincing evidence that the receptors actually "count" the stimulating molecules, each cell being so sensitive that it reacts to a single quantum of odor. (This sensitivity is comparable to that of the rod cells of the vertebrate retina, which respond to single quanta of light.) When a sufficient number of cells (some 40 out of 40,000) have been activated to overcome the background noise caused by the spontaneous firing of receptors in

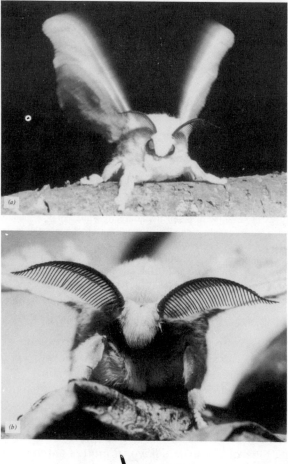

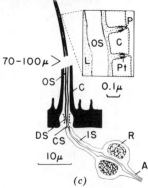

the resting state, a message is received in the moth's brain that the attractant is in the air.

It appears that the spacing and arrangement of moth antennal hairs actually allows them to act as molecular sieves, the width of the mesh they form being small enough so that pheromone molecules, because of their fast thermal movements, cannot pass through without contacting the hairs and being preferentially absorbed. Upon contact, the molecules diffuse through pore openings connected to fine tubules and give rise to an electrical change in the membrane of the basal receptor cells. (For more detailed information about insect olfaction, see Schneider, 1969, 1970, 1971, 1974, and Kaissling, 1971.)

Many receptor organs show interesting parallels to electronic communication receivers, with functional components typically including detectors, decoders, transducers, miniaturized circuitry, and relays. However, their efficiency and sensitivity far exceed any man-made devices (Blum, 1974). Some of the better man-made electronic detection systems currently require input about double the background noise (a signal-to-noise ratio of about 2:1) for reliable efficiency. In contrast, the male silkworm moth can make a positive behavioral response at the incredibly low signal-to-noise ratio of 0.125:1. The sensitivity of male gypsy moths is so great that scientists at the USDA estimate that the 30 grams of disparlure, the synthetic attractant, already on hand will be enough to bait some 60,000 traps per year for the next 50,000 years (Jacobson, 1972).

Because the principal insect olfactory organs are located upon a pair of antennae jutting out from the head, various theories have been advanced to explain this placement (Amoore et al., 1964). One is that use of two such identical sets of receptors may serve to maximize sensitivity and efficiency of the receptor system. Ants following an odor trail characteristically walk a zigzag route in which each antenna appears to be moved alternately in and out of the odor field. The "out" antenna suddenly stops sending signals to the brain; as though seeking to restore

---

Figure 5-3  Olfactory receptor system of the male *Bombyx mori*. (*a*) Wing fluttering and orientation in response to the odor of a female. (Courtesy of D. Schneider.) (*b*) Portrait showing the plumose antennae, each containing 60–70 branches. Each branch is in turn covered with about 17,000 olfactory hairs which average $100\mu$ in length. (Courtesy of D. Schneider.) (*c*) A single olfactory hair (sensillum trichodeum) in longitudinal section. Two receptor cells (R) innervate the hair, and each dendrite is made up of the inner segment (IS), the ciliary segment (CS), and the outer segment (OS), which is bathed in a liquor (L). A portion of the hair wall (C) is enlarged to show the pore-tubule (Pt) system and pores (P) through which odor molecules are adsorbed. The dendrites are enveloped by a dendrite sheath (DS) at the hair base. (Drawing by R. A. Steinbrecht from Schneider, 1971.)

a balance of input from the two sides, the brain causes a steering change which soon results in overcompensation toward the opposite side, and so on. If both antennae remained continuously in the odor field, habituation might quickly occur. By passing in and out of the threshold concentration level, maximal sensitivity to the stimulus would be likely to persist. In addition, such a system imparts a clear orientation axis to

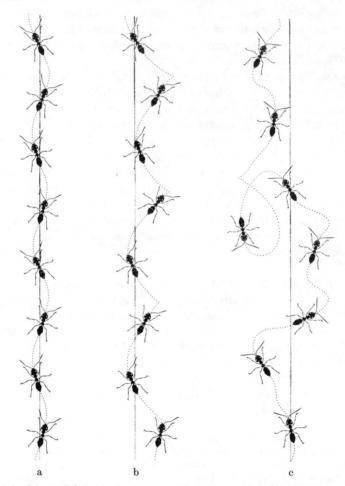

          a                                b                                  c

**Figure 5-4**   Odor trail following by *Lasius* worker ants. (a) Normal trail following in which the ant zigzags evenly first to one side then to the other as the antennae alternately move in and out of the trail's vapor space. (b) With the left antenna amputated, the ant repeatedly overcorrects to the right side. (c) With the antennae crossed and glued, the ant is disoriented and relocates the trail with difficulty; its overall progress in the proper direction is probably mediated by visual cues. (From Hangartner, 1967.)

the signal path. Orientation experiments using ants or bees with amputated or crossed antennae lend support to these ideas (Fig. 5-4).

In concluding this section it is appropriate to consider the odorant chemicals themselves. Wilson and Bossert (1963) have convincingly argued (Table 5-2) that they should have a carbon number between 5 and 20 and a molecular weight between 80 and 300. In addition, they predict, chemicals used for purposes requiring a high degree of specificity (such as sex attractants) should have greater molecular size than those odor signals used to elicit behaviors (such as alarm) for which species uniqueness is less essential. In the main, these predictive generalizations have been upheld by accumulating empirical data; we will refer to them again later.

## TYPES OF CHEMICAL COMMUNICATION

As knowledge about chemical communication systems has grown, the subject has rapidly become more complex. In 1959, the German chemists Karlson and Butenandt and the Swiss zoologist Lüscher proposed a new name for some of these messenger chemicals to replace the contradictory term "ectohormone" then in use. From the Greek *pherein*, to carry, and *horman*, to excite, they derived the word **pheromone**— a substance secreted by an animal that affects the behavior of other animals of the same species.

In 1963, Wilson and Bossert proposed dividing pheromones into two

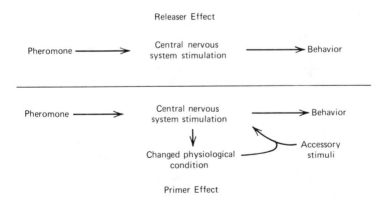

**Figure 5-5**  Pathways of influence of pheromone action. Releaser effects are the classical stimulus–response reactions mediated wholly by the central nervous system. With the primer effect, behavior is usually induced not by the pheromone but by subsequent accessory external stimuli. (Based on Wilson, 1963.)

**Table 5-2** Chemical criteria for airborne communication substances[a]

| Carbon number and molecular weight | Molecular diversity | Olfactory efficiency | Energy expense | Volatility | Related considerations |
|---|---|---|---|---|---|
| Extremely low—carbon number below 5, molecular weight below 80 | Very limited | Low | Low | Very high | Possible difficulty in glandular storage |
| Carbon number between 5 and 20; molecular weight between 80 and 300 | Increasing exponentially; great number of unique compounds possible | Increasing steeply | Intermediate | Intermediate | Differences in diffusion coefficient in this range do not cause much change in properties of active space |
| Very high—carbon number above 20, molecular weight above 300 | Astronomical | Further increases probably confer little or no further advantage | Great expense required to synthesize and transport large molecules | Low | May lead to difficulty in maintaining adequate active space |

[a] Based on Wilson and Bossert (1963).

188

functional groups according to their mode of influence: releasers and primers (Fig. 5-5). **Releaser pheromones** stimulate an immediate behavioral response mediated wholly by the nervous system; these pheromones are thus by definition chemical "releasers" in the terminology of the ethologist (see Chapter 2). They are widespread in insects and serve a great many functions, sex attraction and alarm being two especially important ones. **Primer pheromones**, on the other hand, act to physiologically alter the endocrine and reproductive systems of the receptor animal, reprogramming it for an altered response pattern. In a sense, the receptor's body is "primed" for new biological activity, although such activity may not appear until some future time and may require triggering by another releaser pheromone. These indirect primer pheromones can act either by physiological inhibition or enhancement*. Thus, termite soldier and reproductive castes prevent other termites from developing into castes like themselves by secreting chemicals which are ingested and act at least in part by effects on the corpus allatum. Adult males of the migratory locust, on the other hand, secrete from their skin surface a volatile substance that accelerates the growth and synchronizes the development of young locusts; this plays an important role in the formation of migratory locust swarms.

Next, attention turned from substances that transmit external chemical messages within a species to those that affect individuals or populations of a species *different* from their source; these were termed **allelochemic** (Whittaker and Feeny, 1971). Allelochemic interactions may involve chemical messengers affecting the growth, health, behavior, or population biology of other species. Two categories of allelochemics have been recognized: allomones and kairomones. **Allomones** are chemical agents of adaptive advantage to the organism sending them; **kairomones**, on the other hand, are of adaptive value to the organism receiving them. Figure 5-6 attempts to summarize the diverse roles of chemicals in mediating interactions between organisms. All chemicals produced by one organism that incite responses in another organism are known as **semiochemicals**.

Not all chemicals mediating behavior fit neatly into the above cate-

---

* Perhaps the most novel primer effect in insects is the role played by the enzyme prostaglandin synthetase, transferred from the male to the female house cricket during copulation. Even when provided with moist sand, unmated females will not exhibit oviposition behavior; however, injection of minute amounts of prostaglandin synthetase will subsequently release the full blown oviposition behavior previously lacking. Apparently prostaglandin synthesized after mating acts on the CNS (in the manner of a modifier hormone) altering its responsiveness to environmental stimuli (Destephano and Brady, 1977).

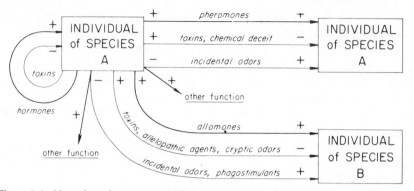

**Figure 5-6** Nonexhaustive summary of the role of semiochemicals in mediating interactions of individuals within and between species. Similar pathways could be constructed for other communication systems such as visual or auditory. The symbols (+) and (−) indicate whether the fitness of emitters or receivers is increased or decreased as a result of transmission. (From Otte, 1974. Reproduced, with permission, from the Annual Review of Ecology and Systematics, Volume 5. Copyright © 1974 by Annual Reviews Inc. All rights reserved.)

gories, however. For example, wasps parasitic on the larvae of certain fruit flies are attracted by compounds released during the fermentation and decay process of the fruit (Greany et al., 1977). Pasteels (1977) has suggested the term **ecomone** to encompass communicative signals which originate either from organisms or the abiotic environment. Duffey (1977) rightly cautions that the process of classifying the diverse responses of organisms to chemicals does not necessarily reflect reality and may obscure thinking about other possible interactions in the spectrum of chemical ecology. As noted in the first chapter, the very labeling of behavioral phenomena tends to color subsequent interpretations of it.

With chemical signals, the same compound may sometimes have two or three roles depending upon context. Nevertheless, these categories are broadly useful, for allomones tend to include compounds having roles in aggressive relationships, such as defensive secretions and repellents (see Chapter 8) and odors mediating mutualistic relationships such as flower scents which attract pollinators (see Chapters 4 and 9). Kairomones include host location cues such as natural plant products that attract herbivorous insects (see Chapter 4) and prey scents that attract predators and/or parasites. Note that from the standpoint of selection, kairomones must be viewed as evolutionary "backfires" for the emitting organism, originally selected to serve either pheromonal or allomonal functions of benefit to it but secondarily used by other organisms (herbivores, predators, parasitoids) to its detriment. Further discussion of the terminology of chemical communication is available in Regnier (1971), Blum (1974), Otte (1974), and Nordlund and Lewis (1976). For a detailed review of behavioral responses to insect pheromones, see Shorey (1973). Birch (1974a), Shorey (1976) and Shorey and McKelvey (1977) cover most aspects of pheromones in depth.

## THE FUNCTIONS OF CHEMICAL COMMUNICATION

Any time an organism gives a signal, there is at least potentially an intended receiver. However, the information being sent may also be intercepted for use by a variety of illegitimate receivers, for example, predators, parasites, and competitors. Furthermore, even for the intended receiver the message which the communicator encodes may differ somewhat from the meaning that the event might have for the recipient. In interpreting communicative emissions, one must be quite careful to distinguish *functions* from *incidental effects*. Functions must increase the fitness of the sender in order to evolve; the same does not necessarily hold true for incidental effects. For example, female rabbits release hormones into their blood which, when picked up by feeding fleas, stimulate flea reproduction. The function of the hormone—rabbit reproduction—is positive for the rabbit; the effect—an increased flea population—surely is not. Otte (1974) further discusses this problem.

### Sexual Stimulation

A diverse assemblage of compounds fall under the term sex pheromone. Though most commonly released by the female, sex pheromones may be produced by either sex. Sometimes, especially among beetles, both sexes may be attracted by the same scent; at some appropriate location such as a suitable host plant, they aggregate and mate. During pheromonal release, certain stereotyped postures called "calling positions" are common among a wide variety of unrelated insects (Fig. 5-7).

Often, the same substance calls in the opposite sex from a distance at low concentrations and elicits courtship behavior at the higher concentrations encountered at close range. Sometimes, however, separate substances are involved. **Aphrodisiacs** are substances produced by either sex (usually by the male and often as only part of a complex pattern of courtship behavior) that facilitate courtship or prepare the opposite sex for copulation *after* the pair have been brought together (Birch, 1974b). Often, attractants are produced by one sex, aphrodisiacs by the other.

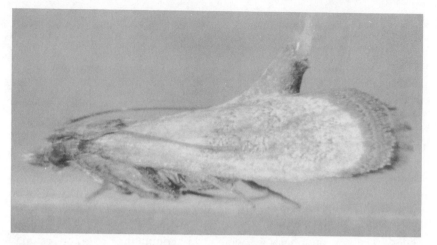

**Figure 5-7** "Calling" position adopted by a female Indian meal moth, *Plodia interpunctella*, during sex pheromone release. During calling the abdominal tip is elevated and glands in the intersegmental membrane between the eighth and ninth abdominal segments are extruded. A similar stance is taken by most calling moths as well as by a number of other unrelated insects. (Courtesy of E. Smithwick and U. E. Brady.)

For example, many butterfly species have virgin females that produce sex attractants, whereas males have aphrodisiac scent glands (Fig. 5-8). Special glandular scales, or **androconia**, are possessed by butterflies such as the male grayling *(Hipparchia semele)*; during courtship, he clasps the female's antennae between his wings, bringing them into contact with these scent scales.

Glands thought to have a similar function have been described in the Neuroptera, Trichoptera, Diptera, and Hymenoptera. The male monarch butterfly and its relatives exemplify a second type of aphrodisiac gland, **hairpencils**; these extrusible organs function as tiny scent-filled brushes (Fig. 5-8).

Some male Orthoptera, such as crickets *(Nemobius sylvestris)*, produce substances on which the females feed during copulation; this is also the case in some cockroaches, where the substance has been given the whimsical name seducin. Some workers feel that the function of these substances is to distract females so they do not eat the spermatophore. Others feel it is more likely that these are aphrodisiacs, perceived by both smell and taste, or that they serve a nutritional function for egg production. Feeding also places the female in the proper copulatory position.

In sexual communication it is especially important for a given species to maintain absolute integrity of its own communication channel. Thus, it is perplexing to find instances where males of two or more species respond to the synthesized female scent of one of them, such as occurs among moths. Such elicitation of cross-attractancy is apparently explained by the fact that most sex pheromones do not act alone. Instead, they are used in combination with other compounds which serve to enhance, inhibit, activate, or mask the primary attractant effects. In phytophagous species, sex odors are sometimes blended with compounds derived from the host plant. The natural occurrence of sex attractant blends also helps explain why crude extracts of sex pheromones often show much stronger attractancy than does purified material.

CASE STUDY: MATE ATTRACTION BY THE SILKWORM MOTH, *Bombyx mori*

*Bombyx mori* (see Fig. 5-3) is the most famous caterpillar in the world— the silkworm whose cocoon spun from a single strand 500 to 1300 yards long furnishes the material for a thread and cloth not yet duplicated by any synthetic fiber. The only truly domesticated insect, it can no longer maintain itself in a natural environment but survives only under cultivation, where most live only until the pupal stage when they are plunged into boiling water and their cocoons unraveled. But in each generation a few moths are allowed to survive to furry, greenish-white adulthood to furnish eggs for new progeny. Neither sex can fly, and therefore the male cannot easily scout the terrain to find a mate. Yet it has long been known that female *Bombyx* in some manner attract males from extraordinary distances.

In 1939, the chemist Butenandt reasoned that a biochemical lure in the bodies of virgin females was probably responsible. Although his available tools would now be regarded as primitive—he lacked such now standard equipment as the gas chromatograph—he began to work on the isolation and identification of the substance.

By 1959, Butenandt and his associates had processed half a million female silkworm moths and had extracted twelve-thousandths of a gram of derivative of the active compound. Using a combination of techniques including chromatography, infrared and ultraviolet spectroscopy, and chemical structural analysis coupled with biological assays, Butenandt obtained a substance which, in minute quantities, was as attractive to male silkworm moths as the most seductive virgin female. It was a primary alcohol with the formula *trans*-10-*cis*-12-hexadecadien-1-ol, which they named "bombykol" (Butenandt et al., 1959). It was the first

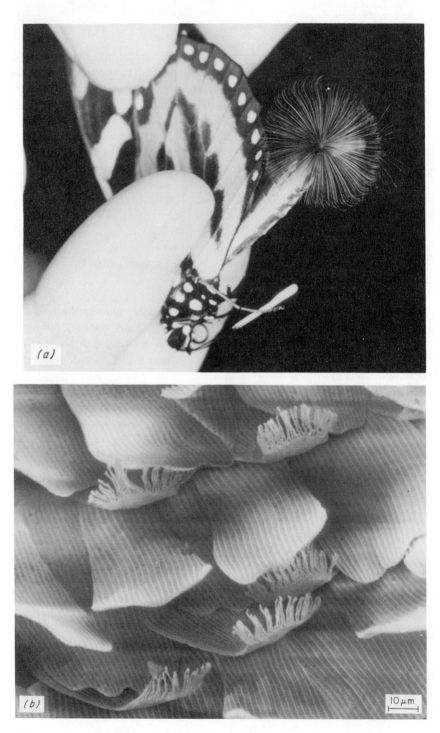

(a)

(b)

10 μm

sex attractant identified in the Lepidoptera and the first chemically characterized insect pheromone.

At about the same time, the olfactory sex attractant ("gyplure") of the female gypsy moth, *Porthetria dispar*, was also identified* and synthesized, and the race was on. The silkworm and gypsy moths are only two of perhaps a hundred thousand species of moths and butterflies making up one of the largest orders of insects, and since 1959 the sex attractants of less than 200 have been identified. However, the means for rapidly identifying others have now been developed, and it is an impressive achievement. Accompanying this has been an exponential expansion of the literature on sex pheromones; much of it has been summarized by Jacobson (1965, 1972). Moulton, Turk, and Johnson (1975) cover many of the increasingly sophisticated techniques used in olfaction research.

Of course, the majority of the research on chemical sex attractants has been economically motivated and practically oriented; the specificity of such compounds makes it possible in theory to single out one particular pest species for detection, monitoring, and/or control. Manipulation of insect behavior by means of sex pheromones is a promising alternative to the use of convential chemical insecticides, although in actual practice it has been fraught with technical problems.

### Assembly and Aggregation

Quietly and quickly lift a pile of logs and you may see a large cluster of crickets beneath it. Look at garden plants and notice growing tips completely obscured by massed aphids or plant lice. Peer into an old well and find, pulsating in unison on its sides, a congregation of daddy-long-legs.

A wide variety of arthropods assemble for various purposes. In its purest sense, **assembly** refers to those situations in which members of a species congregate prior to some activity, such as feeding, mating, or hibernation, with the signals which call them unrelated in any direct

* Ten years later, Bierl and associates (1970) found that the true gypsy moth sex attractant was in fact quite different chemically. Apparently, the attractiveness of the original gyplure was due to contamination with minute traces of "disparlure."

---

**Figure 5-8** Glands of male butterflies whose secretions help induce females to accept them as a mate. (a) Paired extrusible scent organs (hairpencils) at the tip of the abdomen of a hand-held male *Lycorea ceres* (Nymphalidae, Danainae), fully splayed open. (Courtesy of R. L. Silberglied; see Schneider, 1975) (b) Glandular scent scales (androconia) on the wings of a male *Pieris* butterfly. (Courtesy of G. Bergström; see Bergström and Lundgren, 1973.)

manner to the subsequent activity. Usually, except in the social insects, these are temporary groupings, such as feeding aggregations or the mating swarms of mayflies, and pheromones are only one of several ways in which they may be promoted and/or maintained. Among those insects possessing aposematic (warning) coloration (see Chapter 8), pheromonally induced aggregations are particularly prevalent. For example, a male-produced pheromone attracts brightly colored lycid beetles of both sexes to form prominent clusters, sending potential predators very conspicuous advertisements of their distasteful nature, as do the large hibernative aggregations of distasteful ladybird beetles (Coccinellidae). Aggregations and other nonsexual associations are classified in Table 10-1.

In many cases, gregarious behavior clearly functions to bring the sexes together for mating, sometimes in combination with attraction to suitable oviposition sites. Bark beetles are one good example.

One of the most destructive of pests to coniferous forest in North America are the Scolytidae, small cylindrical, bark beetles that destroy millions of board feet of lumber each year. Nearly every time these beetles invade, a very predictable attack pattern occurs. First, a small pioneer group selects a tree, usually one damaged in some way. They initiate an invasion and construct nuptial chambers beneath the bark. Shortly thereafter, an army of attackers of both sexes rapidly converges upon the site to continue the attack. By the time mating has taken place, the number of arriving beetles rapidly declines. But as a result of the mass attack, the tree usually dies. It appears that the initial attraction is due to volatile aldehydes or esters resulting from the abnormal enzyme activity of a subnormal, injured, or cut tree. Beetles of only one sex initially select the individual host tree. In monogamous (see Chapter 9) *Dendroctonus*, the female excavates the entrance tunnel, while in polygamous *Ips* the male performs this function. Then as the pioneer beetles begin to attack, they discharge a pheromone. This simultaneously serves two purposes: attracting additional individuals of both sexes and inducing the opposite sex to enter the nuptial chambers and mate. Because the pheromones are produced only after feeding begins, it has been suggested that a pheromonal precurser may be ingested and metabolized to the attractant, or metabolism of food materials may cause secretory activity in specialized cells (Wood, 1970).

However, while aggregation pheromones in many instances are clearly involved with reproduction, other cases are less straightforward. Many aggregations are composed of immatures, or of individuals of all ages, classes, and both sexes.

One recurrent benefit is to permit more effective exploitation of the environment than would be possible for single individuals. This phenomenon has sometimes been labeled the "group effect." For example, the caterpillarlike larvae of the sawfly, *Neodiprion pratti banksianae*, feed in tight groups upon jackpine trees. When the young larvae were experimentally isolated from their companions (Ghent, 1960), they suffered an 80% mortality, while among those allowed to remain in groups only 53% died. Why? Newly hatched larvae have considerable difficulty chewing holes into the tough cuticle of the jackpine needles. Each larva, even in groups, individually attempts to establish its own feeding site. When finally one successfully cuts through into the inner tissues, other larvae are quickly attracted to the cut, where their feeding widens the breach still further until soon all the larvae are able to feed. (A related strategy is illustrated by a Brazilian sawfly; see Fig. 4-19.) Even in later instars, larvae have the best chance for survival in the largest aggregations, where they synchronize their vigorous defensive reactions into a group display (Fig. 5-9) which is highly effective in discouraging both bird and insect predators.

An example this clear is a rarity, however. A great many more chemically mediated aggregations have been demonstrated than functions discovered. Even when advantages are apparent, the manner in which they accrue is usually unclear. For example, a few years ago in Japan, Ishii (1970a,b) offered a choice of roosting papers to groups of German cockroach nymphs, *Blattella germanica*. Filter paper previously kept in the stock cockroach culture for several days was invariably preferred to controls made of fresh filter paper of the same size (Fig. 5-10). When their antennae were removed, however, the cockroaches aggregated about equally on both roost types. Thus, olfactory cues appeared to be involved. In order to determine the source of the pheromone, Ishii isolated various cockroach body parts, washed them in ether, and impregnated filter paper with the extracts. Abdominal extracts caused the strongest aggregation, and eventually Ishii located the pheromone in the feces, presumably originating from rectal pad cells.

A number of field observations have suggested that certain cockroach species may live together naturally. In general, Ishii found, the aggregation pheromones do not appear species specific; nymphs of several unrelated species in three different cockroach families would aggregate in response to each others' pheromones. However, nymphs appeared more responsive to their own pheromones, and a few species appeared to be repelled by fouling produced by other species. In addition, at least one species produces its aggregation pheromone in its mandibular

**Figure 5-9** A cluster of *Neodiprion* sawfly larvae on a pine branch exhibit their character-istic defensive posture, with the anterior part of their bodies tilted backward and droplets of regurgitated fluids exposed from their mouths. Vigorous jerking movements enhance the effectiveness of this group display. The secretion contains primarily plant-derived sub-stances. (Photograph by the authors.)

glands, so one cannot simply assume that all species of cockroach produce a fecal aggregation pheromone simply because other species collect on paper contaminated by them.

The adaptive significance of such cockroach aggregations is still not entirely clear. However, it has been shown experimentally that German cockroaches reared in isolation grow at a slower rate and suffer greater mortality compared to those raised in groups. The mechanisms behind this differential are still unknown. Additional studies on cockroach aggregation are Bell et al. (1972) and Roth and Cohen (1973).

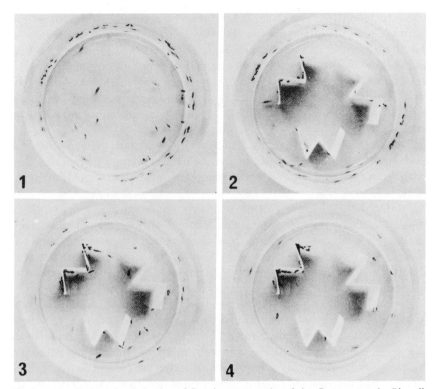

**Figure 5-10**   Aggregation behavior of first instar nymphs of the German roach, *Blattella germanica*. Sixty nymphs are introduced into a glass arena with three folded filter paper roosts: two made of fresh paper, the other (at upper left) of paper left in a stock roach culture for 24 hours prior to experiment. Progressively greater preference for the conditioned roost is shown through time: (1) start of experiment; (2) after 1 minute; (3) after 25 minutes; (4) after 45 minutes. (From Ishii, 1970a.)

Most aggregations are temporary, as we have mentioned. Some, however, are persistent, most notably those forming social insect colonies. Those colonies whose social lives are most highly developed exhibit extensive direct food sharing (trophallaxis, see Chapter 4). Also nearly universal among the social insects is clustering behavior. Workers removed from a nest and placed in bare surroundings will quickly gather in one or several little groups. If the mother queen and/or some larvae are with them, this occurs even more rapidly and the grouping is even tighter.

Of all the types of pheromonally induced insect assembly, perhaps the most dramatic is exhibited by the fertilized social insect queen, continuously surrounded by a retinue of crowding, licking, food-offering attendants (see Fig. 5-14). The effect is least marked in small colonies or where there are multiple laying queens. However, upon the distended bodies of the "physogastric" queens of army ants, fire ants, and termites (see Fig. 10-16), the tightly massed workers may number into the hundreds. Several investigators have shown that pheromonal scents are involved; with army ants, in fact, merely allowing the queen to sit on untreated balsa strips transfers enough scent to make the strips highly attractive to workers.

Harvester ants of the genus *Pogonomyrmex* mainly collect seeds for food, and casual observations in the past have seemed to indicate that foraging was based simply upon individual orientation and initiative. However, some species also carry dead insects back to their nests, and Hölldobler and Wilson (1970) have shown that when individual workers of *P. badius* attack a large and active insect prey near their nest, they discharge an alarm pheromone from their mandibular glands. Just as it does in the presence of dangerous stimuli, the substance both attracts and excites all other workers within distances of about 10 cm, so that the prey is more quickly subdued.

Whenever joint efforts are needed—whether for exploiting a food source, repairing a breach in a nest, or moving to a desirable new nest site—individuals of a great many insects, particularly among social species, are able to chemically summon others of their kind. Although the category is admittedly a loose one, this special case of assembly has been termed **recruitment**, that is, communication that brings conspecific individuals, often nestmates, to some point in space where work is required. It undoubtedly has reached its highest development among the ants, bees, wasps, and termites but is not restricted to them (see Fitzgerald, 1976).

The simplest cases of recruitment appear to border on the unintentional. Honey bee workers, for example, can recognize food odors both from smells adhering to the bodies of successful foragers and by the scent of nectar regurgitated by them. Even in the absence of communicative dancing, workers that have previously encountered a similar odor will search the site for it again. In a slightly more advanced form of recruitment, some evidence suggests that social insects may leave chemical "footprints" to attract others, such as the odor trails laid by both walking honey bees and some wasps in the vicinity of their nests.

CASE STUDY: RECRUITMENT IN THE DOUGLAS FIR BEETLE,
*Dendroctonus pseudotsugae*

In 1969, Rudinsky puzzled over one component of the typical attack pattern of a scolytid bark beetle, *Dendroctonus*, namely that after several hours of mass invasion of the host tree the beetles' attraction always stops very abruptly. It seemed unlikely that this behavioral change was due only to a cessation of pheromone production. It was too sudden, and Rudinsky knew that female frass may remain attractive for days. He also knew that test logs containing only male–female pairs in the bark were not attractive to flying beetles.

Looking for an explanation, Rudinsky was struck by the well-known ability of many Scolytidae to make squeaky sounds (stridulation, see Chapter 7). In general, this sound-producing ability is more characteristic of one sex, the one that is *not* the original invader. Might this yield an answer? Through a series of intriguing field experiments, Rudinsky set out to examine the exact role of stridulation in the Douglas fir beetle.

First, he artificially infested a log section with 30 virgin female beetles. Carefully, he covered their entrance holes with screen so that males could not enter and yet the females could still expel their boring dust, and then he set the test log inside a screen cage. As the females began to bore and produce pheromone-laden frass, beetles flew in to land on the cage, two males being attracted for every female that came. Rudinsky observed no males stridulating.

Then Rudinsky took 30 male beetles from an ice chest and placed a single male on each screened entry to a virgin female. As they warmed up, the males began to stridulate, and suddenly the arrival of flying beetles halted. He removed the stridulating males; flying males again began landing on the cage. Back and forth, on and off, at about 10-minute intervals, he could mask the attraction or renew it at will simply by alternately removing and returning the stridulating males to the screened entries of the attractive females.

Was the tiny chirping sound of the male really the signal responsible for the population-regulating phenomenon Rudinsky was observing? Not satisfied with this simple explanation, Rudinsky attempted to establish more firmly that the stimulus was solely auditory. First, he placed a freshly killed male on each screened female entry hole; consistent with his theory, flying males continued to arrive. In another experiment, he cut the wing covers of 30 live beetles so they could not stridulate. When these silent males rested on the screened holes, they had no effect upon the stream of flying males that continued to arrive. But as soon as the screen was removed so the silent males could enter and join the female, again the flight aggregation stopped. Because the stream of beetles could be halted without stridulation, the auditory signals could not be the only ones involved!

By whacking a hammer on the log bark just above each entry hole, Rudinsky killed the females inside their galleries. Beetles kept arriving, apparently due to residual attraction. Then he added males. Again they stridulated—but flying beetles kept on arriving! Rudinsky removed the females entirely from some logs before adding extra males. Again their stridulation was incapable of halting the immigration.

Now the situation appeared understandable. The arrival of large numbers of beetles quickly leads to the discovery of the infesting females by males, which stridulate as they dig their way into the gallery to join the female. The chirping signals the female, and she responds by releasing a chemical that acts as a "mask" or antiaggregative pheromone, camouflaging the normally attractive odors she has produced in her frass. This in turn serves to inhibit flight response; therefore the number of arriving beetles drops off sharply and suddenly. Such a system would tend to have high adaptive value, promoting an even distribution of available males while both preventing overcrowding with subsequent brood mortality and allowing the mass attack necessary to overcome the host tree's resistance.

More recently, Rudinsky has demonstrated that electronic playback of recorded male sounds has the same effect, thus definitely confirming this novel interaction of sonically and chemically induced behavior. A major fraction of the antiaggregative pheromone has been identified as the ketone 3-methylcyclohex-2-en-1-one (MCH). It was later discovered that this and another isomer of MCH are also released by the entering male when it is stimulated by a clicking signal of the female. Thus, the masking pheromone is released by the male and female together, each as a response to the sonic stimulus of the other sex. MCH is a multifunctional pheromone in which concentration plays a critical role (Fig. 5-11). A dual pheromone system, aggregant and antiaggregant, is now known to exist in other destructive species of *Dendroctonus* and may be a generic characteristic. For additional details see Rudinsky (1969, 1973a,b), Rudinsky and Ryker (1976, 1977) and Rudinsky et al. (1976).

The most elaborate form of chemical recruitment studied to date is the odor trail system of ants, most thoroughly elucidated by Wilson's investigations of the imported red fire ant, *Solenopsis invicta*. Although trail following in ants has long fascinated biologists, the mechanism behind this behavior was unclear until 1959 when Wilson showed that it was possible to artificially induce the complete recruitment process in the fire ant with trails made from glandular extracts or smears containing the trail substance. His subsequent detailed analyses (1962, 1963) have indicated that workers apparently travel through a "vapor tunnel" created by the diffusion of the evaporating pheromone which flows from the Dufour's gland as the trail layer draws the tip of its extruded sting lightly over the ground surface.

Wilson's studies of recruitment in fire ants also provide a good

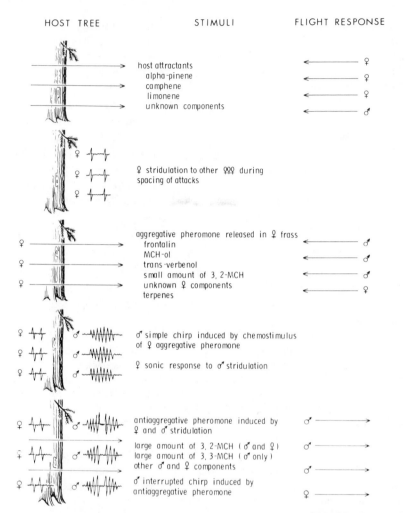

HOST TREE                    STIMULI                    FLIGHT RESPONSE

host attractants                              ⟵————— ♀
  alpha-pinene                                ⟵————— ♀
  camphene                                    ⟵————— ♀
  limonene
  unknown components                          ⟵————— ♂

♀ stridulation to other ♀♀♀ during
spacing of attacks

aggregative pheromone released in ♀ frass    ⟵————— ♂
  frontalin                                   ⟵————— ♂
  MCH-ol
  trans-verbenol                              ⟵————— ♂
  small amount of 3,2-MCH
  unknown ♀ components                        ⟵————— ♀
  terpenes

♂ simple chirp induced by chemostimulus
of ♀ aggregative pheromone

♀ sonic response to ♂ stridulation

antiaggregative pheromone induced by         ♂ —————⟶
♀ and ♂ stridulation

large amount of 3,2-MCH (♂ and ♀)            ♂ —————⟶
large amount of 3,3-MCH (♂ only)
other ♂ and ♀ components                     ♂ —————⟶

♂ interrupted chirp induced by
antiaggregative pheromone                    ♀ —————⟶

**Figure 5-11** Summary of the interaction of olfactory and auditory stimuli in the aggregation response of the Douglas fir beetle and its subsequent inhibition. In low concentrations, in synergistic combination with other host- and beetle-produced chemicals, 3,2-MCH attracts both sexes and causes males to stop and stridulate. At higher concentrations, it prevents flight aggregation of both sexes. (Courtesy of L. C. Ryker; from Rudinsky and Ryker, 1977.)

example of insect **mass communication**—the sharing of information that can be transmitted only by a *group* of individuals. Far from being random, the number of *S. invicta* workers leaving the nest to move along a particular feeding trail is in direct response to the summed amount of trail substance laid down by those workers already returning from the food source. Both quantity and quality are mass communicated. At first, increasing numbers of ants arrive; but as the food becomes crowded, the arriving ants, becoming increasingly unable to contact the food, begin to turn back without laying trails. Thus, numbers equilibrate at a level that is a linear function of the area of the food source. As this area declines, so does the number of trail layers and hence the number of newly approaching workers. Quality of the food source can also be mass communicated, by means of an "electorate" response in which individuals "choose" whether to lay trail or not after inspecting the food find. Through such aspects of mass communication, trail pheromones can provide a control that is more complex than one might predict from individual responses alone (Wilson, 1962).

## Alarm and Alert

In 1609, in what may be the earliest explicit description of an insect pheromone, Charles Butler (in Wilson, 1971, p. 236) commented about an aspect of honey bee behavior already common knowledge among beekeepers:

"When you are stung, or any in the company, yea, though a Bee have strike but your clothes, specially in hot weather, you were best be packing as fast as you can: for the other Bees smelling the ranke favour of the poison cast out with the sting will come about you as thicke as haile: so that fitly and lively did he express the multitude and fierceness of his enimies that said *They came about me like Bees.*"

One of the components of this sting-released pheromone has been identified as isoamyl acetate, which smells like bananas; alone on a cotton ball it continues to attract and anger other worker bees but does not release attack behavior.

Alarm and defensive behavior often go hand in hand, and in some cases a single substance simultaneously releases both behaviors; in others, careful study reveals the simultaneous use of chemicals from different sources. For example, disturbed nymphs of the pyrrhocorid bug *Dysdercus intermedius* secrete a fluid from the third dorsal scent gland; the compound is used simultaneously for defense and for releasing alarm behavior.

Alarm and alert pheromones are alike in that all are produced under conditions of immediate or potential threat, but the responses they elicit grade into several categories. Defense, dispersal, agitation, aggregation, and recruitment are among the more common depending upon which strategy has been evolutionarily more adaptive for the particular stimulus situation. For example, subterranean ants of the genus *Acanthomyops* have populous colonies crowded into relatively small caverns. When stimulated by alarm pheromone, they rapidly converge toward the source of disturbance and meet the danger head-on. Related ants in the genus *Lasius* normally nest in smaller colonial groups in more exposed situations under rocks or logs; their response to disturbance is to scatter and run hurriedly about. Both species employ undecane as the principal component of their alarm pheromones, but *Lasius* is much more sensitive at lower concentrations, in effect, possessing an "early warning system" for rapid evacuation when danger threatens (Wilson, 1971).

A similar concentration effect involves citral, a mandibular product of certain Neotropical stingless bees. In low concentrations droplets of citral are highly attractive to workers of *Trigona subterranea* and serve as a trail pheromone used in recruitment. High citral concentrations in the nest vicinity, on the other hand, release alarm and attack behavior. At least some unrelated stingless bees that have lost the ability to collect their own pollen also use citral to disorient other bees whose nests they raid for pollen. *Lestrimelitta limao* scouts that manage to penetrate nests of susceptible *Trigona* species release citral inside which causes a complete breakdown in the social organization of *Trigona* as well as being a strong attractant to other *L. limao* workers. The same effect was produced by introducing citral-treated blocks into *Trigona* nests (Blum et al., 1970).

Since insect aggregations, whether temporary or permanent, constitute a real jackpot for effective predators, selection for efficient protective behaviors has undoubtedly been especially strong among such species. The existence of alarm-alert pheromones has been demonstrated for many. In gregarious plant-feeding Homoptera, for example, repellents and dispersants have repeatedly evolved. Green peach aphids discharge a sesquiterpene from their abdominal cornicles when disturbed (Fig. 5-12); this acts as a powerful dispersal agent for nearby aphids, even completely unrelated species. Both nymphs and adults of the bedbug, *Cimex lectularius*, discharge an alarm pheromone from thoracic glands in response to irritation; alerted individuals exhibit a response similar to that of aphids (Levinson et al., 1974).

However, the most widespread occurrence of chemical alert systems is among the social insects, particularly in those species forming large colonies. In Europe, for example, Maschwitz (1966) found evidence of alarm-alert pheromones in all of 23 more highly social Hymenoptera

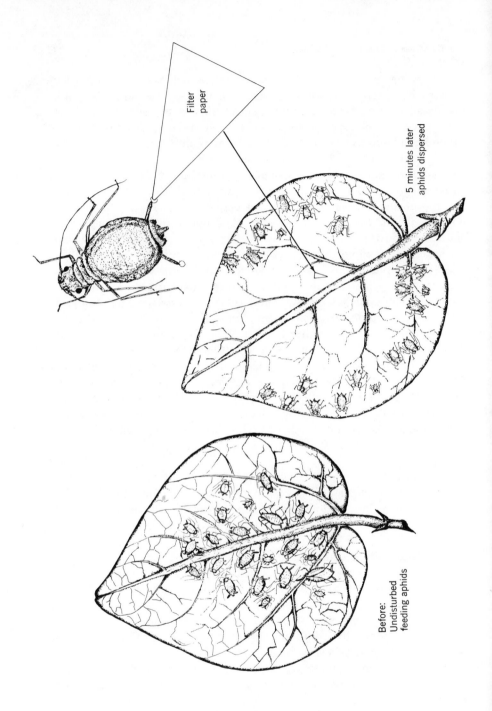

Filter paper

5 minutes later
aphids dispersed

Before:
Undisturbed
feeding aphids

surveyed. The more primitively social Hymenoptera surveyed, particularly bumble bees and *Polistes* paper wasps, showed no evidence of utilizing such pheromones. Likewise, among termites some of the phylogenetically more advanced species produce volatile substances acting as alert signals.

## Spacing

At low population density, adult flour beetles, *Tribolium confusum*, aggregate; at intermediate densities, they apparently distribute randomly. At high population densities, however, they distribute uniformly, with a precision too great to be attributed solely to chance. How do they accomplish this last effect? It appears that this amazingly uniform spacing is due to the secretion of quinones from thoracic and abdominal glands in the beetles. These substances act as repellents above a certain concentration (Naylor, 1959).

Larvae of the flour moth, *Anagasta kuhniella*, secrete compounds from the mandibular gland which affect the dispersion of individuals in the population. In crowded conditions the secretion results in increased dispersal propensity, lengthened generation time of developing individuals, and lowered fecundity of females crowded as larvae (Corbet, 1971).

The increased dispersion resulting from pheromones acting as repellents, territory markers, locomotory stimulants, or deterrents can be highly advantageous under certain circumstances. Consider the parasitoid wasps. Within these groups there appears to have been strong

---

**Figure 5-12** Alarm-dispersant pheromone bioassay for the green peach aphid, *Myzus persicae*. A filter paper triangle daubed with pheromone secreted from the aphid's abdominal cornicles is introduced next to a cluster of feeding aphids. In less than 5 minutes most aphids have left the area and some have dropped off. This alarm dispersal behavior is characteristic of a variety of aphids, but is most pronounced in species not normally attended by ants (such as *M. persicae*). Ant-associated aphids seem to depend more upon their ants for protection than on their dispersive powers. So well integrated is the ant–aphid mutualism that Nault et al. (1976) have shown that a direct alarm communication between ant and aphid occurs, with ants responding aggressively toward artificial sources of aphid alarm pheromone (*trans*-β-farnesene). Moreover, they have shown that the behavior of aphids normally associated with ants varies depending upon the circumstances; those aphids reared without access by ants respond by actively dispersing, whereas those tended by ants remain in intact clusters, which contributes to stabilization of the relationship. (Drawings by J. W. Krispyn from selected frames of a motion picture by C. J. Kislow.)

selection toward "ovipositional efficiency." That is, the females of many of these wasp species are able to make quite fine discriminations regarding those situations in which their eggs, and subsequently their offspring, will have the best chance for survival.

CASE STUDY: HOST-SEARCHING BY THE ICHNEUMONID WASP, *Pleolophus basizonus*

The Swaine jack pine sawfly, *Neodiprion swainei*, a serious defoliator of jack pine in North America, is attacked by a great parasitoid wasp complex, particularly during its cocoon stage. One of these parasitoids is the ichneumonid *Pleolophus basizonus*, a species with the rather slow rate of egg production and oviposition of two to four eggs per female per day. Surprisingly, when Price (1970, 1972) collected sawfly cocoons from different areas, this was consistently the most abundant parasitoid species to be reared from them. How were these "poor layers" able to compete so successfully against 18 other species of parasitoids also competing for the same host? The question was of more than purely academic interest, since this wasp had been purposely introduced several years previously for biological control.

In confined areas at high parasitoid:host ratios, a number of other investigators had noted that several species of wasps showed mutual interference in egg laying. Using replicated caged "arenas" simulating the sandy, lichen-covered forest floor in careful detail, Price began to study the influence of host density on parasitoid oviposition and dispersion of eggs. Six days after placing three mated female *Pleolophus* in each cage with varying numbers of cocoons, Price split open the cocoons to count the number of eggs laid within. At all densities tested, the frequency with which he found only one egg per cocoon was much higher than expected from a random attack. In fact, at the highest cocoon densities, the oviposition pattern was extremely regular, with almost no wastage of eggs through superparasitism. Apparently, searching females could discriminate between parasitized and unparasitized hosts, as had been previously noted in other parasitoids, presumably through some chemical basis.

Performing the obverse experiment, that is, varying numbers of wasps in the cages along with a fixed number of cocoons for two consecutive days, Price watched the cages at hourly intervals and recorded the wasps' positions. As crowding was increased, he observed more females climbing on the cage sides in what looked like an escape reaction.

Placing a barrier across the middle of the arena, he released a single female in one side and allowed her to search for 6 hours while watching continuously and recording her searching pattern. She was then recaptured and released again in the center of the arena with the barricade removed, and the amount of time subsequently spent in each half was recorded. A control was maintained by releasing a female in the fresh arena. Seven of 19 females so tested exhibited a clear recognition and avoidance of areas previously searched by them (Fig. 5-13). Several variables, of course, remained uncontrolled, such as female age, reproductive condition, and degree of possible habituation to the caged conditions. Nevertheless, avoidance was frequent enough to suggest that some "trail odor" deposited on the substrate by females as they searched served later to repel them. (In fact, even different species of parasitoids attacking the cocoons are able to recognize and avoid areas previously searched by *Pleolophus*.) Thus, it was clear that females of *P. basizonus* do not search at random but tend to disperse themselves over an area and search it systematically. Mutual recognition between

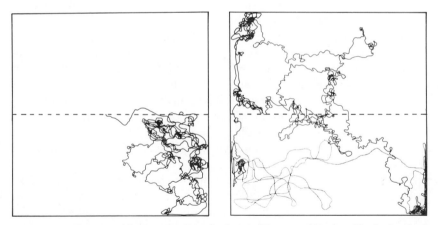

**Figure 5-13** Host-searching movements of a female ichneumonid wasp, *Pleolophus basizonus*, traced following release in the center of a glass covered experimental arena. Dotted line marks position of a removable barrier. In the left diagram, the upper half of the arena had been searched previously for 15 minutes by the same female; her avoidance of it during the subsequent 15 minutes is clearly shown. The right diagram depicts the movements of a female for 30 minutes in a fresh (control) arena. Dashed line depicts instances when the female crawled on the under side of the arena cover. (From Price, 1972; courtesy of P. W. Price and the Entomological Society of Canada.)

females helps to account for the regular placement of eggs, as does the density-dependent interference.

The chemical conditioning of search substrates by searching and ovipositing female parasitoids appears to be mediated by chemical cues called **search deterrent substances**. The adaptiveness of such a system is apparent: individual female efficiency is increased by avoidance of previously searched areas; the parasitoid population more evenly disperses itself throughout the available host habitat; and competition and/ or aggression from other species responding to the repellent odor will be minimized.

## Identification

For the vast majority of insects, recognition of one's own species is necessary only for copulation. Even during mating, another conspecific individual may sometimes still be perceived as food or as potential enemy (see Fig. 9-8). However, in insects that have adopted a gregarious or societal existence, unfailing recognition of conspecific individuals over a wide range of circumstances assumes utmost importance. Olfactory and tactile cues appear to be primarily responsible for this recognition inasmuch as mutual antennal contact seems to be the most universal greeting. At first glance, nestmate recognition appears to be a casual affair, usually no more than a pause and perhaps an exchange of antennal strokes. However, should an alien insect be encountered within the nest, some of the scope of this communication is swiftly and unmistakeably revealed. Even if the intruder is only from a different colony, it normally evokes hostility; if it is of a different species as well, the inspection is swift, the attack violent. Usually the alien is instantly killed or driven from the nest.

For the vast majority of species, however, the actual existence of recognition odors is only a behaviorally supported supposition, for the chemical nature of recognition pheromones is still unknown. For the most part, they appear to be of another class than those involved in attraction, disperson, and other longer-range communication. Instead of being disseminated freely through air or water, they are either absorbed in waxes and other fatty substances on the body surface and released slowly into the atmosphere or are so nonvolatile as to be perceptible only by contact chemoreception. Wilson (1971) has termed them **surface pheromones**: they include not only recognition scents but releasers of grooming behavior, some aspects of courtship behavior, and, in some social insects at least, secretions that stimulate food exchange between individuals. In at least some cases, they may be composed of a combination of species-specific odors and environmentally derived ones (as, for example, from food sources). In some cases, it may be misleading to regard them as pure pheromones in the narrow sense.

A honey bee colony headed by a mated, laying queen normally lives in harmony. Introduce a strange virgin queen, however, and she will be violently killed, as will any workers that happen to be contaminated by brushing against her. If the alien queen is wiped with filter paper and the paper offered to the workers, they will attack it. In other examples, workers of *Polistes* wasps are hostile to both live and dead workers from other colonies but not to the corpses of nestmates. Workers of some bumble bee species will sting anesthetized alien workers of the same species introduced into their nest but not anesthesized nestmates. However, when nestmates are left in an alien nest for an hour or two and then returned to their colony, they are attacked. Such nestmate recognition in most cases is based principally if not exclusively on **colony odor**, perhaps the best known example of surface pheromones. The chemicals involved are typically complex, present only in extremely small quantities, and difficult to separate and bioassay using current techniques. It appears reasonably certain, however, that some of a colony odor's distinctive elements are drawn more or less directly from the environment. For example, several experiments with honey bees suggest that odorants associated with nectar and honey form an important component of bee colony odor; bumble bee nest odors also appear to be acquired at least partly by direct absorption of odorants into the cuticle. The situation in many ants may be similar. Just as ants from another colony are attacked by colony members, so also are ants from their own colony once they have been kept apart for a week or more. Likewise, if colonies are divided and each half fed separately on different diets, the members show hostility when brought together again. Except for some experimental evidence on an apparent pheromonally induced brooding behavior by queen bumble bees (Heinrich, 1974), there is no indication that any colony odor component originates exclusively as an exocrine gland secretion. Thus, colony odor appears to be a constantly changing identification mark with little genetic basis.

Finally, within a given colony, intensity of aggressiveness among aliens varies greatly as a function of other social stimuli. For example, workers of a wide variety of social insects will become more receptive to aliens if deprived of their queens and generally lose most of their hostility when removed from their nest, food source, and nestmates.

For an insect colony to function, of course, the recognition capacity of

**Figure 5-14**   "Retinue" behavior of worker honey bees attracted to the pheromones of the queen bee. (Courtesy of N. E. Gary.)

worker individuals must extend beyond colony odor to include distinguishing, and acting upon, the caste and life stages of nestmates. Honey bee queens are treated differently than workers (Fig. 5-14). Certain paper wasp individuals are given preferences in food exchanges. Ant eggs, larvae, and pupae are segregated into separate piles. In most cases, caste and life stage identification appear to be by antennal contact, suggesting chemoreception; in a few cases recognition appears to be by means of odors transmitted over a distance. However, some ants may distinguish the age classes of their larvae by differences in body hairiness, and the "piping" of young honey bee queens is an example of auditory cues acting to communicate life stage recognition. For further studies on individually distinctive odors in insects, see Barrows (1975) and Barrows et al. (1975).

### "Funeral Pheromones"

Social insects keep the interiors of their nests, and particularly the brood chambers, meticulously clean, removing or covering over disagreeable objects and in some cases using various secretions to form crustlike coats upon the nest walls. Toward corpses they are especially fastidious. Social wasps, bees, and some ants drag their dead from the nest and abandon them. Other ants may eat their dead, consign them to refuse heaps, or pile them in deserted nest chambers or galleries. What stimuli act to release this "necrophoric" behavior pattern?

When *Pogonomyrmex* or *Solenopsis* ant workers meet with a freshly dead nestmate, they will groom it as if it were still alive; its complete immobility and crumpled posture evoke no new response. Upon encountering a corpse that has decomposed for a day or more, however, they antennate it briefly, pick it up, and carry it directly to the refuse pile. But when presented with a corpse which has been thoroughly leached in solvents and dried, the workers usually eat it instead. Wilson et al. (1958) established that bits of paper treated with acetone extracts of the corpses were treated like the corpses themselves; clearly, the cues were chemical. In fact, oleic acid, a common decomposition product in insect corpses, turned out to be a recognition cue. Even living nestmates when daubed with any of several unsaturated fatty acids are picked up and carried, unprotesting, to the refuse pile. Here, the "living dead" scramble to their feet, proceed to clean themselves off, and return to the nest, often only to be mistaken again as corpses and be carried back to the refuse pile time and again until the "scent of death" has finally worn off their bodies.

### THE INFORMATION CONTENT OF PHEROMONES

Consider a single insect sitting on a leaf, releasing a pheromone into perfectly still air. Obeying the law of gas diffusion, the pheromone expands at a calculable rate in all directions surrounding the insect until it forms a sphere at whose outer limits its concentration is zero. Nested within this sphere is a second, more behaviorally interesting one, the **active space**—the zone in which the concentration is at or above the level needed to evoke a biological response from other insects.

Of course, perfectly still air is an oversimplification. If a breeze is blowing, the pheromone expands to form a cone downwind of the insect. Within the cone, the active space becomes not a sphere but a downwind semiellipsoid, rather like the upper half of a giant blimp lying on the ground with the emitting insect sitting at the tip of the airship's nose (Fig. 5-15). The stronger the breeze, the shorter the blimp (or "odor plume") becomes. Even this model does not account for surface drag nor for the fact that most attractant vapors, being heavier than air, will tend to fall before being fully dispersed, thus effectively flattening the

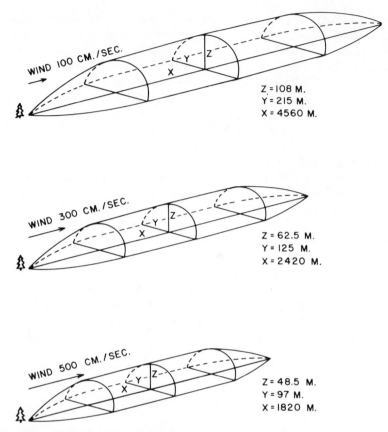

WIND 100 CM./SEC.

Z = 108 M.
Y = 215 M.
X = 4560 M.

WIND 300 CM./SEC.

Z = 62.5 M.
Y = 125 M.
X = 2420 M.

WIND 500 CM./SEC.

Z = 48.5 M.
Y = 97 M.
X = 1820 M.

**Figure 5-15** The active space of the gypsy moth sex attractant, as deduced from linear measurements and general gas diffusion models. Height and width are exaggerated in the drawing. As wind speed increases, there is a contraction of the space within which the pheromone from a single, continuously emitting female is sufficiently dense to attract males. (From Wilson and Bossert, 1963.)

top of the active space. However, the active space concept can be used in combination with linear measurements and gas diffusion laws to generate some predictions and generalizations about the use of pheromones in functionally different communication systems.

### Physiological Adjustments: The $Q/K$ Ratio

Suppose that one knows both the amount of pheromone released and the behavioral threshold for perception for some insect. To quantify the amount of pheromone released, let us consider it to be the number of molecules the insect emits per second (or as a single puff), and call it $Q$. For purposes of brevity, let us call the potentially responding insect's behavioral threshold $K$ and measure it in molecules per cubic centimeter. Then, using predictions developed by Wilson and Bossert (1963), one can estimate three important items: the useful range of communication, the time required for transmission, and the duration of the signal.

The interval between release of the pheromone and disappearance of the active space is called the **fade time**; using Wilson and Bossert's blimplike model, it is the time required for the longitudinal axis of the active space to diminish to below the receiver's perceptual threshold. Behaviorally, variations in the length of time a signal persists would be expected when pheromones serve different purposes. How might this be achieved?

Both intuitively and by studying the model, one can recognize two ways: vary the rate at which the pheromone is sent or vary the receiver's threshold of perception. In quasi-mathematical shorthand, then, **fade time is a function of the ratio of $Q$ to $K$.** The highest $Q/K$ ratio corresponds to a pheromone system with a great signal distance and slow signal fade time. Conversely, low $Q/K$ ratios characterize systems having rapid fade time and a relatively small active space. A theoretically infinite variety of systems exist as intermediates along the continuum; they result from changing the values of $Q$ and $K$ independently of each other (Fig. 5-16).

Because of these relationships, the $Q/K$ ratio is capable of serving as one standard for comparing different communication systems. For example, one might consider the respective information transfer strategies of mate-seeking moths and alarm-sending ants (Table 5-3).

**Table 5-3** The $Q/K$ ratios of three widely differing pheromone communication systems[a]

| Pheromone | $\dfrac{Q}{K} = \dfrac{\text{Natural emission rate, in molecules/sec}}{\text{Behavioral threshold conc, in molecules/cm}^3}$ |
|---|---|
| Imported fire ant $S.$ *invicta* odor trail | 1 |
| *Acanthomyops* ant | $10^3 – 10^5$ |
| Silkworm *Bombyx mori* sex attractant | $10^{10} – 10^{12}$ |

[a] From Wilson (1970).

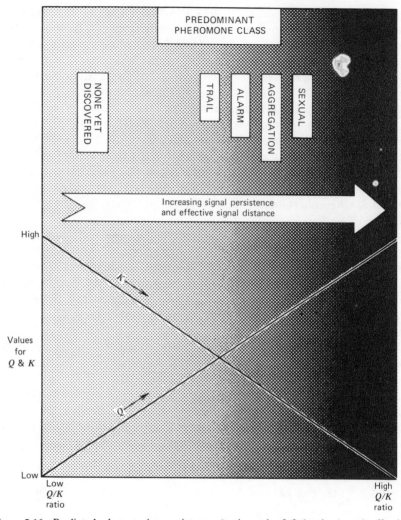

**Figure 5-16** Predicted changes in persistence (reciprocal of fade time) and effective distance of chemical signals along a gradient of Q/K ratios and their relationship to class of pheromone involved.

A virgin female moth may need to broadcast her desirability over a span of miles for several hours before she attracts a male and is mated. Obviously, a slow fade time for her pheromone would be advantageous; it would be desirable also to increase the area in which a potential mate could encounter and respond to the scent, *i.e.* to have a large active space. This necessitates a high $Q/K$ ratio, accomplished by increasing $Q$, decreasing $K$ or both. One would predict metabolic limits on $Q$, how much pheromone the female moth could continuously pump into the environment. (The bombykol content of a single female silkworm moth has been estimated at about one microgram, for example.) Therefore, in order to have a large $Q/K$, the value for $K$ will have to be extremely low; in behavioral terms, the male receiver must have a very low threshold of perception. The *rate* at which information can be transferred would be reduced in this case, however, since signals could not be turned on and off rapidly.

Next, consider the information transfer strategy of an alarm-sending ant. When an ant worker generates an alarm, it is advantageous that other workers be able to locate it sharply in time and space. For this to occur, a signal must have a relatively short fade time; correspondingly, one needs a small or intermediate $Q/K$ value, which can be accomplished either by lowering the emission rate or raising the threshold of concentration, or both. The citronella ant, *Acanthomyops claviger*, uses an alarm pheromone with a $Q/K$ between $10^3$ and $10^5$ cc/second; signals take about 2 minutes to reach an effective radius of about 10 cm and 8 minutes to fade out. The intermediate range and duration are behaviorally predictable. If the citronella alarm pheromone had as high a $Q/K$ as some moth sex lures, a one-ant alarm might theoretically keep the colony in a perpetual unproductive state of tension. On the other hand, some trail-marking pheromones have a $Q/K$ of only 1 cc/second. Were the citronella alarm pheromone to have this low a value, the signals would not travel beyond the distance within which other ants could perceive the danger directly themselves (Wilson, 1970).

So far, of course, we have implicitly limited our discussion to airborne chemical communication systems. While perhaps more unusual in the insects, there also exist cases of pheromones transmitted in water. Here, we may have a very different situation, for the times required to reach the maximum radius of the active space and the fade time are approximately 10,000 times greater in water than in air. To be able to use pheromones in water at all, large increases in the $Q/K$ ratio are necessary. This could be accomplished either by increasing the amount of pheromone, lowering the response threshold, or some combined alteration of both parameters. But such an adjustment would also increase the time to fade out. So, if a reasonably short fade time is required, one can expect to find additional devices at work. For example, the signal might be canceled by unstable molecular structure or enzymatic deactivation. (Such a mechanism is known in the case of the enzymatic deactivation of ingested 9-keto-*trans*-2-decenoic acid by worker bees.)

## Pheromones as Language: Syntax and Lexicon

Because they have tended to be more thoroughly studied and thus simpler to understand, most examples so far have involved single chemicals as pheromones. Often even such "simple" systems are quite complex, communicating different information at different concentrations or in varying contexts. For example, we have mentioned that males of the Douglas fir beetle stridulate when they detect low levels of the pheromone MCH. However, Rudinsky and Ryker (1976) have found that under a high concentration of MCH the chirps of the male beetles change from these of the female attracting calls to a longer, interrupted chirp with quite different acoustic properties (Fig. 5-17); this latter call is characteristic of both courtship with females and rivalry contests with conspecific males. Intuitively, one would expect to find many insects using such multifunctional pheromones; being small animals, it would be particularly advantageous, allowing economy in their receptor systems without sacrificing behavioral diversity. This phenomenon, termed **pheromonal parsimony**, is particularly characteristic of the class of pheromones releasing alarm in social insects. These pheromones evoke an almost astonishing diversity of behavioral responses depending upon the physiological condition of the recipient, concentration differences, duration for which the pheromone persists, and the context in which the signal is presented.

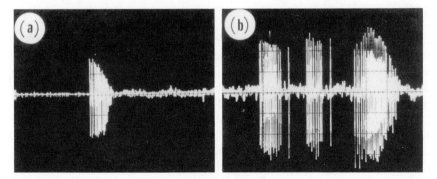

**Figure 5-17** Oscillograms (sound pictures, see Chapter 7) of typical male Douglas fir beetle sounds produced in different contexts. (a) The attractant chirp, made in response to low concentrations of the females attractant. After about 3 minutes the attractant chirp switches to the interrupted chirp. (b) The interrupted chirp is produced in at least four different situations: male in gallery with female; two males in same gallery with one female; two males in same gallery without females; and in response to medium concentrations of MCH. Similar chemically induced sonic signals are also known for many other *Dendroctonus* species. (Courtesy of L. C. Ryker; from Rudinsky and Ryker, 1976.)

Nonetheless, the insect pheromonal language is clearly not composed of an infinite number of chemicals different within each species and unique to each message. In order to achieve and maintain chemical uniqueness with such single chemical systems, insects would need to synthesize increasingly more complex molecules; this strategy obviously would have its upper limits, dependent upon such factors as the biosynthetic capabilities (energy cost) of the organism, the physical characteristics (e.g., volatility) of the molecules used, and the neurological complexity of the receiver. Furthermore, despite the intuitive logic that it would be adaptively advantageous for nearly all insect communication to be species-specific, tests of single chemical compounds under laboratory conditions have indicated that many pheromones may lack specificity even at the generic level. How can this apparent discrepancy be reconciled?

One key probably lies in the nature of the tests themselves. Characteristically, investigations of pheromonal specificity have been made under closely controlled laboratory conditions. Thus, they have been unable to take into account the multitude of factors operative in the field. For example, habitat preferences and seasonal and diurnal cycles may serve to isolate species using the same chemical cues, as for example in the circadian rhythm in responsiveness found in some male moths (Fig. 5-18). Females may also call only during certain time periods of the daily cycle.

Sometimes what appears as a lack of specificity may not be so after all. The odor environment of an insect is enormously complex; an ordinary terrestrial community may contain hundreds of thousands of animal species and scores to hundreds of plant species each producing its own characteristic odors. Out of this vast complex of kaleidoscopically shifting active spaces an insect must select those few signals which are critical to its well-being. One of the most widespread mechanisms for accomplishing this is a simple screening process at the chemoreceptors. That is, the sensory apparatus is very sensitive to relevant stimuli and is able to sort these out from irrelevant ones (chemical "noise"). For example, many ants have a degree of olfactory acuity allowing release of behavior at lower threshold values in the presence of their own odorants than is possible with even closely related compounds; apparently a sophisticated deciphering of closely related molecular forms occurs at the antennal receptor sites. For example, when Blum and associates (Blum and Brand, 1972) studied the alarm-releasing activities of nearly 100 ketones for the ant *Pogonomyrmex badius*, they found many alarm pheromone "mimics" of varying levels of perception and activity but only a few capable, at low concentrations, of transmitting information that could be rapidly decoded and translated into alarm behavior.

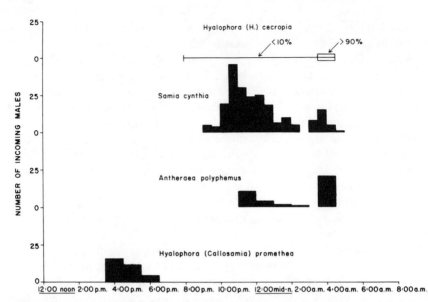

**Figure 5-18**  Periodicity of flight activity for males of four North American species of giant silk moths (Saturniidae); each has a differing time of peak flight activity. Seasonal separation also exists; most cecropia males emerge in May, polyphemus moths, in late May to early June, and cynthia and promethea moths, mostly in June. Thus, while the respective female sex attractant pheromones are known to be very similar, caged females of one species almost never attract males of another species because of these temporal and seasonal differences. In addition, sex pheromone release time can be influenced by ambient temperature during development. Female pupae of the related species, *Antheraea pernyi*, raised at 25°C initiated calling behavior about 6 hours after lights-off, whereas female pupae reared at 12°C advanced the onset of calling to 2 hours after lights-off. Corresponding shifts in flight activity occur in males under these temperature regimes. Thus, temperature, in addition to photoperiod, can also affect calling periodicity and flight activity (see Truman, 1973). (From Wilson and Bossert, 1963, based on Rau and Rau, 1929.)

Some ants produce trail substances in Dufour's gland, others in Pavan's gland, the poison gland, or glands in the hind tibia. Some alarm substances are produced by the anal gland, others by mandibular glands (Fig. 5-19). Insect exocrine glandular systems are extraordinarily diverse. There are a multitude of such glands, with wide variations in ultimate use of a given gland and its products. Furthermore, the products themselves are often complex . For example. the mandibular glands of the queen honey bee contain more than 30 compounds; at least 50 volatile compounds are present in the Dufour's gland of a carpenter ant, *Camponotus ligniperda*.

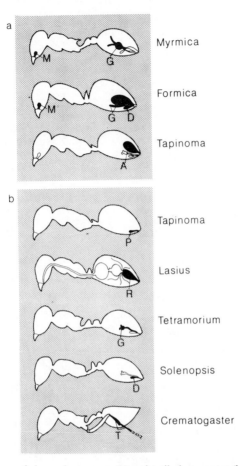

**Figure 5-19**  Sources of alarm pheromones (a) and trail pheromones (b) in various ants. M = mandibular gland; A = anal gland; D = Dufour's gland; P = Pavan's gland; R = rectum or hindgut; G = poison gland; T = tibia. (From Hölldobler, 1973.)

In human language development, the use of a different word for every message in every language has been superceded by the greater efficiency of coding systems based upon combinations of words. While a single word can send a message, adding others produces an almost infinite variety of meanings; consider the different behaviors elicited by "Help!", "Help me," "Help me love," and "Help me kill." An analagous strategy has been evolved by insects and allows them to send messages more efficiently: pheromonal blends.

A so-called "queen substance" is secreted in the mandibular glands

**Table 5.4**   The pheromonal language of insects: message systems

| Nature of chemical messenger | How message may be varied | Advantages | Limits |
|---|---|---|---|
| Unique single pheromone | 1. Alter ecological and/or behavioral context of signal<br>2. Change concentration of pheromone<br>3. Vary physiological condition of recipient<br>4. Change duration of signal<br>5. Modulate signal emission (frequency and/or amplitude) | Economy in receptor system | Need to synthesize increasingly complex molecules to maintain chemical uniqueness |
| Combination of pheromones:<br>1. Simultaneously released<br>  a) By same gland<br>  b) By different glands<br>2. Sequentially released<br>  a) By same gland<br>  b) By different glands | All of above *plus* increased number of messages by releasing same chemicals simultaneously *or* sequentially *or* singly | Greater efficiency in both transmitter and receptor systems; allows wholly unique messages to be more easily generated; more messages possible from same number of chemicals; can take advantage of chemical interactions—synergistic, inhibitory, masking effects, etc. | Increasing complexity of both transmitter and receptor systems |

222

of the queen honey bee. A principal component is the fatty acid 9-keto-*trans*-2-decenoic acid. On its own, it is both an olfactory sex attractant and an aphrodisiac. Combined with 9-hydroxy-*trans*-2-decenoic acid it acts as a primer pheromone and inhibits both queen rearing and oogenesis in worker bees (Gary, 1970, 1974).

It is not unusual to find a chemical blend produced within a single exocrine gland. For example, in 10 *Myrmica* ant species, a variety of natural products were found to be employed for the same communicative purpose—alarm (Blum, 1971). Among eight of the species, the same two chemicals (3-octanone and 3-octanol) were present, but in highly varying ratios. In one, the two chemicals were combined with a third major constituent. Two other species had none of the previously mentioned compounds but shared a single compound characteristic of them alone.

The exploration of chemical coding systems is still in its infancy, but it becomes increasingly apparent that pheromonal blends are probably the rule in arthropods rather than the exception. Such systems appear to have arisen in two ways: by the use of specific blends of chemical produced within a given gland and, perhaps less commonly, by the use of various combinations of glands releasing pheromones simultaneously or in succession.

Chemical combinations have several advantages over single chemical systems (Table 5-4). First and most obvious, the use of blends permits wholly unique signals to be easily generated and thus serves to insulate the signal channel from other species by producing a chemical "fingerprint". Pheromonal blends are of particular significance in maintaining reproductive isolation among certain moths related to the red-banded leaf roller, for example (Fig. 5-20). Also, it has been shown that various isolated fractions of Dufour's gland secretion of the fire ant, *Solenopsis invicta*, will elicit trail following by related species of *Solenopsis*, but the total gland extracts release trail following only in the species from which they are taken (Barlin et al., 1976).

Second, individual components of the mixture may have separate functions capable of being realized in specific, often hierarchial, behavioral contexts. Thus, a highly specific chemical "language" can be developed. Intensive research for over a decade has failed to identify the trail substance of the fire ant species complex. A major impediment has been that Dufour's gland contains an extraordinarily complex blend of components of widely different volatilities (Fig. 5-21) which elicit differing behavioral responses (Table 5-5). Undoubtedly, these interact and serve multiple functions in the trail-following recruitment behavior. For example, workers following a strong odor trail can be diverted by a

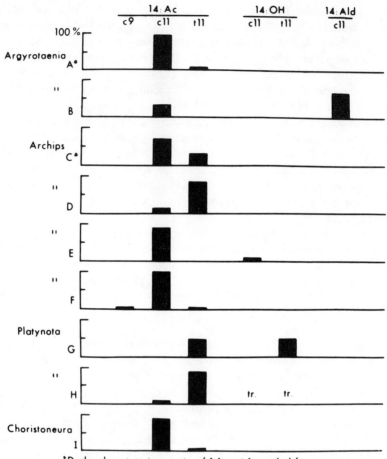

*Dodecyl acetate in a ratio of 1:1 to 4:1 needed for attractancy

**Figure 5-20** Sex attractant pheromone blends in moths of nine species belonging to four genera of the subfamily Tortricinae (Lepidoptera: Tortricidae). All the pheromones are 14-carbon chain compounds, but each species has a characteristic precise blend of components to which it responds most optimally. Some use various mixtures of the *cis–trans* isomers, others utilize acetate–alcohol or acetate–aldehyde mixtures, and some use mixtures of positional isomers. (From Roelofs, 1975.)

freshly laid trail made by a single worker to a new food source. This suggests that a freshly laid trail contains a more volatile fraction functioning particularly for recruitment in addition to the regular components. Other compounds from Dufour's gland extracts fail to elicit any trail-following behavior at all in bioassay tests; possibly these serve as

**Table 5-5**    The relative behavioral response of *Solenopsis richteri* workers to artificial trails made from different fractions of Dufour's gland extracts at three different concentrations[a]

| Fraction | Dilution[b] | | |
|---|---|---|---|
| | 1 ml | 2 ml | 4 ml |
| A | − | − | − |
| B | + | − | − |
| C | + + + | + | − |
| D | + + + | + + + | + + + |
| E | + + + | + + | + |
| F | + + + | + + | + |
| G | + | − | − |
| H | + | + | − |
| I | − | − | − |
| J | + + + | + | − |
| K | − | − | − |
| L | − | − | − |
| M | + + + | + + + | + |

[a] A–M correspond to the fractions marked in Fig. 5-21. Unpublished data courtesy of M. Barlin.

[b] The assignment of + + + indicates that the test substance elicited trail following from most ants. A rating of + + indicates that the test substance elicited trail following from most ants, with hesitation by a few workers. The assignment of + indicates detection of the test substance by only one or two ants which followed the trail in an unnatural way. Assignment of − indicates no trail following at all.

"preservatives" regulating release of more volatile fractions, thereby prolonging perceptibility of the components releasing trail following. The existence of compounds serving these functions has already been demonstrated in some moth sex pheromone blends. Finally, use of chemical "cocktails" can take advantage of synergistic, inhibitory, or masking effects among the different substances. Many examples of these phenomena have been uncovered, particularly among lepidopteran sex attractants.

The other major class of chemical releasers of behavior, the allo-

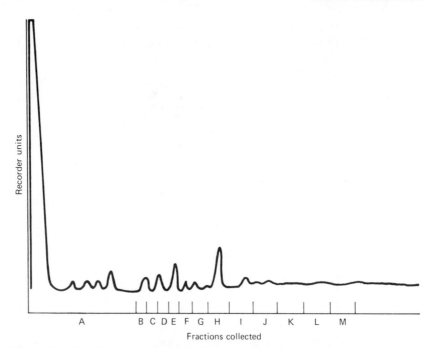

**Figure 5-21** Gas chromatogram of 10 *Solenopsis richteri* Dufour's gland extracts on 10% Carbowax 20 M, temperature programmed from 80° to 200° at 3° per minute; see also Table 5-4. (Unpublished data courtesy of M. Barlin; see also Barlin et al., 1976.)

mones, often also consist of chemical medleys and many of the above considerations apply equally to them. In addition, for defensive secretions, multicomponent allomones may function to simply overload predator receptor systems, thereby effectively hiding the insect chemically from its enemies.

Although a code system of almost any complexity could theoretically be developed by using various combinations of glands simultaneously or in sequence, such a coding system is as yet not well documented. However, many aculeate wasps release simple attractants from the head and sexual excitants from the abdomen. Fire ant workers produce cephalic and Dufour's gland secretions that cause alarm behavior and attraction, respectively, when released singly near workers; when expelled simultaneously by a highly excited worker, they cause oriented alarm behavior.

One additional method of increasing pheromonal information content deserves brief mention. Theoretically, an insect could practice modulation of the frequency and amplitude of pheromone emission, in a behavior analogous to that practiced in a great many visual and acoustic systems. To date, no examples are known of such pheromone modulation, yet the possibility cannot be dismissed. From a statistician's viewpoint, Bossert (1968) suggests two special circumstances where such a means of communication would be not only feasible but highly efficient. One is when transmission occurs in still air over distances of a centimeter or less; the other is in a steady, moderate wind.

## The Chemical Channel: Theoretical Considerations

One of the most striking characteristics of chemical communication systems is certainly their ubiquity, not just in insects but in most groups of animals. Chemical communication is, in fact, such a very general biological phenomenon that many scientists have suggested that it be considered, in some form involving cells or organisms or both, to be one of the fundamental attributes of life itself. How does it compare with other modes of communication?

From the simplest to the most complex, any communication system must fulfill two major criteria—it must be ecologically appropriate and it must fall within the sensory and motor ranges of the species concerned. An efficient system will likely have several of the following features as well: (1) qualitative and/or quantitative specificity, (2) rapid rate of information dissemination, (3) efficiency over considerable distance, (4) directionality of transmission, (5) wide range of information content, and (6) persistence *or* ability to start and stop quickly. Chemical communication systems possess most of these attributes and are certainly as fully versatile and sensitive as acoustic and visual communication. In addition, they possess several unique advantages. For one, they are the only major communicatory mode to have signals capable of lingering in the environment. Second, for chemical signaling, time and space take on a special meaning, for the sender and receiver do not have to be simultaneously coordinated as demanded by visual or acoustic communication. As a third point, like sound, chemical signals are able to go around environmental barriers although, unlike acoustic signals, chemical messages may be difficult to modify on short notice.

In passing, it is worth noting that a major difficulty encountered by nearly all who have attempted to study the olfactory system in most animals is the enormous variety of chemical compounds that may elicit a response. In many cases, the olfactory pheromone–receptor systems of insects may offer unique opportunities for such study, for in them, as in the example of the silkworm moth, we may have a relatively simple olfactory system in which many receptor cells respond identically to only one compound.


## CHEMICAL COMMUNICATION AND INSECT CONTROL

In the past 20 years, interest in insect pheromones has grown steadily, being augmented by the hopes that pheromones might be put to work as lures in the control of insect pests. In some cases this has been done very successfully, as for example in the field trapping research with leaf-roller moths conducted by Roelofs and his associates (see Roelofs, 1975). In at least one case, an insect pest has been completely eradicated in this way; small absorbent cane-fiber squares containing methyl eugenol and a fast-acting insecticide were dropped from airplanes over Rota, a small island in the Pacific Ocean, to completely wipe out the oriental fruit fly there (Jacobson, 1972). Another pheromonal control method which has met with an impressive degree of success has been undertaken by H. H. Shorey and colleagues (see references in Shorey, 1976). Evaporating large doses of synthetic sex attractants in cabbage and cotton fields, they produced sufficient confusion in male cabbage loopers and pink bollworm moths to result in a high percentage of unfertilized females. For additional discussion of the pheromone disruption technique, see Mitchell (1975).

The luring of pest insects with pheromones has in many cases proved valuable in another way. By predicting the outbreak of a large infestation, they have been useful in timing, calibrating, and eventually reducing the application of insecticides, thereby facilitating integrated controls, including natural enemies, etc. For example, consider a program in use against three of the world's most destructive fruit pests—the Mediterranean fruit fly, the melon fly, and the oriental fruit fly. All three are present in Hawaii but not currently on the mainland of the United States; accidental imports have, however, repeatedly occurred. A triple-baited trap containing synthetic attractants combined with a volatile insecticide has proven very effective in detecting new infestations when deployed about ports of entry. USDA officials estimate that between 1958 and 1964 this early warning system saved the government at least nine million dollars in potential eradication costs. In addition, such monitoring made possible significant reductions in insecticide usage; as long as pests are kept from invading, no pesticide is needed to control them. Marx (1973) and Shorey and McKelvey (1977) summarize the use of pheromones in insect control.

## SUMMARY

The field of chemical communication ranks as a vigorous and not so small discipline within insect behavior. Since 1959 when the term pheromone was coined, pheromone study has flourished, the beneficiary of two major technological advances. First was the introduction of gas chromatography, later coupled with mass spectrometry, permitting the identification of secretory products in minute quantities. Second, the development of neurophysiological techniques, notably the electroantennogram, has led to better understanding of chemoreceptor systems in insects. Concurrently, the development of predictive physical models about pheromone behavior has allowed analysis of odor transmission in various media. Several books summarize progress in the field (see Beroza, 1970; Ebling and Highnam, 1969; Tahori, 1970; Moulton et al., 1975; Shorey, 1976; Shorey and McKelvey, 1977; and Jacobson, 1965 and 1972) and a plethora of review papers (Blum, 1969; Blum and Brand, 1972; Hodgson, 1974; Hölldobler, 1973; Kaissling, 1971; Karlson and Butenandt, 1959; Kullenberg and Bergström, 1975; Law and Regnier, 1971; Leonard et al., 1974; Otte, 1974; Pain, 1973; Regnier, 1971; Roelofs, 1975; Schneider, 1971; Shorey, 1973; and Wilson, 1965, 1970).

Pheromone production and release is carried out by specialized glands called exocrine glands whose greatest diversity occurs among the social insects. Chemoreceptors occur abundantly on antennae and also often on mouthparts, tarsi, ovipositor, and cerci. They are extraordinarily sensitive, able to detect and discriminate between only a few molecules of odorant, and release a behavioral response usually involving orientation and locomotion. Some serve as primers, initiating physiological changes that prepare the recipient for subsequent stimuli. Releaser pheromones, on the other hand, mediate a variety of behavioral responses, including sexual behavior, assembly and aggregation, alarm and alert, spacing, and identification. All typically involve orientation and locomotion behavior.

Physical characteristics of most pheromones include a molecular weight of less than 300, fewer than 20 carbon atoms, and varying volatility. Chemical communication systems have evolved to enhance signal specificity and to increase communicative potential. Signal specificity is most often achieved through the use of chemical blends, while communicative potential within a single pheromone system may be increased by physiological adjustments of the senders' emission rate and/or receivers' threshold for response, or by temporal variation in concentration, duration, and context of pheromone emission.

## SELECTED REFERENCES

Amoore, J. E., J. W. Johnston, Jr., and M. Rubin. 1964. The stereochemical theory of odor. *Sci. Amer.* **210:** 42–49 (February).

Barlin, M. R., M. S. Blum, and J. M. Brand. 1976. Fire ant trail pheromones: analysis of species specificity after gas chromatographic fractionation. *J. Insect Physiol.* **22:** 839–44.

Barrows, E. M. 1975. Individually distinctive odors in an invertebrate. *Behav. Biol.* **15:** 57–64.

Barrows, E. M., W. J. Bell, and C. D. Michener. 1975. Individual odor differences and their social functions in insects. *Proc. Nat. Acad. Sci. USA* **72:** 2824–2828.

Bell, W. J., C. Parsons, and E. A. Martinko. 1972. Cockroach aggregation pheromones: analysis of aggregation tendency and species specificity. *J. Kans. Entomol. Soc.* **45:** 414–420.

Bergström, G. and L. Lundgren. 1973. Androconial secretion of three species of butterflies of the genus *Pieris* (Lep., Pieridae). *Zoon Suppl.* **1:** 67–75.

Beroza, M. (ed.). 1970. *Chemicals Controlling Insect Behavior.* Academic Press, New York, 170 pp.

Bierl, B. A., M. Beroza, and C. W. Collier. 1970. Potent sex attractant of the gypsy moth: its isolation, identification, and synthesis. *Science* **170:** 87–89.

Birch, M. C. (ed.). 1974a. *Pheromones.* North-Holland, Amsterdam, Holland, 495 pp.

Birch, M. C. 1974b. Aphrodisiac pheromones in insects. In *Pheromones,* M. C. Birch (ed.). North-Holland, Amsterdam, Holland, pp. 115–134.

Blum, M. S. 1969. Alarm pheromones. *Annu. Rev. Entomol.* **14:** 57–80.

Blum, M. S. 1971. Dimensions of chemical sociality. In *Chemical Releasers in Insects,* Vol. 3, A. S. Tahori (ed.). Proc. 2nd Int. IUPAC Congress, Tel-Aviv, Israel, pp. 147–162.

Blum, M. S. 1974. Deciphering the communicative Rosetta stone. *Bull. Entomol. Soc. Amer.* **20:** 30–35.

Blum, M. S. and J. M. Brand. 1972. Social insect pheromones: their chemistry and function. *Amer. Zool.* **12:** 553–576.

Blum, M. S., R. M. Crewe, W. E. Kerr, L. H. Keith, A. W. Garrison, and M. M. Walker. 1970. Citral in stingless bees: Isolation and functions in trail-laying and robbing. *J. Insect Physiol.* **16:** 1637–1648.

Bossert, W. H. 1968. Temporal patterning in olfactory communication. *J. Theoret. Biol.* **18:** 157–170.

Butenandt, A., R. Beckmann, D. Stamm, and E. Hecker. 1959. Über den Sexual-Lockstoff des Seidenspinners *Bombyx mori.* Reindarstellung und Konstitution. *Z. Naturforsch.* **14:** 283–284.

Corbet, S. A. 1971. Mandibular gland secretion of larvae of the flour moth, *Anagasta kuhniella,* contains an epideictic pheromone and elicits oviposition movements in a hymenopteran parasite. *Nature* **232:** 481–484.

Destephano, D. B. and U. E. Brady. 1977. Prostaglandin and prostaglandin synthetase in the cricket, *Acheta domesticus. J. Insect Physiol.* **23:** 905–911.

Dietz, A. and W. J. Humphreys. 1971. Scanning electron microscopic studies of antennal receptors of the worker honey bee, including sensilla campaniformia. *Ann. Entomol. Soc. Amer.* **64:** 919–925.

Duffey, S. S. 1977. Arthropod allomones: chemical effronteries and antagonists. Proc. XV Internat. Congr. Entomol., Washington, D. C., pp. 323–394.

Ebling, J. and K. C. Highnam. 1970. *Chemical Communication. Studies in Biology.* No. 19. Crane, Russak and Co., New York, 64 pp.

Fitzgerald, T. D. 1976. Trail marking by larvae of the eastern tent caterpillar. *Science* **194:** 961–963.

Gary, N. E. 1970. Pheromones of the honey bee, *Apis mellifera* L. In *Control of Insect Behavior by Natural Products,* D. L. Wood, R. M. Silverstein, and M. Nakajima (eds.). Academic Press, New York, pp. 29–53.

Gary, N. E. 1974. Pheromones that affect the behavior and physiology of honey bees. In *Pheromones,* M. C. Birch (ed.). North-Holland, Amsterdam, Holland, pp. 200–221.

Ghent, A. W. 1960. A study of the group-feeding behaviour of larvae of the Jack Pine Sawfly, *Neodiprion pratti banksianae* Roh. *Behaviour* **16:** 110–148.

Greany, P. D., J. H. Tumlinson, D. L. Chambers, and G. M. Boush. 1977. Chemically mediated host finding by *Biosteres (Opius) longicaudatus,* a parasitoid of tephritid fruit fly larvae. *J. Chem. Ecol.* **3:** 189–195.

Hangartner, W. 1967. Spezifität und Inaktivierung des Spurpheromons von *Lasius fuliginosus* Latr. und Orientierung der Arbeiterinnen im Duftfeld. *Z. Vergl. Physiol.* **57:** 103–136.

Heinrich, B. 1974. Pheromone induced brooding behavior in *Bombus vosnesenskii* and in *B. edwardsi* (Hymenoptera: Bombidae). *J. Kans. Entomol. Soc.* **47:** 396–404.

Hodgson, E. S. 1974. Chemoreception. In *The Physiology of Insecta,* Vol. II, 2nd ed., M. Rockstein (ed.). Academic Press, New York, pp. 127–164.

Hölldobler, B. 1973. Zur Ethologie der chemischen Verständigung bei Ameisen. *Nova Acta Leopoldina* **32:** 259–292.

Hölldobler, B. and E. O. Wilson. 1970. Recruitment trails in the harvester ant, *Pogonomyrmex badius. Psyche* **77:** 385–399.

Ishii, S. 1970a. Aggregation of the German cockroach *Blattella germanica* (L.). In *Control of Insect Behavior by Natural Products,* D. L. Wood, R. M. Silverstein, and M. Nakajima (eds.). Academic Press, New York, pp. 93–109.

Ishii, S. 1970b. An aggregation pheromone of the German cockroach, *Blattella germanica* L. 2. Species specificity of the pheromone. *Appl. Entomol. Zool.* **5:** 33–41.

Jacobson, M. 1972. *Insect Sex Pheromones.* Academic Press, New York, 382 pp.

Jacobson, M. 1965. *Insect Sex Attractants.* Wiley-Interscience, New York, 154 pp.

Kaissling, K. E. 1971. Insect olfaction. In *Handbook of Sensory Physiology,* L. Beidler (ed.). Vol. 4, *Chemical Senses,* Springer-Verlag, New York, pp. 351–431.

Karlson, P. and A. Butenandt. 1959. Pheromones (ectohormones) in insects. *Annu. Rev. Entomol.* **4:** 39–58.

Kullenberg, B. and G. Bergström. 1975. Chemical communication between living organisms. *Endeavour* **34:** 59–66.

Law, J. H. and F. E. Regnier. 1971. Pheromones. *Annu. Rev. Biochem.* **40:** 533–548.

Leonard, J. E., L. Ehrman, and A. Pruzan. 1974. Pheromones as a means of genetic control of behavior. *Annu. Rev. Genet.* **8:** 179–193.

Levinson, H. Z., A. R. Levinson, and U. Maschwitz. 1974. Action and composition of the alarm pheromone of the bedbug *Cimex lectularius* L. *Naturwissenschaften* **61:** 684–685.

Marx, J. L. 1973. Insect control (I): use of pheromones. *Science* **181:** 736–737.

Maschwitz, U. 1966. Alarm substances and alarm behavior in social insects. *Vitamins Hormones* **24:** 267–290.

Mayer, A. G. 1900. On the mating instinct in moths. *Psyche* **9:** 15–20.

McIndoo, N. E. 1917. Recognition among insects. *Smithson. Misc. Coll.* **68:** 1–68.

Mitchell, E. R. 1975. Disruption of pheromonal communication among coexistent pest insects with multichemical formulations. *BioScience* **25:** 493–499.

Moulton, D. G., A. Turk, and J. W. Johnston, Jr. (eds.). 1975. *Methods in Olfactory Research.* Academic Press, New York, 497 pp.

Nault, L. R., M. E. Montgomery, and W. S. Bowers. 1976. Ant-aphid association: role of aphid alarm pheromone. *Science* **192:** 1349–1351.

Naylor, A. F. 1959. An experimental analysis of dispersal in the flour beetle, *Tribolium confusum. Ecology* **40:** 453–465.

Nordlund, D. A., and W. J. Lewis. 1976. Terminology of chemical releasing stimuli in intraspecific and interspecific interactions. *J. Chem. Ecol.* **2:** 211–220.

Otte, D. 1974. Effects and functions in the evolution of signalling systems. *Annu. Rev. Ecol. Syst.* **5:** 385–417.

Pain, J. 1973. Pheromones and Hymenoptera. *Bee World* **54:** 11–24.

Pasteels, J. M. 1977. Evolutionary aspects in chemical ecology and chemical communication. Proc. XV Internat. Congr. Entomol., Washington, D.C., pp. 281–293.

Price, P. W. 1970. Trail odors: recognition by insects parasitic on cocoons. *Science* **170:** 546–547.

Price, P. W. 1972. Behavior of the parasitoid *Pleolophus basizonus* (Hymenoptera: Ichneumonidae) in response to changes in host and parasitoid density. *Can. Entomol.* **104:** 129–140.

Rau, P. and N. L. Rau. 1929. The sex attraction and rhythmic periodicity in the giant saturniid moths. *Trans. Acad. Sci. St. Louis* **26:** 83–221.

Regnier, F. E. 1971. Semiochemicals—structure and function. *Biol. Reprod.* **4:** 309–326.

Roelofs, W. L. 1975. Insect Communication—Chemical. In *Insects, Science, and Society,* D. Pimentel, (ed.). Academic Press, New York, pp. 79–99.

Roth, L. M. and S. Cohen. 1973. Aggregation in Blattaria. *Ann. Entomol. Soc. Amer.* **66:** 1315–1323.

Rudinsky, J. A. 1969. Masking of the aggregation pheromone in *Dendroctonus pseudotsugae* Hopk. *Science* **166:** 884–885.

Rudinsky, J. A. 1973a. Multiple functions of the southern pine beetle pheromone verbenone. *Environ. Entomol.* **2:** 511–514.

Rudinsky, J. A. 1973b. Multiple functions of the Douglas fir beetle pheromone 3-methyl-2-cyclohexene-1-one. *Environ. Entomol.* **2:** 579–585.

Rudinsky, J. A. and L. C. Ryker. 1976. Sound production in Scolytidae: rivalry and premating stridulation of male Douglas fir beetle. *J. Insect Physiol.* **22:** 997–1003.

Rudinsky, J. A. and L. C. Ryker. 1977. Olfactory and auditory signals mediating behavioral patterns of bark beetles. Coll. Internat., Centre National de la Recherche Scientifique, No. 265, Paris, France, pp.195–207.

Rudinsky, J. A., L. C. Ryker, R. R. Michael, L. M. Libbey, and M. E. Morgan. 1976. Sound production in Scolytidae: female sonic stimulus of male pheromone release in two *Dendroctonus* beetles. *J. Insect Physiol.* **22:** 1675–1681.

Schneider, D. 1969. Insect olfaction: deciphering system for chemical messages. *Science* **163:** 1031–1037.

Schneider, D. 1970. Olfactory receptors for the sexual attractant (Bombykol) of the silk moth. In *The Neurosciences: Second Study Program,* F. O. Schmitt, (ed.). The Rockefeller Univ. Press, New York, pp. 511–518.

Schneider, D. 1971. Specialized odor receptors of insects. In *Gustation and Olfaction,* G. Ohloff and A. F. Thomas (eds.). Academic Press, New York, pp. 45–60.

Schneider, D. 1974. The sex attractant receptor of moths. *Sci. Amer.* **231:** 28–35 (July).

Schneider, D. 1975. Pheromone communication in moths and butterflies. In *Sensory Physiology and Behavior,* R. Galun, P. Hillman, I. Parnas, and R. Werman (eds.). Plenum, New York, pp. 173–193.

Sebeok, T. A. 1968. *Animal Communication: Techinques of Study and Results of Research.* Indiana Univ. Press, Bloomington, 686 pp.

Shorey, H. H. 1973. Behavioral responses to insect pheromones. *Annu. Rev. Entomol.* **18:** 349–380.

Shorey, H. H. 1976. *Animal Communication by Pheromones.* Academic Press, New York, 168 pp.

Shorey, H. H. and J. J. McKelvey, Jr. (eds.). 1977. *Chemical Control of Insect Behavior: Theory and Application.* Wiley, New York, 414 pp.

Tahori, A. S. (ed.). 1970. Chemical releasers in insects. *Pesticide Chemistry,* Vol. 3. Gordon and Breach, New York, 240 pp.

Truman, J. W. 1973. Temperature sensitive programming of the silkmoth flight clock: a mechanism for adapting to the seasons. *Science* **182:** 727–29.

Whittaker, R. H. and P. P. Feeny. 1971. Allelochemics: chemical interactions between species. *Science* **171:** 757–770.

Wilson, E. O. 1962. Chemical communication among workers of the fire ant, *Solenopsis saevissima* (Fr. Smith). 1. The organization of mass-foraging. *Anim. Behav.* **10:** 134–147.

Wilson, E. O. 1963. Pheromones. *Sci. Amer.* **208:** 100–114 (May).

Wilson, E. O. 1965. Chemical communication in the social insects. *Science* **149:** 1064–1071.

Wilson, E. O. 1970. Chemical communication within animal species. In *Chemical Ecology,* E. Sondheimer and J. B. Simeone, (eds.). Academic Press, New York, pp. 133–155.

Wilson, E. O. 1971. *The Insect Societies.* Belknap Press of Harvard Univ. Press, Cambridge, Mass., 548 pp.

Wilson, E. O. 1975. *Sociobiology: the new synthesis.* Belknap Press of Harvard Univ. Press, Cambridge, Mass., 697 pp.

Wilson, E. O. and W. H. Bossert. 1963. Chemical communication among animals. *Rec. Progr. Hormone Res.* **19:** 673–716.

Wilson, E. O., N. I. Durlach, and L. M. Roth. 1958. Chemical releasers of necrophoric behavior in ants. *Psyche* **65:** 108–114.

Wood, D. L. 1970. Pheromones of bark beetles. In *Control of Insect Behavior by Natural Products,* D. L. Wood, R. M. Silverstein, and M. Nakajima (eds.). Academic Press, New York, pp. 301–316.

# 6

# Visual Communication

Small black paper balls pulled through the air on a string with sufficient speed will attract male flies, which fly after and grab them. When dragonfly nymphs are presented with various still objects against a background of moving stripes, they will snap at them. More bees settle on flowers on windy days than on still ones. Backswimmers, *Notonecta*, placed in a tank with white sides are unable to avoid being swept downstream in a current.

Salticids are sharp-visioned spiders that stalk their prey and spring upon it. When a mirror is placed in front of a male of the South American species, *Corythalia xanthopa,* the spider holds his body high off the ground, with his palps flexed in front of his face so that the yellow scales on them continue the band of yellow that crosses his clypeus. His abdomen is lowered. His front legs are on the ground, but his other legs are raised, each pair progressively higher than the ones in front of them. He poses motionless in this position. However, if a male has his palps removed and his clypeus shaved to remove the yellow scales, he behaves quite differently. Viewing himself in the mirror, the spider now stands with the forepart of his body slightly raised, his abdomen slightly lowered. He begins to rock from side to side, flexing his legs on one side while extending them on the other, then reversing the movement. He follows this by extending his front legs forward, parallel to each other but pointing upward at about 45° to the horizontal. The spider does not pose in this position but moves (Crane, 1949).

In each of these diverse examples we have a unique glimpse into the fascinating world of insect visual communication; each is exploiting the special properties of light that arthropods can perceive, compounding the elements into complex and specific signals and using visual messages for its own special purposes.

Historically, visual communication has been an unusually appealing subject for behavioral study for at least two major reasons: the tremendous amount of information visual signals may carry and the fact that visual systems play a dominant role in the behavior of man himself. Therefore, it is not surprising that in the theoretical development of

234

ethology over the last half century processes of visual communication have played a starring role. Lorenz, for example, derived many of his theories from the study of fish and bird **displays**—structures and behavior patterns that evolved as communicative visual signals to other conspecific animals. Tinbergen's (1948) landmark paper on social releasers also takes most of its examples from visual signals. At least two important signaling phenomena, ethologists have found, are more readily studied in visual signals than in auditory or chemical signals: the mode of origin of signaling structures and their transformation in the course of evolution, a process known as **ritualization**.

In addition to their central role in the development of ethological theories, visual communication systems have formed a fertile ground for development of two of the behaviorist's most powerful tools—the use of models to elicit behavior and the use of disguise. While the manipulation of relatively few stimulus variables yielded good results in neurophysiology, ethologists discovered very early that the types of behavior of interest to them often could be elicited only by more complex stimulus patterns. The development of ways in which to approach the study of these was one of the outstanding achievements of pioneer ethologists.

The method of using biologically meaningful simple models gained its early impetus in Germany from the use of artificial color patterns to study visual attraction in honey bees. But some responses could not be experimentally studied even by using simplified artificial models. It was just too difficult for an experimenter to effectively imitate all the necessary stimulus configurations, particularly when they might involve more than one sensory mode, as was especially true of some social responses. This limitation could be overcome, however, by the use of disguise; that is, by altering live animals in such a manner as to develop a "model" that looks and acts like the live animal (which it is!) minus certain key characters. For example, the butterfly *Nymphalis io* responds to predation attempts by rapidly lowering its wings, exposing four eye spots that mimic the vertebrate eye rather closely (See Fig. 8-18). Blest (1957) removed spots from some individuals and then compared predation by yellow buntings upon these "models" with that on equal numbers of unaltered individuals. Eye-spotted individuals escaped 76% more often than did unspotted models.

## LIGHT PRODUCTION

For centuries scientific investigators and curious lay observers alike have been fascinated by the production of light by living organisms, or

**bioluminescence.** A great diversity of plants and animals possess such an ability; among the insects themselves, at least eight families in four orders contain self-luminescent species. Bioluminescence has been described in more beetles (Coleoptera) than any other group of insects, of which the widely distributed fireflies (or "lightning bugs")—adult lampyrid and elaterid beetles—are perhaps the most well known. The intensity of luminescence varies greatly from one insect species to another, being in some so low as to be visible to human beings only by the completely dark-adapted eye. (If man had more sensitive eyes, he would probably consider more insects luminous.) But some are exceedingly bright for their size. The common eastern United States firefly, *Photinus pyralis,* has a flash that varies from 1/400th to 1/50th of a candle. The heat set free in this reaction is, however, exceedingly small; in *Pyrophorus* beetles, it has been judged as less than 1/80,000th of that produced by a candle flame of equivalent brightness.

Various insect tissues have become specialized for light production, but the majority of bioluminescent insects utilize the fat body. In most fireflies, the well-innervated luminescent organs are specialized regions of fat body beneath localized areas of transparent epidermis. These may occur in both sexes or in females only and in the larval stage (often called "glowworms"). A layer of light-producing cells with a massive tracheal supply rests upon a reflective layer of cells packed with urate crystals.

The mechanisms behind insect bioluminescence have long been the subject of speculation. The Roman encyclopedist Pliny offered the view that fireflies turn their lights off and on by opening and closing their wings; this statement was repeated again and again down through the Middle Ages. Finally, in 1885 the French physiologist Raphael Dubois attacked the question experimentally. Removing the light organ from *Pyrophorus,* a luminescent click beetle, he ground it up in water and then left it until the light went out. He removed a second light organ from another beetle and ground it in boiling water for a short time until its light was also extinguished. When he combined the two extracts the light reappeared! In this manner, Dubois showed that two substances were required for the light, one of which was inactivated by heat. These were named luciferin and luciferase, both after Lucifer, the bearer of light.

The morphological and biochemical details of insect bioluminescence are extremely complex, variable, and incompletely understood. The review by McElroy, Seliger, and DeLuca (1974) provides a good introduction to the very extensive literature on this subject. The uses to which bioluminescence is put are more clear (see Lloyd, 1971, 1977). In some instances, it is used as a lure for prey. However, in the majority of cases studied bioluminescent communication in insects is directed toward pair formation, mate identification, and location. In insect courtship, the dynamic properties of visual signals are often critically important in eliciting responses from the opposite sex. Certainly one of the most elegant examples studied to date involves the reproductive behavior of fireflies.

In eastern North America, the commonly seen fireflies belong mainly to two beetle genera, *Photuris* and *Photinus*. Courtship typically takes the following sequence: Males initiate flashing during flight at species-characteristic times (often around sunset) in a well-defined habitat area; their flight paths are species characteristic, especially during moments of light emission. Females remain stationary and, upon perceiving a male flash, answer with their own flashes, which follow that of the male after a brief, again species-characteristic delay. Repeated flash–answer sequences soon bring the sexes together, with copulation following (Lloyd, 1971).

As we have noted in the larger *Photuris* beetles, the smaller fireflies belonging to *Photinus* also include several cryptic species, recognizable by consistent differences in flash signals (Fig. 6-1; see also Fig. 1-10) but morphologically distinguishable only in minor details of body color. In many areas two or more *Photinus* species fly together. In other cases related species are geographically isolated. Using electronic devices to produce artificial flashes of known duration and to accurately measure the female's response delay, Lloyd (1966) confirmed that the flash signals of sympatric species differed significantly but that those of species that normally did not occur together were often very similar (Fig. 6-2). In other words, only where there was a possibility of reproductive mistakes being made had refined isolating mechanisms evolved. By varying different signal parameters, Lloyd learned that female *Photinus* were discriminating male light pulse length, the interval between pulses and/or pulse number. Furthermore, males were able to discriminate differing female answer delay times.

Do *Photuris* speak a different flash language from *Photinus*? Female *Photuris* had long been known to be carnivorous, but imagine his surprise when Lloyd observed *Photuris* females attracting and devouring *Photinus* males by mimicking the flash responses of *Photinus* females. As Lloyd (1965) pointed out, this raises a host of new questions: "Is the female *Photuris* predaceous before she has mated? If so, how does her mate avoid the fate of attracted *Photinus* males?. . . Can a single *Photuris* species prey upon more than one *Photinus* species with different signal systems?. . . Is predation on *Photinus* in any sense

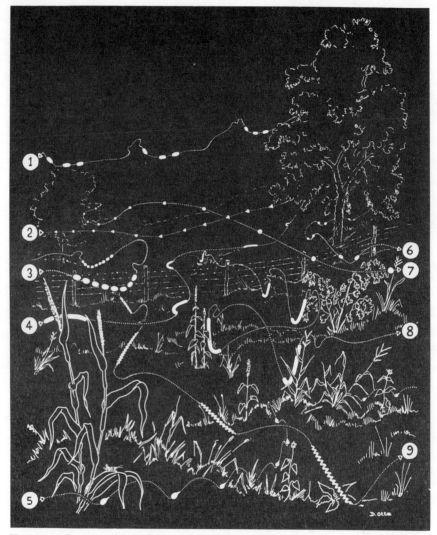

**Figure 6-1** Patterns of light flashes of 9 species of *Photinus* fireflies, illustrating differences in habitat and flash parameters which help to maintain reproductive isolation
between the species. (From Lloyd, 1966; courtesy of J. E. Lloyd.)

obligatory?'' Undoubtedly, this kind of aggressive mimicry (see Chapter
4) has greatly affected the evolution of the signal systems and other
behavior of *Photinus*.

   Lloyd (1975) found that female *Photuris* are indeed versatile and able
to adjust their flashes to successfully attract males of at least four
species with distinctively different flash patterns. The mimicry is quite

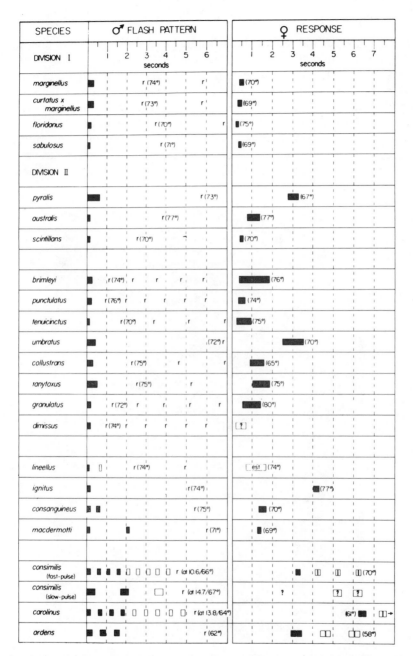

**Figure 6-2** Diagrammatic representation of male flash patterns and female response flashes in selected *Photinus* fireflies. Male flash precedes female; female response time intervals shown are measured from start of last pulse of male flash pattern. Flash pattern repeat interval is indicated by r, and observed variation is depicted by open pulse boxes. (From Lloyd, 1966; courtesy of J. E. Lloyd.)

effective, and females generally succeeded in capturing at least one male for every 10 attempts (Fig. 6-3). Nelson and colleagues (1975) showed that a *Photuris* female becomes a *femme fatale* only after having mated. Mating induces a syndrome of behavioral changes, including locomotor activity, answering postures, predaceous behavior, and response to flashes of males of different species. Conspecific males are not eaten.

While Lloyd (1966) estimates that nothing is known of the communicative behavior of probably 95% of all firefly species, some of the more primitive species are known to have continuously glowing females and in

**Figure 6-3**  A female *Photuris* firefly (*left*) has seized a male of another firefly species in a fatal embrace after attracting him to her by mimicking the mating signal of females of that species, an example of aggressive mimicry. (From Lloyd, 1975; courtesy of J. E. Lloyd. Copyright © 1975 by the American Association for the Advancement of Science.)

others females turn on their glow only in response to glowing males. In other words, the complex flash exchange systems found in *Photinus* and *Photuris* are not necessarily typical of all fireflies but probably represent a quite advanced level of evolutionary development for the group. Many behavioral and ecological adaptations serve to enhance bioluminescent communication efficiency. Females of most species climb up on perches during hours of mating activity. Flashing males assume flight altitudes so that their light is directed toward the ground ahead of them, and many species execute aerial maneuvers which enhance their chances of seeing or being seen by females. In addition, most fireflies exhibit habitat specificity and/or orientation to a mating site; these further restrict the areas that males patrol, reducing "background noise" and the chances of interspecific interaction.

Restricted periods of activity, seasonally or diurnally, have similar advantages. Perhaps the most extreme example of such ecological and behavioral restriction, however, is provided by the "firefly trees" found in parts of tropical Asia. After a river voyage from Bangkok to the sea, a Dutch physician (quoted in Buck and Buck, 1968) in 1680 wrote:

"The glowworms . . . represent another shew, which settle on some Trees, like a fiery cloud, with this surprising circumstance, that a whole swarm of these insects, having taken possession of one Tree, and spread themselves over its branches, sometimes hide their Light all at once, and a moment after make it appear again with the utmost regularity and exactness. . . ."

At times the insects in the trees flash on and off together like this hour after hour, night after night, for weeks or even months. More than 30 similar reports have been published describing these oriental firefly displays, occurring principally from mangrove trees along brackish rivers.

Many explanations have been proposed for these displays, including "a sense of rhythm," "an organic law of rhythmic appreciation," "sympathetic telepathy," and a whole host of other similar nonexplanations with anthropological overtones. One writer even went so far as to attribute it to the twitching of the observer's eyelids, remarking that "the insects had nothing whatever to do with it!"

Through extensive field work with oriental fireflies in Thailand and Borneo and use of photographic and photometric analyses, the Bucks (1968, 1976) showed that synchrony of great numbers of individuals is indeed nearly perfect. They hypothesize that flash synchrony is controlled by an internal resettable pacemaker (Fig. 6-4). Contrary to earlier reports, both males and females occur in these trees, although the

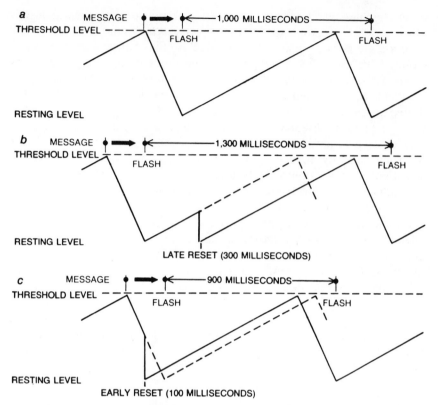

**Figure 6-4** Model of resettable pacemaker hypothesized to be responsible for flash synchronization among oriental fireflies where the normal interval between flashes is 1 second. The underlying mechanism is believed to involve cyclic changes in pacemaker excitation levels. Starting at a resting level, excitation increases steadily for 800 milliseconds to a threshold level at which two events occur simultaneously: a spontaneous triggering of a motor message giving rise to a flash in the light organ about 200 milliseconds later, and a spontaneous decline in excitation of the same duration which returns the pacemaker to its original level and starts the cycle anew. A strong external light signal at any point during the cycle will override the controls and reset the cycle by dropping excitation back to the resting level, after which the new cycle begins spontaneously. (From "Synchronous Fireflies" by John and Elisabeth Buck. Copyright © 1976 by Scientific American, Inc. All rights reserved.)

females do not participate in sychronous flashing. In some instances, aggregations in a given tree may include more than one species, resulting in a complex combination of flashes that is still presumably effective for each species involved.

The Bucks consider synchronous flashing to be part of a complex of behavior patterns (congregation, selection of certain trees, flashing, etc.) for enhancing mating under otherwise difficult conditions. The firefly trees appear to serve as quasi-permanent rendezvous, and synchrony appears to increase their efficiency as beacons. Adult fireflies in these tropical areas live for only a few days and therefore must find mates quickly if they are to reproduce before they die. A single prospective mate is nearly invisible among the thick, tangled plants of these swampy areas, but many males flashing synchronously are easily seen by other fireflies, both male and female.

## LIGHT RECEPTION

In understanding the behavior of a given species, a desirable first step is a thorough examination of the sensory physiology of the species. Vision illustrates this well. How an organism behaves depends greatly upon the way its world appears. When an insect regards a complex scene, it obviously cannot and does not notice every detail. What it sees, or rather, what it responds to, depends on what its evolutionary history has preprogrammed it to "look" for. An insect's response is also determined by properties intrinsic to the photoreceptors themselves and the manner in which they relate to higher centers in the brain. These factors are of particular behavioral interest inasmuch as they are experimentally approachable aspects of vision which relate to the insect's perceptual "self-world," or **Umwelt.**

### Insect Visual Receptors

The most important organs of light perception in the majority of insects are the **compound eyes.** With the exception of certain specially adapted parasitic and cave-dwelling species, most adult insects have a pair of these prominent organs, bulging to varying degrees from either side of their head so as to give a wide field of vision in all directions. Composing each compound eye, and extending inward like narrow columns, is a group of densely packed hexagonal units called **ommatidia.** The number of ommatidia in an insectan eye varies greatly; while the common housefly has about 4000, dragonflies may have as many as 28,000. Each ommatidium in turn consists essentially of an optical, light-gathering part (the lens and crystalline cone) and a sensory part (retinula cells and their differentiated margins, the rhabdomeres) perceiving the radiation and transforming it into electrical energy.

In addition to the compound eyes, adult insects and nymphal hemimetabolous insects typically have three simple eyes called **ocelli**. The ocelli are apparently not involved in form vision but are very sensitive to low light intensities. In certain bees and wasps there appears to be a correlation between large ocellar size and activity peaks at dawn or dusk (Kerfoot, 1967). It is also known that insects can perceive light directly via the brain cells as well as through their compound eyes and ocelli; some have other photosensitive tissues as well.

Larval holometabolous insects such as caterpillars lack both compound eyes and ocelli, but many possess up to six simple ocellilike **stemmata** on each side of the head; with the possible exception of dermal receptors sensitive to light, these are their sole visual organs. Stemmata vary considerably in number and arrangement but apparently are capable of perceiving at least a coarse mosaic, for it is known that caterpillars can differentiate shapes and orient toward boundaries between black and white areas. In some cases, stemmata may also mediate responses to polarized light, as has been shown for larval *Neodiprion* sawflies and tortricid caterpillars. Many larvae have the characteristic habit of moving the head from side to side while advancing, perhaps to compensate for the paucity of visual units in this system; by this behavior, they are able to examine a larger visual field, and the recognition of changes in light intensity is facilitated. General references on insect vision include Horridge (1975, 1977), Mazokhin-Porshnyakov (1969), and Goldsmith and Bernard (1974).

**Perception of Visual Stimuli**

How much of the spectrum can insects see? For man, only the wavelengths between about 400 m$\mu$ (ultraviolet) and 750 m$\mu$ (red) are visible; we cannot perceive the near-ultraviolet portion of the sun's rays that reaches the earth. Spectral sensitivity curves (action spectra) have been plotted for many species of insects. In general, most respond to a range of wavelengths extending from the near-ultraviolet (300–400 m$\mu$) up to a maximum around 600–650 m$\mu$ (orange), with two peaks of greatest activity—one about 350 m$\mu$ (near-ultraviolet), the other about 500 m$\mu$ (blue-green). Near-ultraviolet light is the most effective region of the spectrum in directing phototaxes; the widespread use of ultraviolet lamps in insect traps is a practical application of this. The adaptive significance of ultraviolet phototaxis may be that such light signals "open space." Much of nature, especially green foliage, absorbs ultraviolet wave lengths; the open sky, left as the only extensive source of ultraviolet rays, may signify room for free flight and maneuvering (Mazokhin-Porshnyakov, 1969).

Spectral sensitivity alone does not imply discrimination; however, behavioral evidence supports the interpretation that many insects can and do distinguish between different wavelengths and hence possess true color vision. In his pioneering work on insect color vision in 1914, von Frisch (1967) showed that honey bees were able to differentiate accurately between several major categories of color (Fig. 6-5): yellow, blue-green, blue (including violet), ultraviolet, and "bee purple," a mixture of the spectral extremes, orange and ultraviolet. Von Frisch's techniques employed simple Pavlovian conditioning: marked bees were allowed to feed at a sugar source while simultaneously being exposed to a particular color stimulus, then tested to see whether they would be attracted to the color in the absence of the food. Finally, if this proved successful, they were permitted to choose between the original color and a different, closely similar one to see whether they could discriminate between the two. In recent years these studies have been supplemented with electrophysiological techniques (see also Burkhardt, 1964).

Knowledge of the color sensitivity of insects has significant implica-

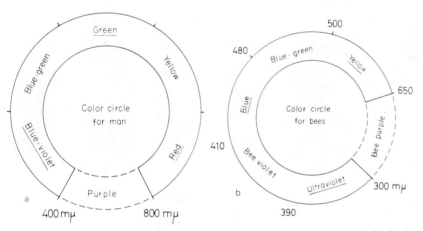

**Figure 6-5** Perception of color by man (*a*) and the honey bee (*b*). Color vision in the bee is trichromatic and closely similar to that of man, but the three primary colors (underlined) differ. In both, complementary colors are opposite to each other, and intermediate colors, produced by mixing the primary colors. A mixture of all wavelengths in the proportions found in sunlight is called white by man. "Bee white" is similarly explained but differs from our white because of the different visual spectrum. (Reprinted by permission of the publishers from *The Dance Language and Orientation of Bees* by Karl von Frisch, Cambridge, Mass.: The Belknap Press of Harvard University Press, Copyright (©) 1967 by the President and Fellows of Harvard College. Somewhat modified, after K. Daumer, Reizmetrische Untersuchung des Farbensehens der Beine, Z. vergl. Physiol. **38**, 413–478 (1956.).

tions for understanding and interpreting visual behavior. For example, to the eyes of a honey bee many flowers glow with color we can see only with the aid of special equipment (see Eisner et al., 1969). Many pale-yellow flowers, such as the evening primrose, *Oenothera,* or the cinquefoils, *Potentilla,* reflect ultraviolet light over the greater part of their petals but have patches at the petal bases that fail to do so; to honey bees, these latter **nectar guides** are sharply differentiated visually (Fig. 6-6). In addition, some red flowers such as poppies reflect ultraviolet light so strongly as to be conspicuous to the bees although the rich red so striking to us is totally invisible to them. Color vision probably also plays a part in the choice of backgrounds by cryptically colored insects, and such choices may be based on an *Umwelt* quite different from our own. As a final example, courting butterflies may be especially sensitive to color patterns on the wings of their potential mates composed of colors that are totally invisible to us (Fig. 6-7). The rhythmic flashing of these wing blotches during flight is of communicative significance during courtship (Silberglied, 1977) but is invisible to

**Figure 6-6**  *Potentilla,* a plain yellow flower in the rose family, photographed through filters equivalent to the three primary color ranges of the honey bee. Seen through a filter with only an ultraviolet transmission band (right), the outer portion of the petals strongly reflect ultraviolet; the nonreflecting center parts contrast strongly to serve as nectar guides pinpointing the food source. Visualize the three photographs superimposed; the composite picture is what the bee actually sees, a "bee purple" flower with a "bee yellow" nectar guide. Because the foliage gives weak and relatively uniform reflectances under all three filters, the green leaves and stems are almost a colorless gray for bees. Therefore, flowers stand out even more conspicuously against this rather drab background, making them easy for scout bees to discover. (Reprinted by permission of the publishers from *The Dance Language and Orientation of Bees* by Karl von Frisch, Cambridge, Mass.: The Belknap Press of Harvard University Press, Copyright © 1967 by the President and Fellows of Harvard College. After K. Daumer, Blumenfarben, wie siedie Bienersehen, Z. vergl. Physiol. **41,** 49–110 (1958). See also Eisner et al., 1969 and Jones and Buckmann, 1974.)

**Figure 6-7**  Three species of pierid butterflies whose wings exhibit regions of high ultraviolet reflectance. The male is uppermost in each photograph and often exhibits a pronounced dimorphism with respect to the ultraviolet reflectance trait (right column, photographed in near-ultraviolet light, 300–400 m$\mu$) which is not apparent to our visual spectrum (left column, photographed in visible light, 400–700 m$\mu$). (A) and (B) are *Colias eurytheme* whose main visible color is orange; (C) and (D) are *C. philodice*, a predominantly yellow species; (E) and (F) are *C. chrysotheme*, an orange-colored species. (From Silberglied and Taylor, 1973; courtesy of R. E. Silberglied.)

vertebrate predators. The combination of flight pattern and wingbeat frequency would seem to have the potential to produce sign signals analogous to the flashing patterns of male fireflies. Patches of closely appressed silvery hairs are commonly present on the bodies of various insects and in at least one case (Deonier, 1974) are ultraviolet reflecting and likely serve as similar communicatory signals between the sexes.

The eyes of an insect are fixed; the insect cannot move them independently of each other. However, as an insect directly approaches the object which it is viewing (or vice versa) the retinal image gradually appears closer toward the inner part of the two compound eyes, thus

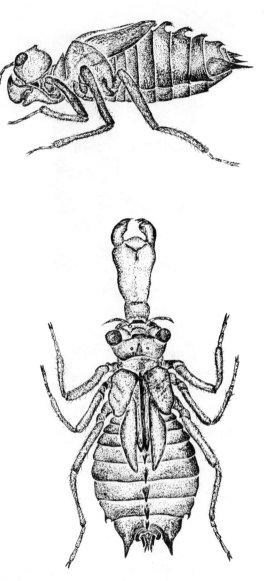

**Figure 6-8** Dragonfly larva (*Epicordulia*) with the mask or labium, in the normal (*above*) and fully extended (*below*) positions. Prey capture is accomplished by shooting out the labium which bears two jaws. The mask is so jointed that it cannot be moved sideways nor used except in the fully extended position. These restrictions dictate that prey can only be caught when they are at one point in space with respect to the head. By facing the prey directly and moving toward it, the larva views the prey with increasing definition. When the prey's image falls on certain inner ommatidial elements it is seen the most clearly and is exactly within proper striking distance which corresponds to the intersection of the optical axes of these ommatidia. (Drawing by J. W. Krispyn after Baldus, 1926.)

affording a method of judging distance. As the visual angle of the ommatidia becomes progressively less over the inner part of the eye, the object comes into sharper vision. Experimental work with predatory insects such as dragonfly nymphs demonstrates this well (Fig. 6-8).

The multiple-unit ommatidial system apparently provides insects with a mosaic image composed of tiny points of light of varying brightness, each provided by one ommatidium. This mosaic will be coarse or fine depending upon the number of facets per unit area, which often varies in different directions in a given eye and in different regions of the eye. For example, one might reason that for many swiftly flying insects acuteness of vision in the vertical axis would be more important than in the horizontal; in fact, the curvature of their eyes lends very different dimensions to the ommatidial angles in these directions.

To what extent can insects see and distinguish shapes? A generalized answer is difficult because form and motion perception are very closely intertwined. However, for stationary objects at a distance, the compound eye universally shows poor resolution. Details of an object often fail to evoke responses at all. With some insects such as the honey bee, the eye receives fuzzy images even when the object is large and nearby.

Much of our present knowledge of insect form vision comes from honey bee behavior during training studies where choice of some shapes has been rewarded with food. However, there are several limitations and difficulties implicit in these studies. Perhaps the most striking has been that all such training studies have been complicated by the bee's spontaneous responsiveness to flicker. Moving shapes of all sorts are more attractive than stationary ones, and while bees could be trained to distinguish solid figures from broken ones, they could not distinguish solid from solid or broken from broken. The adaptiveness of such behavior is apparent if we consider a bee in its natural environment. Not only does wind cause flower movement, but when bees fly low over the ground in search of flowers they experience the passing of a radial cluster of flower petals as a burst of flickering over their ommatidial surfaces. The regularity and high frequency of such flicker patterns identify the stimulus as readily as the total shape does to our eyes.

A second limitation not as immediately apparent is that even these results are quite possibly open to question, for they may be confounded by stimulus filtering appropriate to flower searching but not to other situations. Still a third is that honey bees may not be the appropriate "representative" insects to generalize from, any more than white rats are necessarily "representative" mammals for all purposes. Certainly some other Hymenoptera, such as the predatory wasps *Philanthus* and *Ammophila*, use landmarks for topographic orientation (see Fig. 2-9) to an extent which implies a better ability to discriminate forms than would be expected from theories built upon honey bee observations.

However, even with these limitations, one can safely say that by and large the insect eye is better adapted to perceive motion than static form. In fact, reliance on an object's real or apparent motion (**flicker vision**) is so widespread that it may be regarded as one of the normal concomitants of vision with the compound eye. Because ommatidia recover very rapidly from light impulse stimulation, the insect eye has a remarkable capacity for seeing successively different images at very short intervals and thus for scanning a moving object. As a result, an insect may be able to resolve a finer pattern when it is flying than when it is at rest.

When an otherwise static scene is suddenly interrupted by movement, flicker, or novelty, our attention is immediately attracted. The same is true for insects. Many predatory species, for example, respond only to moving prey; and given choices of stimuli, most insects show a preference for the shapes that cause the most flicker. However, flicker is hardly noticeable to us when it exceeds a rate of about 20 to 30 changes per second. How do insects compare with us in this regard?

When a flickering light is presented to an insect, it is possible to determine the rate of flicker its eye is just unable to distinguish—the **flicker fusion frequency.** If this frequency is exceeded, the light appears continuous. There is a great variation in range of these figures for insects, and the limits are undoubtedly set by retinal mechanisms. There is, however, a strong correlation between insect behavior and rate of flicker fusion. "Slow eyes" with a flicker fusion frequency of as low as 5 to 10 flashes per second are characteristic of relatively slow-moving or nocturnal insects. Rapidly flying diurnal insects such as bees and flies, by contrast, may resolve flicker frequencies as high as to be 10 times as discriminating as the human eye. Male house flies, for example, respond to flickering stimuli during courtship and will court models with increasing responsiveness at flickers of up to 270 stimulus changes per second before beginning to decline; their flicker fusion frequency has been estimated separately through electrophysiological studies at 265 per second. Marler and Hamilton (1966) discuss flicker vision and other temporal properties of light stimuli.

One additional quality of light perception deserves brief mention, and that is perception of the direction of motion of the light waves—**polarization.** It has long been known that light coming from the blue sky is partially polarized, that is, most of the waves are vibrating in the same plane, the extent being dependent upon the position of the patch of sky relative to the sun. The first indication and then the first proof that animals might have the ability to detect this polarization was gained by von Frisch through studies of honey bee communication. The accuracy of the "dance" (see Chapter 7) of a returning forager, which shares information on direction and distance of the food source with other workers, depends in part upon the polarized light pattern perceived from blue sky.

Because the dances are goal directed, the honey bee affords a unique opportunity to experimentally test orientation to polarized light. When the sky is completely cloud covered, bee dances are disoriented. If, however, a small patch of blue sky remains (even when the sun is not visible), bees dancing on a horizontal surface will orient correctly. If the plane of light vibration is altered artificially, as can be done by interposing polaroid sheets, the orientation of the dances changes correspondingly. For example, when the polaroid sheet is rotated clockwise by 30°, the bees will immediately shift the direction of their dances by about the same amount. Perception of the plane of vibration is through the compound eyes, with the plane of vibrations apparently decoded in the rhabdomeres of the sensory cells, whose fine structure reveals the presence of tightly packed microvilli arranged in two planes mutually perpendicular to its long axis. The ability to orient to polarized light has subsequently been demonstrated for a wide range of other insects and arthropods (see von Frisch, 1967).

## FUNCTIONS OF VISUAL COMMUNICATION

Visual systems would seem to have several advantages over other communicatory modes. The range of possible signal variations is theoretically almost limitless; one has only to vary, independently and in combination, such basic signal aspects as color, form or posture, movement, or timing. In addition, the rapid adaptation rates of insect visual receptors could be exploited in development of a wide variety of temporal patterns, developing a system which could be started or stopped immediately. Thus, an insect sensing a predator could freeze and need not communicate its position by any lingering image such as those which might be left by a chemical system. But in situations where it would be advantageous to do so, the insect could make clear its exact position, so that the receiver could respond to it in terms of precise location as well as general presence.

The fact remains, however, that insects do not appear to rely upon visual signals for intraspecific communication to as great a degree as they do upon chemical and auditory ones. The same conspicuous patterns on an insect's body that aid in intraspecific communication can be a hindrance when it must hide from a predator. Thus, permanent intraspecific visual signals tend to be exploited most fully by those insects that are relatively immune to predators. In addition, except in unusual cases such as fireflies, visual signals are useless at night, in dark places, where blocked by the environment, and at long distances. The short visual range and limited capacity for detail possessed by the insect eye also restrict the importance of fine detail in insect releasers. Even within its useful range, a visual signal becomes simpler and bolder, consequently carrying less information, as distance increases. It normally cannot be increased by pumping more energy into it, as is possible with sound or chemical signals.

Functionally, any situation involving face-to-face interaction seems a potential candidate for visual signals. For example, many visual signs are employed in interspecific contexts such as defensive behaviors (especially crypsis, threat displays, and mimicry) and in pollination, functions dealt with in Chapters 4 and 8. In intraspecific contexts, the great majority of currently known examples concern visual signaling associated with reproductive activities. This may, however, simply reflect the disproportionate amount of scientific attention that has been directed toward courtship behavior as compared to other behaviors in which visual cues may be functioning. It is not unreasonable to expect that many additional examples of visual signaling will be forthcoming as attention is directed toward other behaviors where frequent close-range interactions between conspecific individuals are commonplace.

### Aggregation and Dispersion

Behaviors involved in the distribution of members of a species in space have a crucial bearing on their exploitation of environmental resources and thus are of great importance. At close range, such behaviors would seem ideal candidates for visual messages, but aggregation by insects is relatively unstudied in comparison with such behavior in fish, birds, and many mammals, all of which have been demonstrated to utilize prominent visual signals in forming and maintaining such groupings. In bringing together large numbers of widely scattered insects from over large areas, it stands to reason, however, that auditory and chemical cues would be relied upon more heavily than visual ones. The habits of most insects are such that their environment simply does not permit long visual ranges. In addition, the fixed-focus compound eye is not well adapted to long-distance vision.

Dispersal through visual mediation is only slightly better known than insect aggregation using these cues but theoretically might be encouraged by visual signals of several levels of complexity. Insects might simply avoid other conspecific individuals at sight, or they might direct signals to one another, eliciting withdrawal. The signals could be simple ones generated by the insect alone, or compound visual situations involving perception of a complex of environmental factors as well.

Territoriality (see Chapter 9) in dragonflies provides one clear documentation of this type of visual communication. All dragonflies are more or less selective in their breeding sites; and in many cases the male, who generally arrives at the site before the female, localizes his activities over a certain pond or stream area which he will defend against intrusion by other males. Within this area, he will court females. On a smaller scale, damselflies will often establish similar territories closer to vegetation or to the water surface. Sometimes several different species of different size classes will exhibit "stacked territories" over the same pond (Corbet, 1962).

Perhaps the most striking features of the Odonata, aside from their two pairs of intricately netted wings, are their great protruding eyes. Combined with a head that can be rotated readily upon a slender neck, they allow a degree of motion perception (for some up to 40 yards away) that is most unusual for fixed-focus eyes. In addition, the antennae, so prominent in most insects, are miniscule in the Odonata, and their removal appears to make no difference in navigation or prey capture. As these facts might suggest, dragonflies are indeed among the few insects in which the sense of sight is greatly dominant over the other senses. In fact, Odonata behavior is so visually mediated that some have facetiously dubbed them "the bird watcher's bugs."

As with most territorial animals, the dragonfly's territorial behavior centers upon certain ritualized **aggressive displays,** backed up by physical combat as a last resort. In aggressive display, two males of the same species recognize each other as such and then indulge in a formalized ritual which usually concludes with the departure of one of the pair. One example can be seen in one of the most common pond inhabitants in North America, a rather large dragonfly, *Plathemis lydia.* The males have abdomens that are bright silvery white above; and, as with the bright colors found in many other male dragonflies, these play an important role in male interactions. When two male *Plathemis* (Fig. 6-9) encounter each other, one dashes at the other and pursues him. The pursuing male invariably raises the white upper abdominal surface toward the new arrival, who flies away with his own abdomen lowered. Then, surprisingly, after flying 8 to 16 meters, the two males switch roles, the first one now flying with abdomen down, the second pursuing

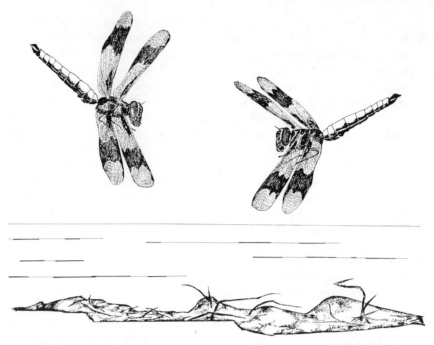

**Figure 6-9** Two males of the dragonfly, *Plathemis lydia,* mutually displaying over an oviposition site, their abdomens raised to display the white upper surface. Success in aggressive display is correlated with abdominal whiteness (which develops gradually with age) and therefore with sexual maturity. Males with abdomens painted black were sometimes ignored by other males. (Drawing by J. W. Krispyn after a photograph in Jacobs, 1955; see also Campanella and Wolf, 1974.)

with his abdomen raised. This pursuit display, alternating in direction, continues until the new arrival finally restricts his movements to the vicinity of another site (Corbet, 1962).

Preying mantises sometimes also defend territories. Upon casual observation it appears as though they are not usually discriminating between conspecifics and prey or enemies but striking out at all alike. However, careful analyses of filmed records of this behavior have shown that in striking out at conspecifics, a small detail of the prey-catching stroke is omitted (Roeder, 1960). The tibia is not closed against the femur. Thus, the "display strike" is nondamaging, an important first step toward "ceremonial" aggression.

Visual displays communicating aggression are often highly variable. This is in marked contrast to the redundant stereotypy of most sexual displays. Aggressive displays often range along a continuum from attack to flight. Such **graded displays** make it possible to communicate slight changes in motivation, an important advantage in facilitating the resolution of territorial and other conflicts.

## Alarm

In marked contrast with auditory and chemical communication systems, specialized visual alarm systems have rarely evolved in insects. One explanation for this may be that most predators place great reliance upon their visual sense while hunting. It is most difficult for a potential prey to emit a visual alarm to its companions without also making itself more conspicuous to the predator. Thus, the commonest visual signals eliciting alertness, alarm, and flight tend to be provided not by specialized systems but unintentionally by the very actions of flight.

One interesting example in which such cues are apparently involved occurs among many Lepidoptera. Adult butterflies of several families will often congregate, sometimes in large numbers, around margins of puddles or animal feces (Fig. 6-10) where they apparently obtain needed

**Figure 6-10** Butterflies drinking at a mud puddle. When disturbed, they will swirl up together to form a confusing mass of colorful forms. Although several individuals of the same species are present, they are not necessarily closely related, and such behavior is better interpreted in terms of survival benefit to individuals who all respond similarly and simultaneously to threat of danger than as some form of visual alarm by one which alerts the group. (Courtesy of R. L. Jeanne; also see Arms, et al., 1973.)

sodium (Arms, et al., 1973). To a predator such concentration of brightly colored butterflies could represent a potential bonanza; in fact, they usually do not, primarily because of their behavior. Upon disturbance, masses of butterflies will suddenly fly up and around, surrounding the predator with a whirling cloud of butterflies moving in unpredictable and chaotic patterns, then gradually settling back again only as the source of disturbance wanes. A predator finds it much more difficult, of course, to single out particular individuals among the swirling cloud than to pursue an isolated individual flying away from the group. This holds true

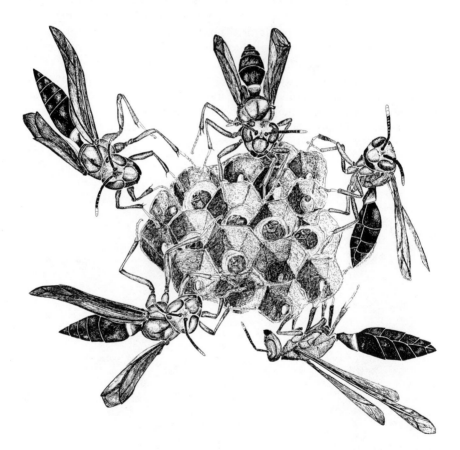

**Figure 6-11** Alarm–defense behavior in the paper wasp, *Polistes annularis*. The approach of an intruder causes resident wasps to assume a characteristic posture with the wings raised and front legs waving in the direction of the source of disturbance. Flight, attack, and stinging will follow if the intruder persists in disturbing the nest. (See also Fig. 10-17. Drawing by J. W. Krispyn.)

whether one or several species are involved. At the same time, greater protection might be expected in a larger crowd. Thus, there would be selective advantage in the convergent evolution of several different species to respond to a generalized visual cue (such as sudden erratic flight) warning of a predator's presence (see p. 351).

True alarm–alert systems are characteristic mainly of group-living organisms such as social insects. Their evolution here forms part of the larger question of the evolution of all types of altruistic behavior, that is, actions that result in self-sacrifice of an individual benefiting others of its kind; this subject is covered in Chapter 10. Even among social insects, though, visual signals are not the most widely used channel for alarm communication, being upstaged by chemical and sound systems. One possible example may be found in paper wasps of the genus *Polistes*. Close approach to a paper wasp nest will immediately alert some of the resident workers, which respond with a graded threat display in which front legs are raised and wings are spread and vibrating (Fig. 6-11). Other workers detect this display and respond by themselves showing an increased state of alertness and patrolling activity, a mobilization to meet the potential threat. Although the possibility of auditory and chemical cues being also involved in this system has not been ruled out, to date no alarm pheromones are reported for any *Polistes* wasps. Moreover, their relatively small colony size and their exposed comb nest environment make it likely that visual and/or auditory alarm communication would be quite effective here.

## Sexual Signals

Perhaps nowhere else is the role of vision in insect communication as well studied and amply documented as in sexual behavior, where visual signals often mediate a chain of stimulus–response interactions between the partners (see Chapter 9). Such signals are often highly redundant and stereotyped, for there is strong evolutionary pressure against making species-identification and sex-identification errors.

It is tempting to assume that Lepidoptera, with their wings of decorative colors and patterns, are primarily visual communicators, making little use of odors and touch in their sexual behavior. The validity of this assumption apparently depends upon what species one is considering. Although there are many intermediates, Lepidoptera may be divided into two groups—those whose distance responses in feeding and courtship are mainly evoked by airborne chemical stimuli and those responding mainly to visual stimuli. The difference may be seen in the behavior of the males; the "chemical type" follows the scent upwind in a gradual zigzag flight to find the female. The "visual type" looks for the female and approaches quickly and directly.

The interplay of signal systems is common. As a case in point, consider the common orange queen butterfly, *Danaus gilippus berenice,* of southern Florida (Brower et al., 1965). In this species, courtship consists of an aerial phase in which the male pursues the female and of a ground phase which begins once the female alights (Fig. 6-12). During

## COURTSHIP OF THE QUEEN BUTTERFLY

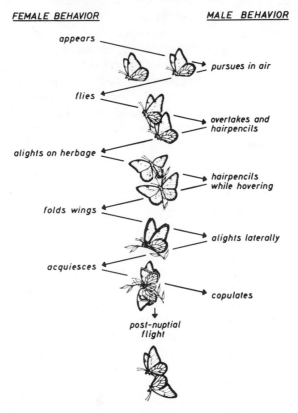

Figure 6-12 Stimulus–response reaction chain in the courtship of the queen butterfly, a species closely related to the monarch. Males may be recognized by the glandular pouch on either hind wing, visible as a conspicuous black dot. Resting males thrust their unsplayed hairpencils into them frequently, but the pouches do not appear to be essential to mating. (From Brower, Brower, and Cranston, 1965; see also Brower and Cranston, 1962. Reprinted from *Zoologica* (Vol. 50, No. 1, 1965), with permission of the New York Zoological Society.)

the aerial phase, males are first attracted visually through a combination of female movement, color, and shape; this attraction is quite general, and males often mistakenly pursue improper objects such as falling leaves. Once the male has overtaken the female, however, a chemically mediated phase begins. Hovering over the female with his brushlike hairpencils (see Fig. 5-8) everted, the male disseminates pheromones which inhibit female flight and wing movement. Males deprived of their hairpencils are capable of courting females but find it impossible to seduce them. Alighting upon herbage, the female signals her sexual receptivity by folding her wings together; in reply, the male first hovers over, then lands next to her and begins copulation. Unreceptive females may land but keep their wings outspread and the male flies off. Successful coupling is followed immediately by a postnuptial flight, which removes the pair to a less conspicuous site and probably helps to ensure that the male's sperm are fully transferred before separation 1–2 hours later.

Another example, the silver-washed fritillary, illustrates how knowledge of the nature and limitations of the insect eye can be profitably combined with the use of models to yield insights into visual communication.

CASE STUDY: THE SILVER-WASHED FRITILLARY, *Argynnis paphia*

The silver-washed fritillary is a spectacular, spotted orange butterfly common in Europe. As with many other butterflies, both male and female fritillaries respond to blue and yellow when seeking nectar and to green when seeking a place to rest. Sexually active male fritillaries, however, actively pursue many different kinds of orange or yellow-brown moving objects. Only when within about 10 cm of the object do they discriminate further, turning away unless encountering odor cues from a female fritillary.

About 20 years ago, Magnus became interested in unraveling the factors behind this male response (Magnus, 1958). Since he knew that many butterflies would court paper models dangled on a string from a wand, he decided to build a more sophisticated piece of apparatus based upon this idea (Fig. 6-13). Using a motor-driven carousel with protruding 6-foot-long arms, he was able to present male butterflies with models "flying" by.

At first, Magnus used models of the same general form and color as females and with flapping wings. Males readily followed the moving dummies and flew off only when they were close enough to discover that the dummy did not smell like a female fritillary. Convinced by this that

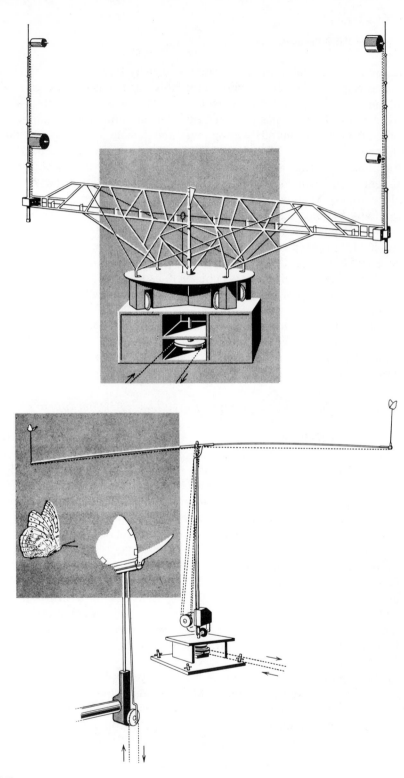

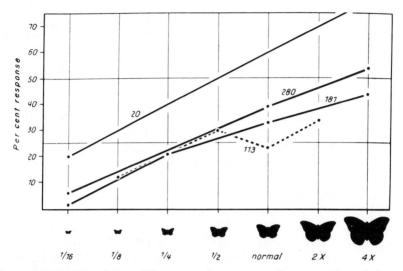

**Figure 6-14** Results of four different experiments comparing approaches of the male fritillary butterfly, *Argynnis paphia,* to nonmoving colored paper dummies of different sizes. The larger than natural size models are clearly more stimulating in each trial. Sizes were presented in different combinations in each experiment; for example, the dashed line trial shows percentages of 113 males which responded to each of four sizes ⅛, ½, normal, and 2×. (Slightly modified from Magnus, 1958.)

his apparatus was satisfactory, Magnus began to vary different aspects of his model so as to determine which qualities were attractive to the males. Tabulation of males' choices between the dummies on different arms of the carousel allowed straightforward comparison of the effectiveness of different models.

By changing the shape of the dummy, Magnus confirmed that shape made no difference; a circle or triangle was just as effective as a "butterfly." Size did matter, however; smaller models were followed less than those as large as a real female. Surprisingly, models twice the area of the female were even more effective; models four times the size were better yet (Fig. 6-14)! Evidently the object had to be moving and colored somewhat like a female fritillary, however.

Soon, Magnus also discovered that the particular type of motion was less important than the rate of flickering perceived by the male butterfly.

**Figure 6-13** Two versions of the motor-driven carousel used to present different stimulus patterns to wild male silver-washed fritillary butterflies, *Argynnis paphia.* The carousel arms carried either a flapping butterfly model or revolving cylinders with alternating colors. (From Marler and Hamilton, 1966; after Magnus, 1958.)

Thus, he was able to use a much simpler type of dummy—a mere rotating spool with alternating bands of color. At first, he used segments of female wings on the spool but found pure orange and black was even more attractive. By tabulating males' choices between spools moving at different rates on each carousel arm, Magnus learned that increasing the rate of flicker also made the model more attractive, with the improved attraction continuing to increase even at flicker rates far greater than those which a flying female is able to attain. Only when the flicker rate climbed above about 140 stimulus changes per second, the upper limit of resolving power of the butterfly eye, did the attractiveness begin to decline.

Thus, Magnus concluded, for a male fritillary the "ideal" female would be up to four times his size, pure orange, and flickering her colors as rapidly as he was able to detect them. Of course, simple physical impossibilities and opposing selective pressures make evolution of such a female unlikely. But such supernormal stimuli (see Chapter 2), which have been demonstrated many times in behavioral studies, can be very helpful in interpreting the underlying physiological mechanisms involved in stimulus filtering.

Several other studies of visual sexual communication in butterflies have agreed in finding the quality universally necessary to elicit male butterfly courtship to be movement, to which the compound eye is particularly sensitive. Butterflies can see colors and tend to be most responsive to the general color of their females' wings. However, complex markings (or their absence) seem to have little effect, and details of patterns very striking to the human eye are often ineffective in influencing male sexual responsiveness. (Color patterns are, however, exceedingly important in another aspect of insect communication— messages sent to potential predators, see Chapter 8.)

SUMMARY

In general, visual communication has been studied earlier, longer, and more thoroughly than other modes, for man himself is a visual animal. A number of important ethological theories have been derived from studies of visual displays as well as from the use of models and disguise.

Light production, or bioluminescence, occurs among a diversity of plants and animals but is particularly widespread in the Coleoptera, where it depends upon the combination of luciferin and luciferase to produce cold light. Bioluminescent communication in insects appears directed toward pair formation, mate identification, and location. Sometimes it may also function in prey capture.

Insect visual receptors include compound eyes, ocelli and stemmata. Most insects respond to wavelengths from near-ultraviolet to orange (300–650 m$\mu$) with peaks of greatest activity in the near-ultraviolet and blue-green. However, while many possess color vision, the insect's perceptual world is quite different from man's. Because their eyes cannot move independently of one another, distance judgment of insects depends upon visual angle, while the multiple-unit ommatidial system provides a mosaic vision in which visual acuteness varies with position and degree of motion. Ability to resolve flicker is much greater than in man and forms the basis for many insect visual behaviors. That light polarization is also discerned is strikingly illustrated in honey bee dance communication.

Visual communication is best known in association with reproductive activities, including not only courtship but such related behaviors as territoriality. Visual alarm systems sometimes also occur.

## SELECTED REFERENCES

Arms, K., P. Feeny, and R. C. Lederhouse. 1973. Sodium: stimulus for puddling behavior by tiger swallowtail butterflies. *Science* **185**:372–374.

Baldus, K. 1926. Experimentelle Untersuchungen über die Entfernungslokalization der Libellen (*Aeschna cyanea*). *Z. Vergl. Physiol.* **3**: 475–505.

Blest, A. D. 1957. The function of eyespot patterns in the Lepidoptera. *Behaviour* **11**: 209–256.

Brower, L. P., J. Van Zant Brower, and F. P. Cranston. 1965. Courtship behavior of the queen butterfly, *Danaus gilippus berenice* (Cramer). *Zoologica* **50**: 1–39.

Brower, L. P., and F. P. Cranston. 1962. Courtship behavior of the queen butterfly, *Danaus gilippus berenice*. Pennsylvania State University Psychological Cinema Register, 16-mm color, sound, 18 minutes.

Buck, J. and E. Buck. 1968. Mechanism of rhythmic synchronous flashing of fireflies. *Science* **159**: 1319–1328.

Buck, J. and E. Buck. 1976. Synchronous fireflies. *Sci. Amer.* **234**: 74–85 (May).

Burkhardt, D. 1964. Colour discrimination in insects. *Adv. Insect Physiol.* **2**: 131–173.

Campanella, P. J. and L. L. Wolf. 1974. Temporal leks as a mating system in a temperate zone dragonfly (Odonata:Anisoptera) I: *Plathemis lydia* (Drury). *Behaviour* **51**: 49–87.

Corbet, P. S. 1962. *A Biology of Dragonflies.* H. F. & G. Witherby, London, 247 pp.

Crane, J. 1949. Comparative biology of salticid spiders at Rancho Grande, Venezuela. Part IV. An analysis of display. *Zoologica* **34**: 159–214.

Deonier, D. L. 1974. Ultraviolet-reflective surfaces on *Ochthera mantis mantis* (DeGeer) (Diptera: Ephydridae). Preliminary report. *Entomol. News* **85**: 193–201.

Eisner, T., R. E. Silberglied, D. Aneshansley, J. E. Carrel, and H. C. Howland. 1969.

Ultraviolet video-viewing: the television camera as an insect eye. *Science* **166**: 1172–1174.

Frisch, K. von. 1967. *The Dance Language and Orientation of Bees.* Belknap Press of Harvard University Press, Cambridge, Mass., 566 pp.

Goldsmith, T. H. and G. D. Bernard. 1974. The visual system of insects. In *The Physiology of Insecta*, Vol. 2, 2nd ed., M. Rockstein (ed.). Academic Press, New York, pp. 166–270.

Horridge, G. A. (ed.). 1975. *The Compound Eye and Vision of Insects.* Oxford Univ. Press, New York, 595 pp.

Horridge, G. A. 1977. The compound eye of insects. *Sci. Amer.* **237**: 108–120 (July).

Jacobs, M. E. 1955. Studies on territorialism and sexual selection in dragonflies. *Ecology* **36**: 566–586.

Jones, C. E. and S. L. Buchmann. 1974. Ultraviolet floral patterns as functional orientation cues in hymenopterous pollination systems. *Anim. Behav.* **22**: 481–485.

Kerfoot, W. B. 1967. Correlation between ocellar size and the foraging activities of bees (Hymenoptera: Apoidea). *Amer. Nat.* **101**: 65–70.

Lloyd, J. E. 1965. Aggressive mimicry in *Photuris*: firefly femmes fatales. *Science* **149**:653–654.

Lloyd, J. E. 1966. Studies on the flash communication system in *Photinus* fireflies. Misc. Publ. Mus. Zool., Univ. Mich., No. 130, 95 pp.

Lloyd, J. E. 1971. Bioluminescent communication in insects. *Annu. Rev. Entomol.* **16**: 97–122.

Lloyd, J. E. 1975. Aggressive mimicry in *Photuris* fireflies: signal repertoires by femmes fatales. *Science* **197**: 452–453.

Lloyd, J. E. 1977. Bioluminescence and Communication. In *How Animals Communicate*, T. A. Sebeok (ed.). Indiana Univ. Press, Bloomington, 1344 pp.

Magnus, D. 1958. Experimentalle Untersuchunger zur Bionomie und Ethologie des Kaisermantels *Argynnis paphia* L. (Lep. Nymphalidae). *Z. Tierpsychol.* **15**: 397–426.

Marler, P., and W. J. Hamilton, III. 1966. *Mechanisms of Animal Behavior.* John Wiley and Sons, New York, 771 pp.

Mazoklin-Porshnyakov, G. A. 1969. *Insect Vision.* Plenum Press, New York, 306 pp.

McElroy, W. D., H. H. Seliger and M. DeLuca. 1974. Insect bioluminescence. In *The Physiology of Insecta*, Vol. 2, 2nd ed., M. Rockstein (ed.). Academic Press, New York, pp. 411–460.

Nelson, S., A. D. Carlson, and J. Copeland. 1975. Mating-induced behavioural switch in female fireflies. *Nature* **255**: 628–629.

Roeder, K. D. 1960. The predatory and display strikes of the praying mantis. *Med. Biol. Illus.* **10**:172–178.

Silberglied, R. E. 1977. Communication in the Lepidoptera. In *How Animals Communicate*, T. A. Sebeok (ed.). Indiana Univ. Press, Bloomington, 1344 pp.

Silberglied, R. E. and O. R. Taylor. 1973. Ultraviolet differences between the sulfur butterflies, *Colias eurytheme* and *C. philodice*, and a possible isolating mechanism. *Nature* **241**: 406–408.

Tinbergen, N. 1948. Social releasers and the experimental method required for their study. *Wilson Bull.* **60**: 6–51.

# 7
# Mechano-
# Communication

In Kansas, great numbers of female cicadas are attracted to a tractor with its motor running. In Malaya, natives attract another cicada species by clapping their hands rhythmically. In France, a third species responds to whistling. Female corixid bugs of certain species come to the ultrasonic sound of a frequency generator. Spiders have been reported to be attracted to a wide range of musical instruments, including the bagpipe, harpsichord, lute, and violin. What is happening? A simple answer easily comes to mind. These invertebrates are hearing, and responding to, sounds. Probably the sounds are very similar to those which would normally serve some communicative function for them.

But if these examples seem to clearly involve hearing, what should one make of the following ones? When a noise is sounded near some types of caterpillars, they react by rearing up either the anterior third of the body or sometimes the tail end; decapitated caterpillars, or even isolated body pieces, show the same reaction. Whirligig beetles normally swim about on the water surface film in freely moving swarms; when their antennae are altered, individuals collide. Migratory locusts have hairs on the fronts of their heads; stimulation of these always results in flying. Parasitic wasps that lay eggs within wood-boring larvae can locate their prey with great accuracy even through an inch or more of bark. How many of these should be called "hearing"? Any? Some? All?

Help in answering these questions can be obtained by considering the whole of **mechanoreception** (Table 7-1), the perception of any mechanical distortion of the body. The range and sensitivity of all this equipment strikingly illustrates what an immense amount of information an insect actually has about the outside world; indeed, mechanical stimuli are involved in more behavioral activities than any other type of external stimulus.

Mechanoreception in its broadest sense encompasses not only hearing but touch and the detection of forces exerted by gravity. A number of different receptors are involved, but all share a common concern: the oscillations of various media and reactions to mechanical pressures of

**Table 7-1** A simplified functional classification of insect mechanoreception[a]

| Mechano-receptive subsense | Temporal pattern of stimulus energy | Producer of temporal pattern | Information content | Receptors | Remarks |
|---|---|---|---|---|---|
| Gravity and pressure | Constant or slowly changing forces | Activity of the living insect | Spatial relations | "External stato-cysts"; displacement of various body parts | Cooperative with vision |
| Movement | Forces of inertia | Surroundings as well as insect's own movements | Stabilization against disturbing forces, especially rotation | Halteres; mass of ear, head and hair plates in the neck | |
| Current | Movement of the surrounding medium | Air or water movements relative to the insect | Orientation to currents | Johnston's organ; sensory hairs on head and antennae | Often linked with activation of sense of smell |

| | | | | | |
|---|---|---|---|---|---|
| Touch | Contact with solid structures | Active or passive contact with environmental structures | Temporal and spatial distribution | Single hair sensilla | Most primitive expression of mechanoreception |
| Vibration | Oscillations of the substrate | Shock to an elastic structure or forced vibrations produced by insect's rhythmical movements | Alert; social messages | Sensory hairs on tarsi; chordotonal sensilla of legs; subgenual organs | Closely related to hearing, frequently using same receptors; very slow vibrations may be considered tactile |
| Hearing | Oscillations of air and water | Displacement of air or water molecules | Phonotaxis, phonokinesis; sexual behavior; alerting, localization, communication | Auditory hairs and sensory spines; Johnston's organ; tympanal organs | Hearing and sound production often linked |

[a] Based on Schwartzkopff (1974).

different sorts. In theory, they fall into two general groups, those yielding information about an arthropod's position in space (the subject of Chapter 3) and those involved in communication with other living organisms. In practice, however, the divisions are less clear-cut. Touch receptors, for example, function in such diverse activities as avoiding obstacles, fighting, and copulating. Since communication is the major concern of the present chapter, however, let us concentrate upon the structure and function of the receptors involved in the last three subsenses in Table 7-1.

All three of these mechanocommunicatory subsenses, one notes, often share the same types of receptors. But in actuality, *most* of the sense organs of insects are quite similar in form, namely, small structures with a single sense cell and a single nerve fiber. In fact, many of the tiny bristles or hairs that occur to one degree or another on most insects are sense organs. These "little sense organs," or **sensilla**, provide the armorencased insect with sensitive points of contact through its quite insensitive cuticle. On the front of an insect's head, sensilla perceive air movement. Between head and thorax, they perceive gravity and position. On the antennae and other appendages, many sensilla are touch receptors. On the tail filaments of some insects, they clearly serve as hearing organs. In places they may join together to form a loose field or dense pad, functioning together as sense organs of higher order. Inside the body, homologous sensilla may also join to form still other types of mechanoreceptors.

While the basic structure of the cuticular mechanoreceptors may tend to follow a single plan, their shape and their mechanical and physiological properties vary considerably (for a review, see McIver, 1975). However, all mechanoreceptors fall into two functional classes: pressure sensitive and velocity sensitive. Pressure sensitive hairs show a repetitive neural discharge during a static deformation. In simpler language, they continue to fire all during the period of time in which they are bent. Most common on those body areas where position is important, they are the **proprioceptors**. Because they adapt slowly, proprioceptors are not suited to registering sudden stimulus change. However, they give an accurate measure of stimulus intensity, transmitting information about the state of muscles even when contracted for long periods of time. Proprioceptors help the insect to maintain its position, both the relation of various body parts to each other and the relation of its whole body with respect to gravity (see also Chapter 3, especially Fig. 3-6). Other receptors, particularly visual and tactile, assist in this capacity. Mill (1976) provides an up-to-date treatment of invertebrate proprioceptors.

Velocity-sensitive sensilla, on the other hand, fire only while the stimulus is changing, such as when a sensory hair is deflected, moves back to its original position, and then is deflected again. Such waves of deflection are produced most commonly by oscillations—vibrations which cause ongoing alternate compression and expansion of the adjacent medium. Not surprisingly, tactile and auditory reception depend upon this class of receptors. Velocity sensitive mechanoreceptors usually have a very rapid adaptation rate and thus provide less accurate information about differences in stimulus intensity. They are well suited, however, to record temporal patterns of stimulation.

Hearing and touch differ in the types of pressure alternations involved, sound stimulation having a phasic nature while touch stimulation has an unfluctuating or irregularly fluctuating nature. More commonly, however, we separate the two on the basis of distance. Touch refers to mechanoreception involving contact; hearing, to mechanoreception involving a distance between sender and receiver. The differences thus are akin to those between taste and smell in chemical communication. More specifically yet, the term "hearing" is usually restricted to the detection of airborne sounds by specialized receptors, relegating all other sound reception to a "vibration sense." Though we will follow this restricted definition here, detection of "sounds" and "vibration" overlap more often than not in the insect world. For example, between the second segment and the rest of the antenna, called the flagellum, almost all insects have a group of sensilla known as **Johnston's organ**. In culicine mosquitoes where it was first discovered (and in midges as well), the Johnston's organ is enormously developed and has a clearly auditory function. In most other insects studied, it appears to act primarily as a tactile organ relaying several sorts of mechanical information. In aphids it is used for the control of flight. Cutting off the antennal flagellum beyond the Johnston's organ causes an aphid's flight to become erratic; When an artificial antenna is reattached, normal flight resumes. Schwartzkoff (1974) thoroughly reviews the subject of mechanoreception.

## TOUCH AND TIME: SEMATECTONIC COMMUNICATION

The very large and complex nests of many social insects present an enigma. No one colony member can oversee more than a small piece of the construction work or envision the nest in its entirety. Sometimes, a nest may require a number of worker lifetimes to complete, and each new part must be brought into balance with the old. How can the workers communicate so effectively over such a long time period? Who has the nest blueprint?

The first to answer such questions in any detail was Grassé (1959), studying nest building in various termites. The key process, he suggested, was that it is the product of previously accomplished work rather than direct communication among nestmates that induces the insects to perform additional labor. Even if one should constantly renew the work force, the nest would be completed "according to plan," for the nest structure already finished determines what further work will be done.

Several subsequent workers have challenged the completeness of Grassé's explanation, but it does contain an important insight. The most durable signals which communicate information between insects of the same species are those incorporated in structures built by the insects, such as nests. Wilson (1975) has called such communication **sematectonic**, from the Greek words for sign and builder, and defined it as "the evocation of any form of behavior or physiological change by the evidences of work performed by other animals, including the special case of the guidance of additional work."

Sematectonic communication is not necessarily limited to social species. For example, many solitary bees and wasps construct nests in hollow twigs in which cells are arranged end to end in a linear series (see Fig. 10-8). At emergence, adults always chew outward in the direction that they are facing, which is almost invariably toward the nest entrance. In general, the nest tunnel is not broad enough to permit the adult to turn around inside should it emerge from its cocoon facing the wrong way. How does the mature larva correctly orient itself with an accuracy that far exceeds that expected by chance alone?

As the wasp nest is constructed, the mother builds partitions between successive cells. As a result of the techniques of construction, the inner and outer faces of each partition differ in texture and concavity. The mother wasp works from the outside, and the partition tends to become concave on the outer surface. It is also smoothed on the outside, while the inner surface has an irregular, bumpy texture on its convex face. By testing independently the effects of the four possible cues on larval orientation, Cooper (1957) was able to clearly show that concavity was the primary cue used by the spinning larvae (Fig. 7-1), which always oriented the head away from the concave surface. Moreover, Cooper found that completely unrelated wasps, including some parasitic species, orient correctly by the same cues. In twig-nesting solitary bees, where the cell partitions tend to be rather amorphous walls of resin or chewed leaves, the significant orientation cues seem to lie in the placement of the pollen mass. Spinning larvae always orient cocoons facing away from the pollen mass.

Except for the special case of sematectonic cues, however, touch

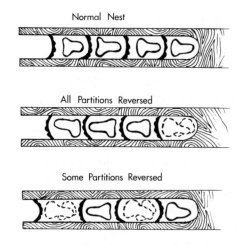

**Figure 7-1**  Sematectonic communication in solitary wasps. Larval direction when spinning cocoon depends upon cues received indirectly from the mother via concavity and smoothed texture on mud partitions between the linearly arranged cells. (The swollen end of each cocoon contains the head.) (Based on Cooper, 1957.)

communication as a whole among insects is quite poorly known. Theoretically, touch could form the basis for a complex and important communication system, transmitting a great variety of messages by varying frequency, pressure, and time of contact. However, tactile communication appears to be relatively unimportant in comparison with the other communicatory modes. Most tactile systems have one overriding limitation: the sender must be in contact with the receiver. As a result tactile methods are usually restricted to close-range situations such as courtship and mating and occasionally alarm (see also Fig. 4-21). Furthermore, the sense of touch is one of the most difficult modalities to investigate in a communicatory context because it is hard for an observer to interpose himself or his instruments in such a way as to register the signals in the form that they are actually received without disrupting the system. And because of this difficulty, even in cases where tactile signaling is involved, other communicatory modes tend to receive the first attention.

## THE ACOUSTIC CHANNEL

Auditory communication has apparently evolved hundreds of times and now occurs in tens of thousands of species. Yet clear demonstrations of "hearing" have been restricted to five orders producing sounds of relatively strong intensity—Orthoptera, Homoptera, Lepidoptera, Coleoptera, and Diptera. And even though the songs may be clearly audible, their true nature is seldom fully appreciated by man. A major reason is simply that for insect sounds the human ear has little to recommend it as an analyzing instrument. Its frequency response range is too small, its time constant too long. To compare, contrast, or even properly describe insect songs, the use of various instruments is essential.

The advent of high quality recording equipment and sophisticated sound-analyzing instruments has greatly helped to breach the gap between the known occurrences of insect sound and the relatively little known significance of them. For more information, the reader may wish to begin with the bibliography of Frings and Frings (1960) and the books and reviews of Haskell (1961) and Alexander (1967).

**Sound Production**

The majority of insect species produce communicative sound at some stage of their life cycle, using an enormous variety of mechanisms (Table 7-2). Frictional methods predominate, particularly among the Orthoptera, Hemiptera, and Coleoptera. Given the durable elastic cuticle of insects, initial chance sound production could occur during locomotory, feeding, or cleaning movements; many of the present mechanisms do in fact appear to have arisen after specilaization in just such ways. In some insect or other, almost every body part has become modified to produce sound.

**Stridulation**, the rubbing of one body part against another, is so common that some authors have broadened their definition of this useful term to include "any sound produced by an insect." One body part, the file, is usually a series of pegs or teeth; the opposing part, the scraper, is generally a single edge or ridge. Associated with many frictional sound production devices are various types of resonating systems which impart distinctive features to the resultant sounds.

Frequency and pattern may vary tremendously. Crickets produce the purest sounds known for any insects and, in general, the simplest patterns; each movement of their elytra produces one pulse of sound (Fig. 7-2). In other insects with complex songs, such as tettigoniid grasshoppers, each elytral movement can produce many sound pulses. DuMortier (1963) and Pierce (1948) provide a well-illustrated review of insect frictional mechanisms.

Sounds produced in another way, by a vibrating membrane driven

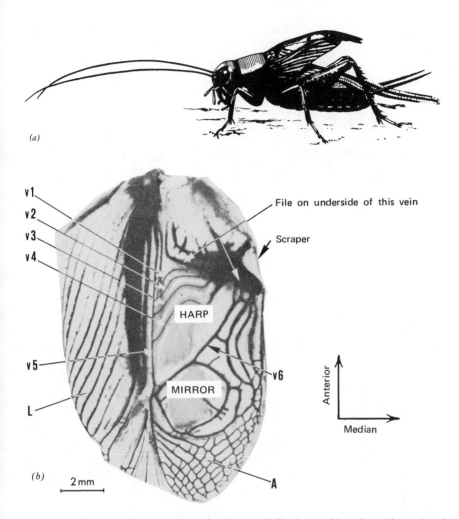

(a)

v1
v2
v3
v4

File on underside of this vein

Scraper

HARP

v5

L

v6

MIRROR

Anterior

Median

(b)

2 mm

A

**Figure 7-2** Sound production by a male cricket. (*a*) Singing position. Sound is produced by scissoring the elevated front wings back and forth, so that a sharp scraper near the base of one front wing rubs along a ridged file on the underside of the other front wing. Only the closing wing stroke (homologous to the downstroke of flight) produces a sound. (Courtesy of R. D. Alexander; see also Davis, 1968.) (*b*) Dorsal view of the left front wing illustrating the positional relationship of the file and scraper. The harp region functions to boost the acoustic power of the chirp; calculations based on its triangular shape predict it to produce a resonance frequency close to 4 kilohertz, which corresponds to the predominant frequency of the calling and rivalry sounds (see Fig. 7-8). (Modified from Michelson and Nocke, 1974.) For discussion of the neurobiology of cricket song see Bentley and Hoy (1974).

**Table 7-2** A simplified behavioral classification of insect sounds by mechanism[a]

| Production mechanism | Selected examples | Use in communication |
|---|---|---|
| I. Byproduct of another activity | | |
|   Feeding | Woodboring beetle larvae | Spacing |
|   Emergence | Megarhyssa wasps | Sexual attraction |
|   Flying | Heliothis moths, Aedes mosquitoes | Sexual attraction |
| | Schistocerca locust and other Orthoptera | Aggregation maintenance? |
| II. Impact of body part against substrate | | |
|   Head | Termites | Alarm? |
| | Deathwatch beetles (Xestobium, Anobium) | Sexual? |
|   Abdomen | Booklice; some Plecoptera | Mating call of female; unknown |
|   Tarsi | Several Orthoptera | Unknown |
| III. Frictional rubbing (stridulation) | | |
|   Almost every body part utilized by some, but mechanism in all essentially similar | Of widespread occurrence; associated especially with Orthoptera, Hemiptera, Homoptera, Coleoptera | Sexual and aggressive behaviors |
| IV. Vibration of a membrane | | |
|   Tymbals driven by muscles | Restricted to Hemiptera, Homoptera, and Lepidoptera; especially cicadas, pentatomid bugs, arctiid moths | Sexual and aggressive behavior, aggregation, predator confusion |
|   Thoracic sclerites | Virgin queen honey bees | Identification |
| V. Pulsed air stream | | |
|   Epipharynx and pharynx | Death's head hawkmoth (Acherontia) | Disturbance |
|   Spiracles | Gromphadorrhina cockroaches | Disturbance, courtship |

[a] Based primarily on DuMortier (1963) and Haskell (1974).

directly by muscles, are most common among Homoptera but also occur in some Hemiptera (Pentatomidae) and Lepidoptera (Arctiidae). These special membranes, called **tymbals**, are extremely elaborate in cicadas, where they are usually restricted to the males. All tymbal mechanisms appear to act in a similar way. When the tymbal muscle contracts, the tymbal produces a single sound pulse; when it relaxes, the tymbal produces another, a mechanism analogous to a rounded tin can lid pressed inward with one's finger and then released. The insect's tymbal muscle contracts with such rapidity (170 to 480 contractions per second have been recorded) that to human ears the sound output appears continuous.

Other insects vibrate other body parts. Virgin queen honey bees, for example, produce a "piping" sound by thoracic sclerite vibration; other queens still in their larval cells respond with their own piping in short pulses at a lower frequency. In other cases, noise is produced by striking some body part against the substrate. An example is the deathwatch beetle; sexually mature beetles bend their head down and bang it against the floor of their wooden burrows seven or eight times a second. In addition, many insects produce sounds in the process of defensive behaviors, such as the explosive anal discharges of bombardier beetles (see Fig. 8-27) and the buzz of stinging bees and wasps; the acoustic

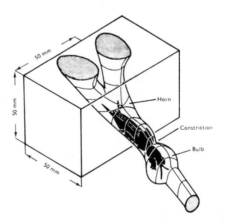

**Figure 7-3**  The burrow of a mole cricket, *Gryllotalpa vinae*, constructed in the form of a double horn which serves to orient the song in space and acts in a manner analogous to a loudspeaker. Songs sung outside the burrows have only about 4% as much acoustic power as those sung inside the burrow. The intense song, which attracts both mole cricket sexes, may be heard by man up to 600 meters away; it lasts almost an hour each evening at dusk. (From Bennet-Clark, 1970; see also Ulagaraj and Walker, 1973, and Prozesky-Schulze, et al., 1975.)

element may be an important addition to the effectivness of these mechanisms. One of the few insects known to modify its surroundings for acoustic purposes is the mole cricket (Fig. 7-3).

## Sound Reception

Sound and touch reception have already been mentioned in our introductory overview of mechano-communication, where we noted that all rely upon a basically similar plan of tiny innervated sensilla joined in various groupings. The chordotonal sensilla, unique to the insects but found in every insect in which they have been sought, illustrate this well. They are apparently not exclusively audioreceptors as originally thought. Many are proprioceptors, and at least some have a mixed function. Subcuticular, with no external evidence of their presence, they are widely distributed in the insect body, occurring in mouths, legs, wing bases, halteres, antennae, abdomens, and tracheal systems. Upon the antennae, they are grouped into the Johnston's organ, mentioned previously, which detects movements of the shaft of the antenna.

Generally, insects also have chordotonal organs at four locations along their legs, located in bundles in such a way as to sense vibrations of the surface upon which they are standing. One set of particular importance are the "below the knee" or **subgenual organs**. These are extremely sensitive at their optimal frequencies, for the amplitude of displacement required to stimulate the organ is very small. Since they can apparently be used to localize a source of vibration not in contact with the insect, they can properly be termed organs of sound perception as well as touch perception. Since the subgenual organ is one of the most sensitive receptors available to insects, it is of interest that termites have been shown to use it to detect the sounds of their fellows (see Howse, 1964).

The propagation of sound through media other than air has received relatively little study to date. Given the wide variety of wood-boring insects, for example, and the considerable acoustic range obtainable through the use of the subgenual organ, one might predict that many cases of stridulatory communication transmitted as substrate vibration will be turning up in the years ahead. (One such example, the wood-boring Douglas fir beetle, was considered in Chapter 5.) Wilcox (1972) and Jansson (1973, 1976) provide detailed analyses of sound communication in aquatic Hemiptera, and Ryker (1976) and Van Tassel (1965) have studied acoustic behavior in certain water beetles.

A third type of specialized auditory organ has arisen in insects, also from organs serving originally for proprioception. In this case the chordotonal sensilla have increased in number and become attached to a thin, taut cuticular membrane which is set into vibration by waves of airborne sound. This membrane, a sort of "eardrum" or **tympanum**, in turn has a closely appressed internal tracheal sac and associated "tympanic muscles" of unknown function. The air rather than tissue backing reduces inertia of the membrane making it very sensitive to sound. The tympanic organs are paired and often interconnected by a series of tracheal chambers. Most often they occur on each side of the abdomen just behind the thorax or on the last segment of the thorax itself, but in some cases they are found on the tibiae of the fore legs (Fig. 7-4). Tympanic organs occur in a wide variety of families in the

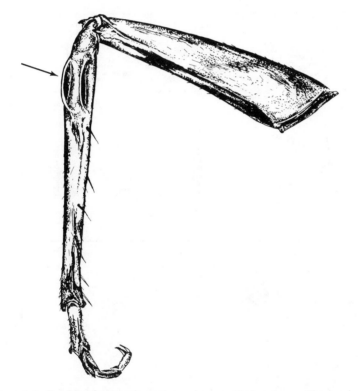

**Figure 7-4** Foreleg of a long-horned grasshopper, *Tettigonia* sp. The hearing organ is concealed behind slits in the tibiae (arrow). By swinging its forelegs as it walks, this insect can scan its surroundings and localize sound sources. Recent evidence suggests that the preferential sites of sound entry to the tettigoniid tympanal organs are the prothoracic spiracles, however, rather than the tibial slits as one might suppose. (Drawing by J. W. Krispyn; see also Nocke, 1975.)

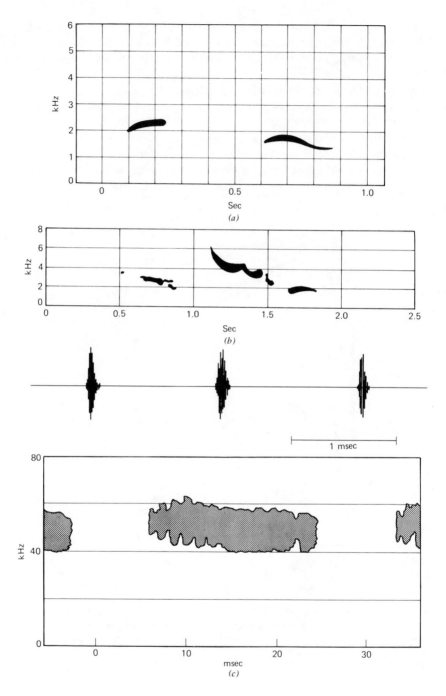

Orthoptera, Hemiptera, and Lepidoptera. The tympanic organ of moths appears to be used primarily for predator evasion and was discussed in that regard in Chapter 2 (see Fig. 2-4); there is also some evidence that these organs may be involved in navigation during flight.

Many insects that lack tympanic organs are nonetheless quite sensitive to sound waves in air. (Most if not all caterpillars, for example, can detect sound and will react by "freezing" or thrashing as a protective response). It does however appear that those sensilla types most highly developed for airborne sound reception differ from tympanic organs in several ways. First, they respond over much lower frequency ranges and fatigue fairly rapidly. Second, they do exhibit a limited frequency discrimination, for they will respond synchronously with stimulus frequency over certain ranges. Finally, they tend to equilibrate, that is, as a sound continues through time, the amplitude of the massed spikes in the responding nerve declines. Such hair sensilla often serve for tactile purposes and react best to high-intensity, airborne sounds. They appear in all insect orders.

### Parameters of Insect Song

Despite the strong reliance on sound exhibited by arthropods, even the most complex of insect sounds have fewer dimensions than most vertebrate sounds. One reason stands out: insects cannot carry a tune. Rather, sound intensity and timing form the basis of insect acoustic communication, in a nested pattern of time-related groupings of individual sounds at various volumes. However, the scale on which these elements occur is normally hidden to us, for as we have already noted, the time constant of human ears is too long to effectively analyze insect sound without electronic technology. For example, when we hear a buzz, the individual behavioral song components (**phonatomes**) and their first- and second-order groupings cannot be resolved by our ears. But

---

**Figure 7-5** Sonagrams and oscillograms of various sounds. Recorded either as a pen trace on paper or an electron beam on film, these graphic representations enable one to appreciate song characteristics and differences between songs at a glance. Sonagrams show the carrier frequency on the vertical axis (ordinate) expressed in kilohertz and time on the horizontal axis (abscissa). Oscillograms also represent time on the abscissa, but the ordinate expresses the pattern of sound intensity (loudness) fluctuation, without specifying the frequency. Degree of blackening correlates with sound intensity on sonagrams. (*a*) Sonagram of a wolf whistle, whose form can be easily recognized. (*b*) Sonagram of the typical song of the eastern meadowlark. (*c*) Song of the long-horned grasshopper, *Phlugis*, displayed on an oscillogram (*top*) and a sonagram (*bottom*). ((*c*) from Michelson and Nocke, 1974, adapted from Suga, 1966.)

when the phonatome rate is slow enough, we can hear the individual phonatomes as countable "ticks" or "smacks" or, when they occur slightly too rapidly to count, as a "rattle." Because temperature has a substantial effect on phonatome rate, the same song component that is identified as a rattle at one temperature may be called a tick sequence at a lower temperature. (Sometimes this relationship to temperature is quite precisely known. For example, snowy tree crickets are well known to naturalists as "thermometer crickets," for by counting the number of chirps heard over 13 seconds and adding 40, one can fairly accurately calculate the Fahrenheit air temperature.)

By definition, a phonatome includes all the sound produced during one cycle of wing, leg, or body movement during stridulation (i.e., one complete stroke of the scraper over the file and return). Because it is a behaviorally defined term, it is difficult to apply precisely in cases where the behavioral basis of sound production is not known. In contrast, a **pulse** is the simplest element of amplitude convenient to recognize—a wave train isolated or nearly isolated in time by a substantial amplitude modulation (frequently, an interval of silence). Sometimes, as in a single cricket chirp or one squeak of a restrained velvet ant, this is equivalent to a phonatome. In other cases, such as the meadow katydid, *Orcheli-mum*, the simplest sound element is a pulse lasting a fraction of a millisecond, and the travel of the scraper along the file generates a whole train of pulses which collectively are equivalent to a phonatome.

In the laboratory, many insects can apparently distinguish sound frequencies between about 100 and 800 cycles per second, but most insects produce sounds that are in the upper part of the auditory range detected by their sound reception organs, where frequency discrimination appears to be rudimentary or absent. In addition, most insect sounds are noiselike, covering a broad band of many nonharmonically related frequencies. These sibilant sounds are commonly below 18 kHz, but many range into the ultrasonic. (The chirp of the domestic cricket is comparatively unusual, occupying a narrow band of frequencies having a nearly pure tone which we recognize as a definite pitch.)

Just as the colors in light can be separated out by a prism, the component pitches in a sound can be separated out by special electrical equipment. The resultant sound pictures are of two types, spectrograms and oscillograms (Fig. 7-5). A sound spectrogram, or **sonagram**, records the frequency spectrum of a sound as a function of time and is commonly used to study bird songs because it permits one to distinguish pitch differences. Because an insect song lacks melody (in other words, the dominant frequency stays at a more or less common pitch), a sonagram is less useful to insect sound study than is an oscillogram, however. An **oscillogram** shows volume; the greater the deflection, or amplitude, of the tracing above and below the abscissa baseline, the louder the sound. This is important because insect songs often show considerable amplitude or intensity modulation, which results in a rhythmically alternating loud–soft pattern through time. Equally important, oscillograms are able to resolve the phonatome sequences which occur too rapidly for unaided human ears to separate (Fig. 7-6).

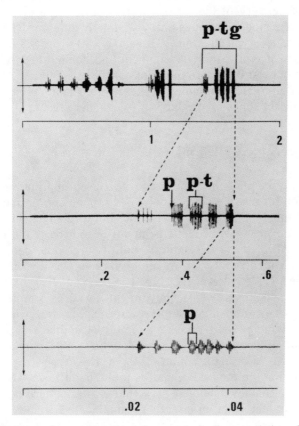

**Figure 7-6** Diagrammatic representation of successively finer resolution of the components of insect sound. Abscissa shows time in seconds; ordinate indicates volume (relative amplitude). Top, Analysis of a hypothetical stridulatory signal composed of a sequence of three pulse train groups (PTG). Middle, Expanded oscillogram of only the last pulse train group reveals it to be composed of several pulse trains (PT) separated by brief silent intervals. Bottom, Detailed analysis of only the last pulse train reveals that it in turn is composed of several individual pulses (P) or wave trains each of which is the equivalent of one tooth strike as the scraper runs over the file. (From Jansson, 1973; see also Morris and Walker, 1976.)

There seems to be no universally accepted terminology for these nested groupings of pulses in time. We prefer the term **pulse train** (PT) for a first-order grouping of two or more pulses, preceded and followed by a period of silence substantially greater than any of the time intervals between pulses. A second-order grouping of pulses is then a **pulse train group** (PTG); and a third-order grouping, consisting of a series of closely spaced pulse train groups, is a **mode** (Morris and Walker, 1976). Most buzzes, rattles, ticks, etc., heard by the unaided human ear are modes. The rate of delivery of the pulses in a mode is not a criterion for the definition. The mode continues until a distinctly different PT is produced or the pause is atypically long. Obviously, PTs and PTGs are never identical, and it is a matter of judgment and common sense as to how different they must be before they are considered distinct. Elsner (1974) employs an alternate terminology. Older, often more subjective, terms applied to insect songs are defined by Alexander (1967).

## FUNCTIONS OF INSECT SOUND

It might truly be said that one could devise an appropriate functional classification for insect sound only by devising an appropriate classification for all functions of animal behavior. But a relatively simple system (Table 7-3) covers most acoustic behavior in four basic but overlapping categories. All of these fall naturally into two groups, depending upon whether the sound shows any repetition of some basic element, that is, whether the song is **patterned**.

Unpatterned sounds, while they might serve as simple behavioral releasers, cannot transmit complex information. Interspecific signals tend to be characterized by their high intensity and lack of pattern; they are the sort of sounds often released upon tactile stimulation. Most "alarm" and "warning" sounds are of this type (see Fig. 7-12). Patterned sound, on the other hand, is found in all cases where complex intraspecific behavior mediated by sound exists; pattern is, in fact, the most constant feature of an insect's communicative sounds.

### Congregation

Of all the acoustic responses an arthropod may give, the most widely observed and easily demonstrated is the **phonoresponse**—upon hearing a noise the insect replies by making one. Acoustic stimulation may produce two different types of phonoresponse, depending upon the species. In some, the responding insect alternates its emission with that of the stimulus. Such alternation is frequently displayed between two males, either at the emission of the calling song or of the rivals' song; it is also observed in species where a male and female emit an "agreement song". Phonoresponses involving two or more individuals of the same sex are the most common type. Among other insects, particularly species living closely packed together, the first sound emission may set off a collective and synchronous song of all the population. This **chorusing**, although the subject of many observational data, remains poorly understood. Such behavior has several levels of possible complexity, and in some cases males may synchronize or alternate calling phrases for indefinitely long periods of time, as do Cicadidae and some Orthoptera.

Chorusing may result in large numbers of insects coming together in dense aggregations. This has been most thoroughly studied in periodical cicadas *(Magicicada)*, where a complex of three species show extraordinary chorusing behavior. In each, a calling or aggregating song produced individually but sung in chorus activates and assembles both males and females. Not only do the songs of each species differ in their acoustic parameters, but each species has a different time of peak chorusing activity as well (Fig. 7-7). These two factors result in a slow but clearly defined grouping of each species in its habitat. Thus, in their first few days of adult activity the three species may be intermixed; but after a week or two, the grouping has become so intense that a tree may contain hundreds of cicadas of one species and only one or two of the other two species.

Acoustic signals would also theoretically be useful to aggregations of crepuscular or night-flying insects. Little is known in this regard, but it is suspected that swarms of mosquitoes and midges maintain their

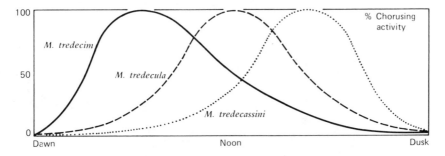

**Figure 7-7** Approximate diurnal times of maximal chorusing activity in the three sympatric species of "13 year locusts", *Magicicada*, found in southeastern United States. (From Alexander and Moore, 1962.)

**Table 7-3**  A simple functional classification of insect communicative sounds[a]

| | Song type | Equivalent names | Signal characteristics | Response or purpose |
|---|---|---|---|---|
| **CONGREGATIONAL** | Aggregational | Congregational song | | Heterosexual gathering, sexual or social |
| | Calling | Ordinary, usual, normal, wonted, spontaneous, indifferent, solitary, common song, *gewöhnlicher Gesang*, *chant d'appel* | Loud, specifically patterned, long range; has perfectly defined physical characteristics | Attracts females; aggregation, pair formation. Some female Orthoptera produce a stridulatory copy of male's calling song |
| **SEXUAL** | Courtship | Serenade, *Werbegesang*, *chant de cour*, nuptial song | Low intensity, short range, often with weakened frequency spectrum | Insemination timing; insemination facilitating; pair maintenance |
| | Courtship interruption | | | Pair maintenance; pair reforming |
| | Copulatory | | | Maintains female passivity |
| | Postcopulatory or inter- copulatory | | | Pair maintenance |

284

| | | | | |
|---|---|---|---|---|
| AGGRESSIVE-ALARM | Aggression | | | |
| | Rivals duet | *Rivalengesang, chant de rivalité*, answering song | Often appears derived from calling song; Alternated by two individuals (anaphony) | Second male often replies with same song; separates rivals, establishes dominance; territorial, sometimes also pair-reforming |
| | Warning song | | | Intruder withdraws; territorial |
| | Fighting song | | | Intruder withdraws; territorial |
| | Protest and alarm | Call of distress, complaint cry, cry of agony, reflex cry | High intensity, lack of pattern | Repulsion of predators; conspecific warning |
| SOCIAL | Recognition signals | Contact cry | | Limited to subsocial and social species; pair- and family-maintaining |
| | Food-finding and nest site directives | | | Limited to social species |

[a] Based on Haskell (1974), Alexander (1967), and Dumortier (1963).

coherence through reactions to one another's flight tones. For further discussion of the evolution of synchronous and alternating phonoresponses in insects, see Alexander (1975) and Otte (1977).

## Sexual Signals

Most people have at some time heard the chirping of crickets on a midsummer's evening. Cricket song, which at first seems like a random sequence of chirps and trills, is actually a communication system of considerable complexity which consists of senders (always male, for females are silent) and receivers (which may be of either sex) and a variety of signals conveying vital information (Fig. 7-8).

Acoustic sexual signals span a wide range of behaviors including attraction, courtship, copulation, and postcopulatory pair maintenance;

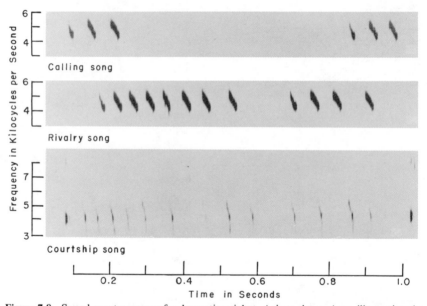

**Figure 7-8** Sound spectrograms of a domestic cricket, *Acheta domesticus,* illustrating the acoustic repetoire. The loudest and most commonly heard song is the calling song sung by males to attract sexually receptive females to them. Once the sexes are together, a new song—the courtship song—facilitates mating. A quite different song, the rivalry or aggression song, is sung when two males encounter one another and serves to determine the relative dominance of an individual. In other species the male may sing a postcopulatory song after mating, called by some the "triumphal song"; it serves to keep the sexes together for a period following sperm transfer and may help the male to "protect" his sperm investment. (Courtesy of R. D. Alexander).

in their broadest sense they include aggregation and chorusing as well. In practice, it is usual to make a distinction between long-distance acoustic signals (pair formation and calling sounds) and those used after the two sexes have come together (courtship and copulation sounds). In general, the evolution of these rather different aspects of sexual signaling appears to have proceeded quite independent of each other.

In acoustic pair formation, we usually think first of a direct locomotory response of a silent female to a single sedentary male's calling song, such as occurs in most crickets. Such oriented movements in response to acoustic signals are termed **phonotaxes** (Fig. 7-9). Many variations in pair formation exist, however. Some may involve elaborate visual/acoustic displays by both sexes, or they may depend upon alternating and distinctive songs between males and females. Habitat ecology is undoubtedly a major determinant of the type of system that can be employed. For example, diurnal desert grasshoppers can use visual–acoustic systems with ease in their open habitats, but many crickets and tettigoniid grasshoppers must rely more heavily upon acoustic elements because of their visually restricted habitats and nocturnal and/or cryptic habits (Otte, 1970).

The acoustic behavior of more than half of the 1000 or so species of Orthoptera and Homoptera in North America and Europe has been studied to some extent, and pair-forming or calling signals have been experimentally demonstrated more frequently than any other kind of intraspecific acoustic signal. In no case has identical or confusingly similar sound signaling been discovered between sympatric species singing at the same time and place. However, closely related but geographically and/or temporally isolated species often have identical or very similar sounds (Alexander, 1968).

It is worth noting that one American field cricket species has been shown to have an abruptly lowered auditory threshold at just the frequency (pitch) of its song, about 4.5 kilocycles per second. If other crickets also have their auditory organs so precisely "tuned" to their own frequencies, the songs of different genera might be rendered almost inaudible. Then rhythm patterns of each could evolve without reference to one another, and songs could be rhythmically quite similar without confusion. Some weight is given to this theory by the observation that ground crickets (Nemobiinae), which often live with field crickets, have songs usually pitched quite differently from them (about 8–10 kilocycles/second). In a region such as North America, where 40 or 50 insect species often may be stridulating simultaneously in the same locality, such a development would obviously be a great advantage.

One usually thinks of each species stridulating its own specific sex-

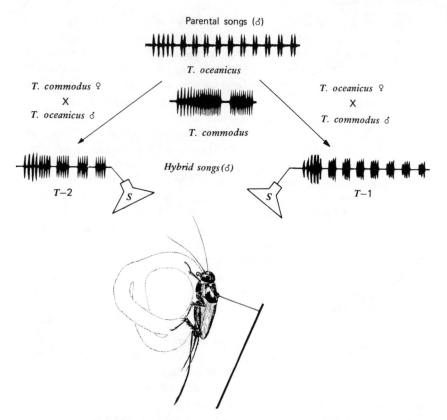

Parental songs (♂)

T. oceanicus

T. commodus ♀
X
T. oceanicus ♂

T. oceanicus ♀
X
T. commodus ♂

T. commodus

Hybrid songs (♂)

T–2

S

S

T–1

Hybrid female phonotaxic response
♀ (T–1 or T–2)

**Figure 7-9** Positive phonotaxis response used in experimental studies on the inheritance of song in hybrid *Teleogryllus* crickets. Walking along a featherweight styrofoam Y maze, tethered receptive hybrid adult females listen to different hybrid male calling songs played simultaneously by equally distant speakers, and reveal their preference at fork points in the maze. (Slightly modified from Hoy et al., 1977. Copyright © 1977 by the American Association for the Advancement of Science.)

related messages; if they do not, one immediately assumes some other clear-cut isolating mechanism. But sometimes the case is less well defined than this. Two British grasshoppers, *Chorthippus brunneus* and *C. biguttulus*, are distinguishable morphologically only by a very few small, nonoverlapping characters; although they have different songs, they are not separated ecologically. Surprisingly, cross-mating readily occurred in the laboratory (Perdeck, 1958), with the production of viable offspring, although very little hybridization occurred in the field. How could this be? Careful field study revealed that the major difference between the species lay in the *degree* of stimulating and orienting influence their songs possessed. In the presence of its own specific song, a male moved faster and made more copulatory attempts; a female was also more ready to copulate in the presence of the song of her own species. These behavioral factors were apparently sufficient for effective isolation of the two species, another illustration of behavioral distinctiveness preceding morphological isolation.

Courtship signals include those occurring after pair formation and before copulation. Under this definition, special acoustic courtship signals are definitely known for some male Orthoptera (Acrididae, Gryllidae, Tettigoniidae) and for cicadas, and there is suggestive evidence for their occurrence in males of numerous other arthropods from crabs to *Drosophila* (see Bennet-Clark and Ewing (1970). Responses to courtship signals have not been widely confirmed experimentally. Often, they seem merely to take the form of females "allowing" the male to copulate. The presence or absence of such readiness needs to be tested using normal and silenced males and deafened and nondeafened females; such tests have as yet not been performed on any scale.

In the "classic" case, a male insect stridulates sex-related messages and the female receives them through her tympanal organs. However, sound production may also arise as a by-product of other activities such as flight, and in some species the sounds used in communication fall within the sensitivity range of hair sensillae and antennal receptors. Predictably, the significant properties of such sounds are rather different than we found in orthopteran songs. The mosquito is a case in point.

CASE STUDY: SEXUAL ATTRACTION IN THE YELLOW FEVER MOSQUITO, *Aedes aegypti*

Many mosquitoes produce an audible hum during flight, often so distinctive that some workers have claimed to be able to identify mosquito species by their sound alone. Astute observers have long known that mosquitoes react to sounds as well, for swarms of various species have been seen to alter their behavior following acoustic stimuli as varied as pistol shots, locomotive whistles, musical instruments, and the human voice (Fig. 7-10). Yet until the perceptive research of Roth, (1948), the basis of mosquito sexual communication was a matter of dispute. For example, one researcher claimed that males were drawn to females by odor. Another postulated that the female's motion alone was sufficient for male recognition and attraction. A third had suggested that

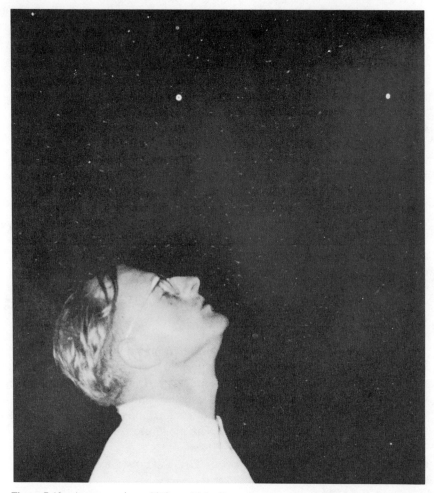

**Figure 7-10**  An opera singer hitting a high glissando gets a mouthful of male mosquitoes! (From Nachtigall, 1974.)

knoblike projections (halteres) behind the wings of the female produced attractive "birdlike" sounds both at rest and in flight. Roth had his own hunches, of course. He felt that male mosquitoes were attracted by the buzzing sounds produced by a female's wings during flight and that males received this sound through their enlarged antennal bases which have conspicuous Johnston's organs. But how best to prove this?

Beginning with careful behavioral observations, Roth established that once male *A. aegypti* became sexually responsive 15–24 hours after emergence, they remained in a constant mating state through the rest of their lives (Fig. 7-11). When he forced caged females to fly continuously, males attempted mating with them repeatedly; after a time, the females tended to resist further mating attempts but as long as they continued to fly, males continued to be attracted. However, when he killed female mosquitoes and presented them, the same males which readily mated with flying females were indifferent to these freshly dead bodies. In addition, even though a resting female might be surrounded by males with some so near that they touched her, never did a resting female induce a male to mate. It therefore appeared unlikely that odor or constantly produced sounds were involved.

Next, Roth exposed mosquitoes of both sexes to some mechanically produced sounds. He found no evidence that female *A. aegypti* were attracted to any, though they sometimes gave shock reactions to certain frequencies. Males, however, showed their characteristic phonotaxis and mating response to any sound in the range between 300 and 800

**Figure 7-11**  A tethered female mosquito, *Aedes aegypti*, attracts a male soon after she begins to fly who then mates with her. Note the feathery antennae characteristic of the male. (Courtesy of T. Eisner; from Roth et al., 1966.)

vibrations per second. This was true whether he produced the sound by a recording of another mosquito, by an audio oscillator, or by a tuning fork. His hunch further strengthened by this, Roth fastened fine wire loops about the necks of female mosquitoes, then suspended them in a cage full of males. As long as a female hung with motionless wings, the males flying near her remained indifferent. Even when Roth swung the female to and fro upon her tether, they did not respond. As soon as she began to vibrate her wings, however, the males flying nearby immediately seized her and began to mate. To put the theory of attractive "birdlike" sounds from the halteres firmly to rest, Roth repeated the experiment again using females with halteres removed. Again, males responded as the females vibrated both wings. However, males ignored wingless females which still had their halteres.

Some questions arise. Why does the male respond to such a range of frequencies? Upon reflection, one can postulate that this behavior ensures the "attractiveness" of females which may produce sounds of different frequencies because of certain uncontrollable factors such as size, wing damage, extent of distension of abdomen with food, etc. A possible disadvantage of a wide response range might be that males could be attracted to females of other species and attempt to mate with them.

What sets the limits of the range of sounds within which the male is attracted? Roth found the key in the sexual maturation of the mosquito: mature females and freshly emerged males beat their wings at about the same rate and are pursued by older males, but young, freshly emerged females have a different wingbeat rate and are not pursued. By response limits such that only mature females are pursued, the likelihood is increased that the female has ingested a blood meal and that a mating will produce viable progeny. Young males avoid sonic misidentification and pursuit by older males by simply remaining quiescent until they complete their sexual maturation. By the time a male begins to fly, the sound he makes in flight is sufficiently high in pitch to be beyond the range that will stimulate older males.

What is the source of the attractive humming sound of the females? When Roth cut off greater and greater portions of their wings, the sound progressively decreased in volume while the pitch rose progressively. Yet males were attracted to and mated with females which had only part of their wings, even stumps, vibrating. This suggested that some basal organ might be the principal sound producer. Morphological studies have in fact confirmed the presence of a stridulating organ at the base of the female's wings.

The above discussion covers only a few aspects of the complex, wide-ranging investigation Roth undertook; the interested student may wish to peruse his original study (Roth, 1948) for further detail. A surface pheromone has since been discovered which mediates the final stage of species recognition.

## Disturbance, Alarm, and Aggression

After an encounter with the stridulating carabid beetle *Cychrus,* the common shrew will not attack this insect again. When threatened, both sexes of the velvet ant *Dasymutilla occidentalis* (Fig. 7-12) stridulate with enough volume to be readily detected by the human ear a meter away. Small insectivorous birds called tits will not approach stridulating *Necrophorus* carrion beetles. Bats will avoid palatable mealworms shot into the air if the sounds of unpalatable arctiid moths are played concurrently.

Insects of almost every order are known to react with some noise when picked up, pinched, probed, restrained, or otherwise disturbed. In fact, more descriptive papers have been published on this general kind of sound communication than on any other aspect of arthropod acoustical behavior. For most arthropods, the only known acoustic communication is such a **protest sound**, a high-intensity unpatterned sound of a broad frequency spectrum. These signals are elicited primarily upon tactile stimulation. Their function is still unknown, and whether such signals are even primarily acoustic is open to question. It has been

.1 sec

**Figure 7-12**   Oscillogram of the disturbance stridulation produced by a harassed unrestrained adult female velvet ant, *Dasymutilla occidentalis.* The unpatterned sound is produced by an in-and-out movement of the abdomen segments which rubs a series of transverse striations (the file) on the dorsal anterior margin of certain abdominal segments across raised ridges (scrapers) on the posterior dorsal margins of the preceding segments. Top recording represents one direction of abdominal movement; lower trace represents the opposite direction. Greatest sound intensity falls in the range of 1–10 kHz. (Courtesy of J. O. Schmidt; see Schmidt and Blum, 1977.)

suggested that the effective stimuli are tactile or vibratory communicative signals; in this view, the air-transmitted portions of the signal which man hears may be incidental.

The emission of protest sounds is so widespread in the Arthropoda that it is most tempting to consider these as predator-escape mechanisms. However, after actual observations one finds it at least as easy to persuade oneself that such signals do not upset predators at all; the literature is certainly full of documentary notes on insects being eaten while emitting such sounds. Of course, failure in a certain proportion of cases does not necessarily mean that it cannot have a defensive function. Some startled predators may still release the prey. Even if it only worked to the prey's advantage some of the time, such behavior would still have positive selective value.

It has also been postulated that protest sounds in at least some instances may be analogous to aposematic color patterns (see Chapter 8). Many arthropods emitting so-called protest sounds have defenses such as noxious secretions, an offensive taste, or possession of chelicerae or a sting. A predator might, after one or several experiences, associate protest sounds with earlier painful feelings. Once thus conditioned, the predator would treat the stridulation as a warning to keep its distance. The ultrasonic clicks of certain moths appear to serve as such a device (Dunning, 1968).

Another suggestion has been that these protest sounds should be regarded merely as displacement activity, an outlet for the intense nervous excitement involved in capture or cornering. A variant of this theory states that these sounds have no meaning at all. Why, it is asked, does the possible meaning of this acoustic display even have to be considered when no one thinks of giving equal consideration to the role and efficacy of the cries of a mouse captured by a cat? Of course, on this subject one is truly "unencumbered by the facts." Until some perceptive experimental work is undertaken, the situation will remain so. It is likely that both hypotheses are valid in different situations, and that in others the very novelty of an unfamiliar acoustic signal may elicit predator avoidance.

Actual **alarm signals** differ from protest sounds in some important ways. They are *intra*specific signals characteristic of social insects and show a wide range of intensities. Their sounds are almost always patterned, from simple to often complex. Releasing stimuli are often elaborate. At present, they are poorly researched, but one experimental investigation is provided by the work of Howse (1964) on *Zootermopsis angusticollis*. This termite uses substrate vibrations to send alarm signals, producing these by tapping its head on the roof and floor of its tunnels. Both soldiers and workers produce such vibrations under disturbance conditions such as sudden exposure to bright light or puffs of air. They often also produce these upon meeting other termites and will respond to an alarm vibration by making one themselves. The vibrations are received through their tibial subgenual organs, which appear to be tuned somewhat to the frequency and pulse rate of their particular disturbance response. Stuart (1977) has reexamined head banging behavior in a related *Zootermopsis* and concluded that it does not function in the transmission of alarm by sound or vibration. He suggests instead that head banging may be a mechanism serving to lower the termite's response threshold to other stimuli. The alarm distress sound of another social insect, the leaf cutting ant, *Atta cephalotes,* has been studied by Markl (1965).

Much experimental work on disturbance sounds is needed. However, it is worth noting that on theoretical grounds an effective alarm should have a very wide frequency range and be of high intensity, and indeed these characteristics do typify most sounds in this category. Such sounds are both easily heard and readily pinpointed as to source.

A second type of intraspecific disturbance sound is the patterned aggressive signal, or **rivalry sound**, mediating competitive interactions between conspecific individuals. An example among bark beetles was mentioned previously (see Fig. 5–17). However, the role of acoustic signalling in aggression has been most thoroughly studied in crickets.

CASE STUDY: AGGRESSIVE SINGING IN CRICKETS, *Acheta domesticus* AND *Gryllus pennsylvanicus*

Ever since ancient times, cricket fights have been a popular sport in China. If two adult male crickets are caged together with just one burrow for hiding, they fight over it. The winner and loser are established through heated battles, after which actual physical combat occurs less frequently because the subordinate cricket avoids encounters with the dominant one. Perched alone at his burrow entrance, a male gives the calling song which we recognize as his familiar "chirp." But when two males approach each other or engage in battle, they produce an aggressive rivalry song quite unlike their sounds in other situations (see Fig. 7-8). The amount of such aggressive stridulation the crickets display is positively correlated with winning encounters. In fact, the subordinate male rarely chirps at all after an encounter. Do these songs cause the subordinate male cricket to maintain his lessened aggression? Phillips and Konishi (1973) decided to attack the question experimentally.

First, Phillips and Konishi marked adult males of the common house cricket, *Acheta domesticus*, and the field cricket, *Gryllus pennsylvanicus*, with various combinations of silver enamel dots. Then, two previously unacquainted conspecific crickets were introduced at a time into a very small cage. Under these conditions, they encountered each other frequently and fought or demonstrated clear dominance. The procedure was repeated many times until a statistically sufficient number of encounters was recorded. Then, the losers in these matches were anesthetized and the tympana in their forelegs carefully torn. By electrophysiologically monitoring the responses of the tympanic nerve to recorded chirps, the investigators satisfied themselves that the crickets were deaf, their tympanal organs unable to respond to sounds of the frequency involved in aggressive chirps. Monitoring of the cerci and the subgenual organ in the foreleg showed these unable to register the stridulatory sounds as well. While the subgenual organs in the middle pair of legs responded to the chirps when they were played at supernormal intensity, control experiments ruled out any behavior effects of such responses under normal situations.

Having thus carefully established the subordinates' deafness, the researchers proceeded to rematch them with the previous winners. In all cases, the deafened losers immediately became extremely aggressive! They initiated combat and engaged in sustained battles, and often came to clearly dominate their former superior rivals. In a second series of experiments, the researchers chose sets of three or four crickets which had already established a dominance hierarchy in an empty aquarium. The lowest-ranking individual in these sets was then deafened and rematched in tournaments with all the higher-ranking individuals. In every rematch, the deaf cricket significantly increased the proportion of winning encounters.

How might one explain the results of Phillips and Konishi? One can assume that the acoustic signals of the dominant cricket's song inhibit the aggressive tendencies of the subordinate cricket and that inability to hear these signals frees the deaf cricket from this inhibition. This would explain the increased aggression which deafened individuals displayed. However, in the rematches, the deaf individual did not *always* come out on top. Therefore, it seems likely that the probability of a cricket *winning* an encounter is additionally influenced by the existence of some other, as yet unspecified factors. In fact, tactile antennal stimuli are also important; other research has shown that a dominant male can be "defeated" by repeatedly exposing him to artificially produced aggressive sounds while lashing him with bristles simulating cricket antennae (Alexander, 1961).

## Social Sounds

Large wingless cockroaches of the gregarious genus *Gromphadorrhina*, native to Madagascar, frequently hiss when their culture boxes are disturbed. *Pogonomyrmex* harvester ants trapped in a small landslide stridulate loudly, attracting other workers who dig them out. Hornets produce and are sensitive to a variety of sounds inside their nests (Ishay, 1977). Honey bees initiate swarming in part in response to a contagious buzzing sound, the *Schwirrlauf*, started by a few workers running excitedly through the hive.

Group-living insects range from those existing in simple aggregations to complex societies (see Chapter 10). To be social necessitates methods of communication smoothly integrating the cooperative behaviors of all group members. Most often such integrative communication is chemical; several instances were noted in Chapter 5. However, sound can also serve as an effective means of communciation in social insects. Social communication is rarely restricted exclusively to one particular mode. The communication system of the Douglas fir bark beetle discussed in Chapter 5, an interlocking combination of odor, sonic, and tactile signals, is a case in point. However, one of the classic and best-studied examples of the interaction of communicatory modes is provided by the classic social insect, the honey bee.

### CASE STUDY: THE INTERPLAY OF COMMUNICATORY MODES IN THE HONEY BEE, *Apis mellifera*

In the late 1940s the Austrian biologist von Frisch excited the scientific community with the results of his thorough observations of honey bees, which indicated that a returning bee was able to communicate to hivemates the quality, direction, and distance of a food supply with such accuracy that bees left the hive after receiving the message and flew directly to the food. Through careful study, von Frisch had learned that a forager bee that found a good food source would bring back a full load of pollen and nectar to the hive. It would then fly out again to the food, thus establishing the most direct outward route. Back in the hive upon its second return, the successful forager communicated its vital information to others by performing specific, repeated, stereotyped behaviors, which von Frisch termed **dances**, upon the vertical face of the honeycomb inside the hive.

The forager's message was a complex one. One aspect of it was distance. Distance may be communicated by several methods but appears to be described by bees in terms of the energy expended in covering it. Thus, if a bee must fly into a headwind or uphill or if it is forced to walk, it will report a distance farther than actually occurs. The distance to a food source can be correlated with the type of dance the forager displays (Fig. 7-13). With the waggle dance in particular, any of a number of features can be correlated with food source distance: the speed or length of the wagging run, the duration of the semicircular run, the duration of a single tail-wagging phase, or the number of tail wags during the straight run. While the last two of these are generally favored at present as showing the best correlations, which of these features if

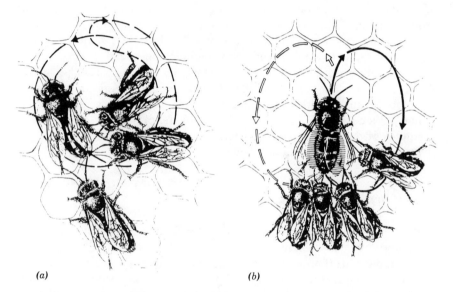

(a)                                             (b)

**Figure 7-13** The dances of the honey bee. A forager locating a source of food close to the hive (within a distance that varies among strains) moves in circles of alternating direction, performing a **round dance** (a). Followers of the round dance obtain information that food is nearby without having the direction specified. As the distance to the food increases, the round dance performance gradually changes to an open figure-8 (termed a sickle dance) which is transitional to the **waggle dance** (b) performed by foragers who have discovered food sources far from the hive. As the bee passes through the straight run portion of the dance, she "waggles" her abdomen rapidly from side to side, then circles back to the beginning of the straight run alternating turning right, then left after each repetition of the straight run. Information about the distance and direction of the food find is acquired by followers during the straight run. (Reprinted by permission of the publishers from *The Dance Language and Orientation of Bees* by Karl von Frisch, Cambridge, Mass: The Belknap Press of Harvard University Press, Copyright © 1967 by the President and Fellows of Harvard College.)

any are used by the recruited bees is not clear. What is clear is that humans observing the dance can relate these features to distance.

The direction of the tail-wagging run in the waggle dance gives an indication of food source direction. The angle of this direction with respect to the vertical is equal to the angle formed between the sun's position and the food source (Fig. 7-14). Such communication is theoretically complex, for the dance is performed out of sight of the sky and with reference to gravity rather than directly to the sun itself (see Fig. 3–6).

The dance of the honey bee has become perhaps most famous as an

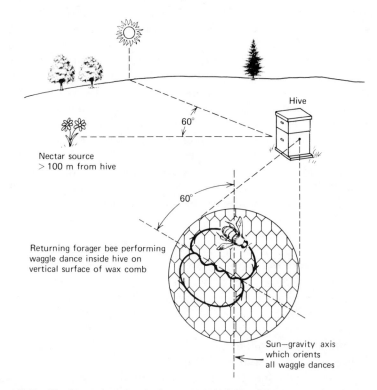

**Figure 7-14** The honey bee waggle dance communicates food source direction: the angle of the straight run with respect to vertical equals the angle between the sun's position and the food source. Thus, a crossrun angle of 60° to the left of the upward vertical axis as the bee dances on the vertical face of the comb means that the food was on a path 60° to the left of the sun–hive line, going toward the sun. Interestingly, the honey bee sometimes dances at the hive entrance on sunny days; in this horizontal position, the tail-wagging run of the dance points directly to the food source. (Redrawn after von Frisch, 1967.)

example cited in discussion of visual communication. It is, however, actually quite difficult to classify this remarkably complex communication. Esch, who spent some time studying bees with von Frisch, was puzzled by the visual interpretation given to the bee dance language. How, in the total darkness of the hive, could bees observe the dance and visually interpret its tempo? Seeking an answer with the aid of modern instrumentation, Esch built an artificial motor-driven bee capable of doing a waggle dance (Esch, 1967). When it was introduced into the hive, bees clustered around it just as they would a genuine forager. No bees, however, left the hive to search for food. Why?

Perhaps, reasoned Esch, the dummy wasn't sounding right, since it did not vibrate as he knew a live dancing bee did. So Esch placed a microphone in the hive to record actual dancing bee sounds. To his surprise, not only did the dancing bees produce strong sound signals with vibrations of their wings, but the period of sound production coincided exactly with the duration of the straight run of the dance. Since von Frisch had already found a correlation between the latter and the distance to the food source, a logical conclusion was that the bees might equally well be using sound as the cue to distance rather than using the visual or tactile cues of the waggle itself. Wenner (1964) independently discovered that sound was produced during the bee's dance. The sound, mainly at 250 cycles/second, is produced in bursts whose number may be directly correlated either with the distance or with sugar concentration of the food. The pulse rate of these sounds is directly related to the rate of waggling and is approximately 2.5 times as fast.

The primary sensory modality involved in honey bee dance communication is still open to question, for an important lesson illustrated by the studies of von Frisch and Esch is that communicatory episodes may involve more than one sensory modality simultaneously. Failure to consider such possibilities when analyzing communication systems can be a pitfall.

With the honey bee, sound may indeed be the most important channel for distance indication. But since the followers can be seen to have antennal contact with the dancer, other researchers have felt it is also likely that tactile cues are important. The speed of waggling could be sensed this way as a measure of distance. Contact could also be necessary in another way. By following the dancer and closely imitating the path of its dance, as many of the bees do, information on direction might be derived from proprioceptors monitoring this imposed movement.

As if this were not enough, some scientists have questioned the entire concept of bee communication. Wenner (1971) claimed that bees are recruited solely by a conditioned response to food or odor, without using the coded dance information at all. He felt that bees lacking experience with a particular food source have great difficulty finding it. Only a relatively few seem to be successful, and these take much longer to arrive than would be expected if they were flying a straight line between the hive and the source. Inexperienced foragers, Wenner suggested, locate the source simply by dropping downwind of the hive and then searching for the right combination of odors from food, locality, and other bees. This scientific controversy, including von Frisch's defense of his work, has appeared in a number of scientific journals and is far from settled. Gould has performed a variety of experiments aimed at resolving the dilemma (Gould, 1975, 1976). He concluded that, depending upon conditions, honey bee recruits use either the dance language and odor information or odors alone. He further suggests that von Frisch and Wenner were both studying situations involving exploitation of an abundant food source but focusing on different stages of this process: von Frisch on the early stages of discovery, Wenner on later stages. Future imaginative experimentation is clearly invited and will surely be attempted, but for the present the honey bee dance language remains one of the most striking, best-studied examples of mixed channel communication and a highly complex transfer of information. Von Frisch shared a Nobel prize in 1973 for his studies in honey bee communication behavior.

In an attempt to understand ways in which this system may have evolved, attention has turned to related tropical stingless bees (Meliponini) (Esch, 1967; Lindauer, 1971; Lindauer and Kerr, 1960). The 11 species of these social bees which have been studied all show some means of alerting their colonies to food. However, the efficiency and degree of development of these communication systems vary greatly. In the simplest case, *Trigona iridipennis* workers that have found a food source simply fly back to the comb and excitedly run about, thus alerting other workers to the existence of food but not to its location. In the most complex case, a *Trigona postica* worker constructs a trail system of marks left at short intervals by rubbing her mandibles on convenient objects such as leaves to deposit secretions of a special mandibular gland. Back at the comb, the bee then runs noisily about, giving out food samples. When a number of other bees have been acoustically and chemically alerted, the forager leads these bees along the scent trail. Such a system is of special advantage to these forest-dwelling species, for here vertical movements may be as important as horizontal ones and a single-coordinate directional system may thus be inadequate.

Elements of the bee dance have been uncovered in many diverse insects since investigators began searching for evolutionary antecedents to this remarkably complex behavior. A number of nonsocial insects perform stereotyped movements after feeding or flying. For example, blowflies will perform a crude sort of dance after eating, with a vigor and persistence related to the distance the fly has flown after feeding. Some saturniid moths rhythmically sway upon settling down after any movement; the number of oscillations is closely correlated with the distance flown before settling. Of course, there is no evidence that these activities have any communicative function, perhaps being simply by-products of the physiological state of the insect concerned. But they do share the important characteristic of persistence, that is, they continue to be performed for a considerable time after the cessation of the activity which produced them. Such persistence is of paramount importance in communication, especially in recruitment of foraging bees. If, as seems true with all other animal language, such by-products may acquire signal value as indicators to another animal of the state of the sender, then they are ideal materials for the construction of a language.

Those interested in more details of the bee language story will find the books of von Frisch (1967, 1971), Lindauer (1971), and Wenner (1971) especially useful. Additional important papers are those of Gould et al. (1970), Gould (1975, 1976), von Frisch (1974), Esch (1967), Krogh (1948), and Wenner (1964).

**Sound as Communication Method: Theoretical Considerations**

With the development of sophisticated instrumentation, researchers are beginning to appreciate more fully the immense range and variety of insect sound. The widespread utility of auditory signals is not surprising, however. Acoustic messages form an unusually complete and efficacious mode of imparting information. They are easily diffused, are resistant to disturbances, and exhibit the potential for creating a vocabulary. Sound may be distinguishable at very low levels; it may be raised above environmental noise by pumping energy into the system. In many cases, an insect would not even need to suspend other activities in order to transmit or receive sound signals.

Of the varied communicatory modes, chemical messages have perhaps the widest range, visual messages probably the shortest. Acoustical messages have a theoretically greater range than optical ones if one considers the distance at which signals are no longer physiologically perceptible. However, the distance at which they cease to release a reaction is undoubtedly much shorter than this.

Primer effects have yet to be demonstrated in acoustic communicative systems involving invertebrates, though examples have been found in birds and mammals. This is the only communicative mode for which this is true; tactile, chemical, and visual signals functioning as primers all clearly exist among the insects.

Acoustic signals seldom initiate and orient behavior patterns by themselves. But unlike specializations associated with chemical and visual input or output, specialized acoustic behavior is always a communicative phenomenon. This is true whether the signals function intraspecifically or interspecifically, or even if they are of use only to the individual producing them, as in echolocation. To examine them in terms of selective action, therefore, they must be studied as communicative units.

Song is a species-characteristic feature very much like morphological features. While external factors, especially temperature, may considerably modify the song, these modifications are only simple responses to changes of environment much like changes in pigmentation or ornamentation. No experiment has yet suggested that learning is involved; in arthropods, the song appears to be totally hereditary. What plasticity is present is preformed; it cannot be modified through learning. Song may also serve as an important reproductive isolating factor. When new arrangements in its parameters arise (a rare but not impossible situation), they are capable of producing instant isolation of the individual from its species. Hence, with this barrier, progress toward speciation has already begun. This situation has been encountered in several situations, with species indistinguishable by their morphology being found to be separated by their song (see Walker, 1964).

What would an ideal communication system be like? It should exhibit a broad spectrum of information content and be precise in both the nature and emphasis of the messages it encodes. It should be able to reach the intended receiver rapidly, traveling over long distances when necessary. It should be able to either start and stop quickly if needed or to persist in the environment when that strategy becomes advantageous. And it should be difficult to exploit.*

Obviously, no system meets all these criteria. The function the signal

---

* Interestingly, examples of predators breaking the communicative code of their prey are known in all the major communicatory modes. For examples see the aggressive mimicry by *Photuris* fireflies (Chapter 6), interspecific group living (Chapter 4), and papers by Sternlicht (1973) and Cade (1975).

must fulfill, the anatomy and physiology of the organisms sending and receiving it, and the nature of the immediate environment all must play a part in determining which qualities are to be emphasized and what communicatory compromise is to be reached. Sebeok (1968, 1977) provides very comprehensive coverage of animal communication.

## SUMMARY

Mechanoreception includes the perception of any mechanical distortion of the body, and as such encompasses hearing, touch, gravity reception, and a host of specialized subsenses dealing with an organism's position in space. Detection of sound and vibration commonly overlap in the insect world; the Johnston's organ provides a good example. Hearing and touch, both of which depend upon velocity-sensitive receptors, are communicative senses. Pressure-sensitive sensilla, or proprioceptors, help the insect maintain its position with respect to gravity and the relation of various body parts to each other.

Tactile communication, limited to various close-range situations, is difficult to document without disrupting the system. Additionally, it mandates the interplay of signal systems at least through time, for other methods of communication are required to bring individuals close enough for tactile contact to occur. Sematectonic communication involves durable signals in the form of evidence of work which evokes behavior or physiological change in the receiver; a special case, guidance via tactile cues, is illustrated by nest building in both social and solitary Hymenoptera.

Sound production uses a great variety of mechanisms and body parts, but frictional methods (i.e., stridulation) predominate. Vibrating membranes (tymbals) driven directly by muscles are also common. Sound reception mechanisms include various modifications of groupings of tiny innervated sensilla located subcuticularly in various body parts and regions. Tympanic organs and subgenual organs are most common.

An unusually fruitful subject for study in almost all of animal behavior, acoustic communication has been equally profitable in insects. Unlike most vertebrates, however, insects sing a monotone; their songs lack variations in pitch and for their distinctiveness depend instead upon differences in sound level, stereotyped volume changes, the presence or absence of silent pauses between song elements, quality differences, and the rate of delivery of the "notes." The nested nature of the song components has led to a great many terminologies; one straightforward system includes pulses, pulse trains, pulse train groups, and modes. An unaided human ear usually discerns only the modes.

The functions of insect sound are many, but most are included by four basic but overlapping categories—congregational, sexual, aggressive-alarm, and social. The simultaneous interplay of sensory modalities in insect communication is particularly well illustrated in the case of honey bee communication. Useful overviews of insect mechanocommunication include those of Alexander (1967, 1968), Bennet-Clark (1975), Frings and Frings (1958), Haskell (1961, 1974), Michelson and Nocke (1974), and Moore (1973).

## SELECTED REFERENCES

Alexander, R. D. 1961. Aggressiveness, territoriality and sexual behavior in field crickets. *Behaviour* **17:** 130–223.

Alexander, R. D. 1967. Acoustical communication in arthropods. *Annu. Rev. Entomol.* **12:** 495–526.

Alexander, R. D. 1968. Arthropods. In *Animal Communication: Techniques of Study and Results of Research,* T. A. Sebeok (ed.). Indiana Univ. Press, Bloomington, pp. 167–216.

Alexander, R. D. 1975. Natural selection and specialized chorusing behavior in acoustical insects. In *Insects, Science, and Society,* D. Pimentel (ed.). Academic Press, New York, pp. 35–77.

Alexander, R. D. and T. E. Moore, 1962. The evolutionary relationships of 17-year and 13-year cicadas, and three new species (Homoptera, Cicadidae, *Magicicada*). Misc. Publ. Mus. Zool., Univ. Mich., No. 121, 59 pp.

Bennet-Clark, H. C. 1970. The mechanism and efficiency of sound production in mole crickets. *J. Exp. Biol.* **52:** 619–652.

Bennet-Clark, H. C. 1975. Sound production in insects. *Sci. Progr.* **62**(246): 263–284.

Bennet-Clark, H. C. and A. W. Ewing. 1970. The love song of the fruit fly. *Sci. Amer.* **233:** 84–92 (July).

Bentley, D. and R. R. Hoy. 1974. The neurobiology of cricket song. *Sci. Amer.* **231:** 34–44 (August).

Cade, W. 1975. Acoustically orienting parasitoids: fly phonotaxis to cricket song. *Science* **190:** 1312–1313.

Cooper, K. W. 1957. Biology of eumenine wasps. V. Digital communication in wasps. *J. Exp. Zool.* **134:** 469–514.

Davis, W. J. 1968. Cricket wing movements during stridulation. *Anim. Behav.* **16:** 72–73.

Dunning, D. C. 1968. Warning sounds of moths. *Z. Tierpsychol.* **25:** 129–138.

DuMortier, B. 1963. Morphology of sound emission apparatus in Arthropoda. In *Acoustic Behaviour of Animals,* R. G. Busnel (ed.). pp. 277–345.

Elsner, N. 1974. Neuroethology of sound production in gomphocerine grasshoppers

(Orthoptera: Acrididae). I. Song patterns and stridulatory movements. *J. Comp. Physiol.* **88**: 67–102.

Esch, H. 1967. The evolution of bee language. *Sci. Amer.* **216**: 96–104 (April).

Frings, H. and M. Frings. 1958. Uses of sounds by insects. *Annu. Rev. Entomol.* **8**: 87–106.

Frings, H. and M. Frings. 1960. *Sound Production and Sound Reception by Insects: A Bibliography.* Penn. State Univ. Press, University Park, 108 pp.

Frisch, K. von. 1967. *The Dance Language and Orientation of Bees.* Harvard Univ. Press, Cambridge, Mass., 566 pp.

Frisch, K. von. 1971, *Bees, Their Vision, Chemical Senses and Langauge,* rev. ed., Cornell Univ. Press, Ithaca, New York, 161 pp.

Frisch, K. von. 1974. Decoding the language of the bee. *Science* **184**: 663–669.

Gould, J. L. 1975. Honey bee recruitment: the dance language controversy. *Science* **189**: 685–693.

Gould, J. L. 1976. The dance language controversy. *Quart. Rev. Biol.* **51**: 211–244.

Gould, J. L., M. Henerey, and M. C. MacLeod. 1970. Communication of direction by the honey bee. *Science* **169**: 544–554.

Grassé, P.-P. 1959. La reconstruction du nid et les coordinations interindividualles chez *Bellicositermes natalensis* et *Cubitermes* sp. La theorie de la stigmergie: Essai d'interprétation du comportement des termites constructeurs. *Insectes Sociaux* **14**: 73–102.

Haskell, P. T. 1961. *Insect Sounds.* Quadrangel Books, Chicago, 189 pp.

Haskell, P. T. 1974. Sound production. In *The Physiology of Insecta,* Vol. II, 2nd ed., M. Rockstein (ed.). Academic Press, New York, pp. 353–410.

Howse, P. E. 1964. The significance of the sound produced by the termite *Zootermopsis angusticollis* (Hagen). *Anim. Behav.* **12**: 284–300.

Hoy, R., J. Hahn, and R. C. Paul. 1977. Hybrid cricket auditory behavior: evidence for genetic coupling in animal communication. *Science* **195**: 82–83.

Ishay, J. 1977. Acoustical communication in wasp colonies (Vespinae). Proc. XV Internat. Congr. Entomol., Washington, D.C., pp. 406–435.

Jansson, A. 1973. Stridulation and its significance in the genus *Cenocorixa* (Hemiptera, Corixidae). *Behaviour* **46**: 1–36.

Jansson, A. 1976. Audiospectrographic analysis of stridulatory signals of some North American Corixidae (Hemiptera). *Ann. Zool. Fenn.* **13**: 48–62.

Krogh, A. 1948. The language of the bees. *Sci. Amer.* **179**: 18–21 (August).

Lindauer, M. 1971. *Communication Among Social Bees.* Rev. ed. Harvard Univ. Press, Cambridge, Mass., 161 pp.

Lindauer, M. and W. E. Kerr. 1960. Communication between the workers of stingless bees. *Bee World* **41**: 29–41, 65–71.

Markl, H. 1965. Stridulation in leaf-cutting ants. *Science* **149**: 1392–1393.

McIver, S. B. 1975. Structure of cuticular mechanoreceptors of arthropods. *Annu. Rev. Entomol.* **20**: 381–397.

Michelsen, A. and H. Nocke. 1974. Biophysical aspects of sound communication in insects. *Adv. Insect Physiol.* **10**: 247–296.

Mill, P. J. (ed.). 1976. *Structure and Function of Proprioceptors in the Invertebrates.* Chapman and Hall, London, 686 pp.

Moore, T. E. 1973. Acoustical communication in insects. In *Syllabus. Introductory Entomology*, V. J. Tipton (ed.). Brigham Young Univ. Press, Provo, Utah, pp. 307–323.

Morris, G. K. and T. J. Walker. 1976. Calling songs of *Orchelimum* meadow kadydids (Tettigoniidae). I. Mechanism, terminology, and geographic distribution. *Can. Entomol.* **108**: 785–800.

Nachtigall, W. 1974. *Insects in Flight*. McGraw-Hill, New York, 153 pp.

Nocke, H. 1975. Physical and physiological properties of the tettigoniid ("grasshopper") ear. *J. Comp. Physiol. A, Sens. Neural Behav. Physiol.* **100**: 25–28.

Otte, D. 1970. A comparative study of communicative behavior in grasshoppers. Misc. Publ. Mus. Zool., Univ. Mich., No. 141, 168 pp.

Otte, D. 1977. Communication in Orthoptera. In *How Animals Communicate*, T. A. Sebeok (ed.). Indiana Univ. Press, Bloomington, 1126 pp.

Perdeck, A. C. 1958. The isolating value of specific song patterns in two sibling species of grasshoppers *(Chorthippus brunneus* Thunb. and *C. biguttulus* L.). *Behaviour* **12**: 1–75.

Phillips, L. H. and M. Konishi. 1973. Control of aggression by singing in crickets. *Nature* **241**: 64–65.

Pierce, G. W. 1948. *The Songs of Insects*. Harvard Univ. Press, Cambridge, Mass., 329 pp.

Prozesky-Schulze, L., O. P. M. Prozesky, F. Anderson, and G. J. J. van der Merwe. 1975. Use of a self-made sound baffle by a tree cricket. *Nature* **255**: 142–143.

Roth, L. M. 1948. A study of mosquito behavior. An experimental laboratory study of the sexual behavior of *Aedes aegypti*. *Amer. Midl. Natur.* **40**: 265–352.

Roth, M., L. M. Roth, and T. E. Eisner. 1966. The allure of the female mosquito. *Nat. Hist.* **75**: 27–31.

Ryker, L. C. 1976. Acoustic behavior of *Tropisternus ellipticus, T. columbianus,* and *T. lateralis limbalis* in western Oregon (Coleoptera: Hydrophilidae). *Coleop. Bull.* **30**: 147–156.

Schmidt, J. O. and M. S. Blum. 1977. Adaptations and responses of *Dasymutilla occidentalis* (Hymenoptera: Mutillidae) to predators. *Entomol. Exp. Appl.* **21**: 99–111.

Schwartzkoff, J. 1974. Mechanoreception. In *The Physiology of Insecta*, Vol. II, 2nd ed., M. Rockstein (ed.). Academic Press, New York, pp. 273–352.

Sebeok, T. A. (ed.). 1968. *Animal Communication. Techniques of Study and Results of Research*. Indiana Univ. Press, Bloomington, 686 pp.

Sebeok, T. A. (ed.). 1977. *How Animals Communicate*. Indiana Univ. Press, Bloomington, 1126 pp.

Sternlicht, M. 1973. Parasitic wasps attracted by the sex pheromone of their coccid host. *Entomophaga* **18**: 339–342.

Stuart, A. M. 1977. Some aspects of communication in termites. Proc. XV Internat. Congr. Entomol., Washington, D.C., pp. 400–405.

Suga, N. 1966. Ultrasonic production and its reception in some Neotropical Tettigoniidae. *J. Insect Physiol.* **12**: 1039–1050.

Ulagaraj, S. M. and T. J. Walker. 1973. Phonotaxis of crickets in flight: attraction of male and female crickets to male calling sounds. *Science* **182**: 1278–1279.

Van Tassell, E. R. 1965. An audiospectrographic study of stridulation as an isolating

mechanism in the genus *Berosus* (Coleoptera: Hydrophilidae). *Ann. Entomol. Soc. Amer.* **58:** 407–413.

Walker, T. J. 1964. Cryptic species among sound-producing ensiferan Orthoptera (Gryllidae and Tettigoniidae). *Quart. Rev. Biol.* **39:** 345–355.

Wenner, A. M. 1964. Sound communication in honeybees. *Sci. Amer.* **210:** 116–124 (April).

Wenner, A. M. 1971. *The Bee Language Controversy: An Experiment in Science.* Educational Programs Improvement Corp., Boulder, Colo., 109 pp.

Wilcox, R. S. 1972. Communication by surface waves: mating behaviour of a water strider. *J. Comp. Physiol.* **80:** 255–266.

Wilson, E. O. 1975. *Sociobiology. The New Synthesis.* Belknap Press of Harvard Univ. Press, Cambridge, Mass., 697 pp.

# 8

# Defense: a Survival Catalogue

A great variety of moths spend their days resting safely upon tree trunks, so perfectly matching the mottled bark that their very invisibility forms their defense. To repel honey thieves, some stingless social bees erect walls of sticky resin in front of or around the nest entrance, while other species smear a repugnant liquid there. Taking a more active and direct approach, pentatomid bugs earn their common name of "stinkbug" from the odorous, distasteful chemical they discharge when disturbed. The puss moth caterpillar, *Cerula vinula*, is even more elaborate in its defense. When it rests upon a branch it resembles a curled poplar leaf with a blackened margin, even to holding the two prongs of its tail together like a leaf stalk. When a threat appears, however, the caterpillar quickly throws off its passive protection, rearing its head to display the startling crimson front of its prothorax with its two eye spots. Simultaneously, the caterpillar spreads its forked tail, everting red whiplike threads from its ends and flourishing them in the air. Finally, when the attack persists, the caterpillar forcibly ejects from its prothoracic gland a burning, colorless fluid containing 40% formic acid.

A continual evolutionary battle of wits between predator and prey is being waged, and defensive maneuvers are not equally effective against all enemies. For example, *Eleodes* beetles react to threats by performing a headstand and firing a stinking irritant spray into the faces of their small vertebrate assailants. Some mice have evolved a way of thwarting this defense: quickly grabbing the beetle, they stuff its abdomen into the sand and proceed to calmly devour their prey from the head down.

A familiar estimate states that a single pair of houseflies are capable of producing 125 billion great-great grandchildren were all their offspring to survive. Obviously this does not occur. When averaged over a span of many years, the hundreds or thousands of eggs a female insect may lay usually result in only a few adults which reproduce again. This high percentage of losses is due to a variety of factors, but a major one is predatory attack. Almost all insects face a nearly constant threat of death from predators. Under such severe pressure, selection greatly favors those individuals which are able to reduce the risk of attack or injury with the most effectiveness while simultaneously interfering the least severely with their performance of other necessary activities.


DEFENSE MESSAGES

There are many aspects to defense, and widespread convergence in various structural and behavioral adaptation occurs. Table 8-1 classifies one subset of such defensive adaptations, the basic types of visually perceived anti-predator adaptations (excluding defensive secretions) based on the predator's response.

A more comprehensive approach to classifying insect defenses is to consider the prey-to-predator message implicit in a particular behavior or structural adaptation. Approximately half a dozen such messages will handle the bulk of the variety of insect defenses.

Crypsis: "I Am Not Here"

A whole series of insect larvae including geometrid moth "loopers," noctuid caterpillars, and sawfly larvae have come to resemble pine needles, with alternate stripes of pale and dark green and a resting behavior in a straight line parallel to the needles themselves. Stick insects often resemble the substrate upon which they rest with striking accuracy (Fig. 8-1). Katydids may mimic leaves to such an extent as to include copies of blemishes, fungal spots, or bird droppings in addition to reproducing the proper leaf tint and venation (Fig. 8-2).

By far the commonest method by which insects evade potential predation is through camouflage, or **crypsis**, that is, by imitating certain environmental background features. Crypsis involves at least shape, color, and color pattern. In some instances it probably also involves scent and sound matching although human sensory apparatus may not be equal to the task of discerning it. A wide variety of diverse unrelated insect groups often come to simulate the same inedible (to a carnivore) object—grass stems, tree bark, bird droppings (see Fig. 8-12), etc.

Even the most exact reproduction will be of little value in concealment without appropriate behavior, such as resting on the proper background in the right orientation and attitude, moving seldom and/or slowly and in such a manner as to attract as little attention as possible. Crypsis, properly considered, is not solely a matter of coloration or morphology but a type of behavior as well. For example, many moths have wing stripings resembling grooves in bark; those moths with

**Table 8-1** Visually mediated systems of passive defense and their correlates[a]

| Adaptation Category | Correlates | | | Predator Response |
|---|---|---|---|---|
| | Morphological | Behavioral | Populational | |
| I. Crypsis<br>Background imitation | Countershading; disruptive coloration; homochromy; dorsoventral flattening; body flanges; often possess secondary startle mechanisms | Capacity to remain immobile for prolonged periods; feed principally at night; proper orientation (often via dorsal or ventral light reactions); proper background selection; special resting attitudes; movement by dash and freeze or slow locomotion | Genetic plasticity and/or polymorphism for color forms; low population density; dispersed distribution | Predator fails to discriminate prey from substrate |
| Protective resemblance to inedible objects: | Appropriate shape, color, and color pattern; finely detailed patterns favored; often possess secondary startle mechanisms | Same as above, plus mimetic attitudes, movement by rocking or teetering | Dispersed distribution; low population density | Predator confuses prey with the inedible |

311

**Table 8-1** *(Continued)* Visually mediated systems of passive defense and their correlates[a]

| Adaptation Category | Correlates | | | |
| --- | --- | --- | --- | --- |
| | Morphological | Behavioral | Populational | Predator Response |
| II. Bizarre Forms | Markedly unusual in appearance, style, or general character; may possess secondary startle mechanisms | Often nonpredictable; unusual and/or startling behaviors favored | Dispersed distribution; low population density | Predator fails to recognize prey as food (or is startled into letting prey escape) |
| III. Simple Aposematism | Possession of weaponry (stings, bites, sprays, etc.); poison synthesis and/or utilization of host poisons acquired during feeding; conspicuous coloration and/or structure; bold vivid patterns favored; often yellow, orange, or red on black background | Usually diurnal; may aggregate; conspicuous behavior, warning displays | Clumped distribution; high population density | Predator learns to recognize prey as distasteful and/or dangerous |

IV. Mimicry

| | | | | |
|---|---|---|---|---|
| Müllerian mimicry: | Same as above; coloration and structure superficially very similar to others in the complex; usually aposematic | Conspicuous behavior very similar to others in the complex; may aggregate | Clumped distribution; high population density | Predator learns that a particular *Gestalt* is distasteful and/or dangerous |
| Batesian mimicry: | Conspicuous coloration and/or structure similar to model(s) | Similar to model(s) | Mimetic polymorphisms favored; dispersed distribution; low population density | Predator confuses prey with another it has learned to avoid |
| Wasmannian mimicry: | Similar to predatory host, at least in releasers; chemical, auditory, tactile mimicry also favored | Similar to the predatory host, which serves as both model and selective agent | Low population density; associated with social insects having large colonies at maturity | Predator allows prey to approach without molestation; predatory species is usually exploited and may become prey itself |

[a] Based in part on Robinson (1969a) and Rettenmeyer (1970).

**Figure 8-1**   A cryptic tropical stick insect (Phasmida) from Panama resting on a moss- and lichen-covered branch. The body and legs of the insect possess lichenose outgrowths which enable it to merge with its background. A notch on the inside of each front femur near the base permits the forelegs to be held together anteriorly and fit snugly around the head. (Courtesy of C. W. Rettenmeyer; see also Robinson, 1969b.)

vertical stripes rest on tree trunks with their heads pointed up or down, but horizontally striped moths orient themselves at right angles to the trunk.

To be maximally camouflaged, a cryptic individual has other problems. A major one is contour. Probably the most widely used solution is **disruptive coloration**, a visual breaking of the insect's outline so that parts of it appear to fade separately into the background (see Fig. 1-3). Effective even when used alone, such patterns are often combined with other cryptic features and help the animal blend in with its background (Fig. 8-3). A second way of minimizing clues as to contour involves actually or apparently reducing any telltale shadows. This may be accomplished through a dorsoventral flattening, as occurs in many bugs, often in combination with lateral flaps or various irregular body protuberances which bridge the gap between body and substrate (Fig. 8-4).

**Figure 8-2**   A well-camouflaged mosslike katydid resting on a mossy branch. (Courtesy of C. W. Rettenmeyer.)

Shadow may be minimized by proper body alignment also. (It has been suggested that many cryptic butterflies perch relative to the sun so that their wings throw the least shadow, but the observation is complicated by the fact that butterflies also control body heat by soaking up or avoiding sunlight.) Still another widely encountered method for eliminating shadows is **countershading**, that is, a compensatory deepening or lightening of body color to counteract for apparent color changes due to light intensity. Insects that have countershading invariably rest with the darker surface directed toward the light, a behavior critical to the success of countershading (Fig. 8-5).

Some insect species have the chameleonlike ability to change their color to match temporary backgrounds. Many locusts, by varying the quantities of orange, yellow, and black pigments they form, can adapt their color from dirty white to yellow, brown, or black. Thus, even adults kept for a few days on burnt ground can darken. Only when the hoppers are kept amid green vegetation and fed with abundant moist food in a very humid atmosphere do bright-green locust hoppers appear. Likewise, certain caterpillars are able to change color and markings to

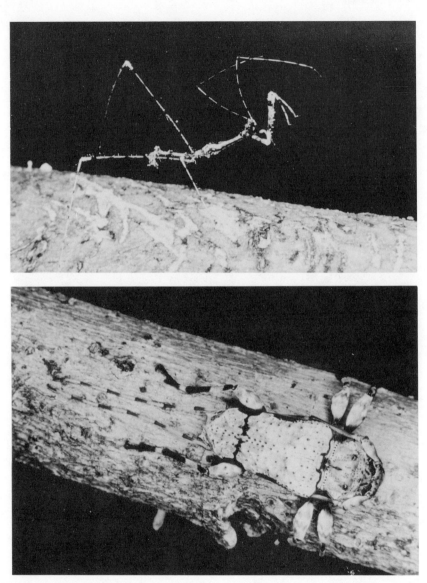

**Figure 8-3** Two unusual examples of disruptive coloration. *Above,* a threadlike reduviid bug (Emesiinae) from Panama that is so thin it "disappears" except when viewed from the side. The banded legs are an example of disruptive coloration, which further enhances its ability to merge with its background. (Courtesy of R. E. Silberglied.) *Below,* a resting Neotropical long-horned beetle (Cerambycidae) whose black- and white-banded antennae and hind legs could further serve to divert a predator by creating the illusion that the beetle is going in the opposite direction. (Courtesy of R. E. Silberglied; from Silberglied and Aiello, 1977.)

316

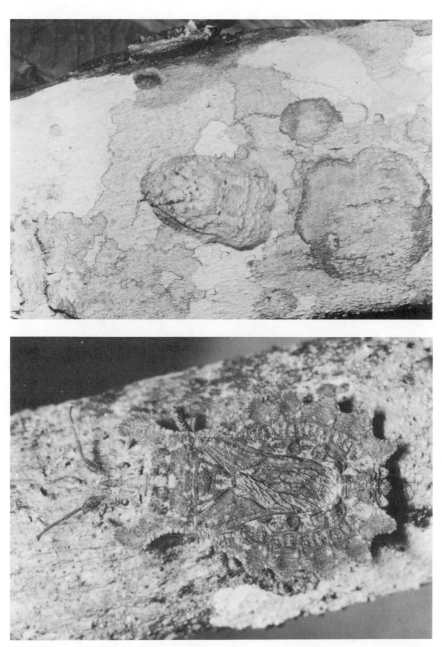

**Figure 8-4** Two examples of dorsoventrally flattened cryptic bugs. *Above,* dorsal view of a toadlike fulgorid with laterally extending wings which cause it to resemble a flat blotch on the tree bark. *Below,* an aradid bug with lateral projections upon its thorax and abdomen enabling it to blend into the bark upon which it rests. (Courtesy of C. W. Rettenmeyer.)

**Figure 8-5** Countershading. Above, a hornworm caterpillar (*Manduca*) in normal resting position. Nearly every green caterpillar which rests in the open has upper parts (in this position) that are a progressively deeper shade of green. This shading, which compensates exactly for the shadow on the lower parts, produces an apparent flattening. Below, when the caterpillar is inverted (or the light source reversed), the illusion is not only destroyed but the addition of countershading to natural shadow renders the caterpillar highly conspicuous. (From Cott, 1940; see also de Ruiter, 1955.)

match the background of their varied diet. Larvae of peppered moths, for example, are smooth and purplish brown with darker "lenticels" resembling a birch twig when feeding upon this host. On oak, however, they are brownish green with the different markings suitable to these trees. On oak covered with lichen, they become mottled in such a way as to mimic a lichen-covered twig!

How do insects "copy" their background in this manner? The most plausible hypothesis at present is that such insects are genetically polymorphic and that perception of the general pattern of the insect's surroundings provides a "switch mechanism" bringing into action the appropriate genetic color form. The pupa of the cabbage butterfly *Pieris*, which can adapt to its background by obscuring its basic green by white or black pigments in the cuticular surface layers, loses this adaptive ability if the caterpillar's eyes are cauterized before the chrysalis is formed (see Wigglesworth, 1964). A number of other examples of crypsis are covered in Cott (1940), Portmann (1959), and Edmunds (1974).

Genetic variations in cryptic coloration also occur. One of the most thoroughly studied cases of rapid directional selection in progress involves crypsis in a common English insect, the peppered moth.

CASE STUDY: MELANISM IN THE PEPPERED MOTH, *Biston betularia*

Peppered moths, like many others, fly at night and rest quietly by day on the sides of tree trunks. Naturalists of the nineteenth century knew them well and noted that they were usually found on lichen-covered trees and rocks, against which their pale mottled coloring made them practically invisible. Prior to 1845, all museum records of *B. betularia* were light-colored or "typical" specimens, but in that year, near the growing industrial center of Manchester, one black or "melanic" moth of that species was captured, so distinctive that it was given the name *B. betularia* f. *carbonaria* (Fig. 8-6). As the years passed, black individuals turned up with increasing frequency, mainly from the vicinities of industrial towns. Eventually, by the middle of the present century, the peppered moth population around Manchester came to consist of nearly 99% *carbonaria*.

Why were melanic moths becoming predominant? In 1952, Kettlewell undertook the first field experiments on the phenomenon. Smoke from the expanding industrialization of England had polluted the surrounding countryside, he noted, blackening rocks, ground, and even the tree trunks. It seemed plausible that upon this darkened background the dark moths were better protected from predators, particularly birds, but Kettlewell had little support for this hypothesis. Many entomologists and ornithologists vigorously objected that no one had ever seen any bird eating a *B. betularia* of any color, nor was there any published support for such predation on any of a variety of related cryptic moth species.

Thus, Kettlewell first needed to determine whether in fact birds actually ate peppered moths and whether they did so selectively. Into a

**Figure 8-6** The peppered moth, *Biston betularia,* and its black form *carbonaria* at rest on a lichen-covered tree trunk in a nonindustrial area of England. The typical light form illustrates the principle of disruptive coloration as it blends with the background. The dark form, *carbonaria,* however, rests in conspicuous contrast against the same background. Genetic studies have shown that the dark coloration is under control of a single dominant gene that arises spontaneously by mutation from the "typical" color gene. (Courtesy of H. B. D. Kettlewell.)

large outdoor aviary containing a nesting pair of great tits with young, he placed both light and dark tree trunks and boughs of equal surface area. into the aviary he released 10 moths—five melanic and five typical *B. betularia*. Based upon their appearance against these backgrounds at a distance of 2 yards, resting moths were classed as "conspicuous" or "inconspicuous".

For the first 2 hours of the first day, the birds failed to recognize the moths of either form as food. But in the next hour, all five of the "conspicuous" forms and two of the "inconspicuous" moths had been taken. In a repeat experiment the next day, Kettlewell released 18 moths; 16 were eaten in the first half hour, leaving only two "inconspicuous" survivors. Subsequent replications followed the same pattern, suggesting that the birds required a period of contact with the prey before recognition but that once recognition was established, all the moths were under risk of predation. Satisfyingly, the data seemed to indicate that the most conspicuous were preyed upon first.

Encouraged, Kettlewell turned to field experiments in two study sites, a polluted area and an uncontaminated woodland like that which must have prevailed 200 years before. Setting up camp, which included sheds to house some 3000 *B. betularia* pupae to raise for release, Kettlewell began round-the-clock studies. During the day, he released melanic and typical moths in known numbers, carefully marking each with a paint spot beneath the wings where it could not be seen by a predator, and watched wild bird behavior toward the moths. During the night he operated light traps, capturing the peppered moths of both forms attracted to the ultraviolet bulbs. (The use of such a method was essential because of the moths' crypticity. When Kettlewell attempted to visually discover and count his released moths again immediately after their release in an area where his continual presence precluded predation, over a third of the cryptic forms were already "missing," that is, so well hidden that they eluded even Kettlewell's experienced eye.)

With Tinbergen's help, Kettlewell watched bird predation on the moths, both with binoculars and with a movie camera from behind camouflaged blinds. In this way he was able to document at least five bird species actually selecting and eating the moths, which they do with such alacrity that it is surprising they were not previously observed. More important to his hypothesis, on the majority of occasions all the bird species at both locations took all of the more conspicuous moths before any of the inconspicuous phenotypes. Further proof of the selective advantage of having coloration appropriate to one's background was obtained by recapture of marked moths at night in the light traps. In unpolluted woodland, survival rate of light moths was about twice that of dark ones, while in soot-darkened woodland the ratio was reversed.

Do moths discriminate among a choice of possible resting substrates? As a pilot study, Kettlewell outfitted a barrel with a lining of alternate black and white stripes (of identical texture to eliminate tactile cues). Each night he released three moths of each form into the barrel; in the morning, he recorded their resting positions. Of 110 moths so tested, 65% chose the "correct" background. However, the artificiality of such tests disturbed Kettlewell. So he selected a number of lichen-covered tree trunks and then carefully removed the bryophytes from one side. Over this denuded half he painted a soot suspension and then placed the trunk upright and covered it with a muslin tent. At the midline of the two sides, he released equal numbers of females of each form and recorded their final resting position. Of 31 *carbonaria* individuals, all (100%) chose the black background; of the light forms, 75% selected the "correct" background. These highly significant results provide quite conclusive evidence that female *Biston betularia* are able to selectively choose their background with respect to their own coloration, probably through visual cues.

Several important additional implications arose from Kettlewell's studies. One major revelation was the speed with which evolution can occur at the level of the single gene. Geneticists had previously considered a selective advantage on the order of 0.1% as being "normal"—Kettlewell illustrated an advantage on the order of 50%! This selection, intense enough to cause a significant change in gene frequency within a single generation, when extended over tens of generations could result in a nearly complete reversal in the favored gene.

Now that English industries are beginning to control their soot emission, the trees are becoming lighter; in the last several years, Kettlewell has obtained evidence that the light phenotype of the peppered moth is now on the increase. It is too soon, however, to know whether a complete reversal will occur.

The now classic "*Biston* affair" and numerous other aspects of melanism are nicely summarized in Kettlewell's (1973) book, where references to the original papers may be found. Interested readers may also wish to refer to Kettlewell's earlier (1959, 1961) reviews of industrial melanism. Bishop and Cook (1975) consider recent developments.

**Aposematic Defenses: "I Am Dangerous"**

The Carolina locust, *Dissoteira carolina*, combines a vivid yellow and black coloring with a loud crackling sound. Orange lycid beetles and various other insects (Fig. 8-7) gather into large aggregations. Brightly colored bees and wasps sound a warning buzz sufficiently effective to be mimicked by a wide variety of nonrelatives. Caterpillars that possess stinging hairs often undulate conspicuously as they crawl about. Other insects swing rhythmically, produce rattling or rustling noises, or adopt various bizarre display postures.

With bright, contrasting colors and shapes and conspicuous behaviors, a wide variety of insects literally flaunt their presence in front of predators. Why? Simply, when one is poisonous, noxious tasting, or

**Figure 8-7** Orange and black coreid bug nymphs in a normal aggregation on a twig; such behavior serves to "pool" their aposematic warning signals. These bugs possess potent chemical repellents secreted from abdominal glands. (Courtesy of R. E. Silberglied.)

otherwise disagreeable to dangerous, it pays to advertise. Standing out brightly not only differentiates oneself clearly from cryptic palatable prey; it also may jolt potential predators out of any nondiscriminating "state of mind." In addition, the mere fact of being bright and different can afford some protection. Many birds, for example, have been shown to avoid novel food items (Coppinger, 1970). Such conspicuous coloration, structure (and in its broadest sense, behavior) indicating special capacities for defense is termed warning, or **aposematic**.

Aposematic coloration in insects tends to follow definite patterns. Most frequently, these tend to be combinations of red, yellow, or orange on a contrasting background, often black. The color pattern of a bumble bee is an example. Such patterns contrast sharply against backgrounds that are generally green, brown, gray, or blue.

It seems clear that aposematic coloration evolved in response to vertebrate predators, particularly birds. Insects perceive light at the red end of the spectrum poorly, if at all (see Chapter 6); red presumably appears gray or black to most and would therefore be of little significance as a signal to other insects. On the other hand, birds readily perceive bold patterns of red or orange. That the vivid patterns are so limited in variety suggests that inborn avoidance tendencies on the part of predators are being exploited. Limiting the number of patterns also

**Figure 8-8** Two examples of Batesian mimicry of social wasps. *Above*, a wasp-mimicking katydid about to jump from its perch. The wings of this Neotropical species are orange with black tips, resembling many "tarantula hawk" wasp species. *Below*, a diurnal Central American sessiid moth with reduced wing scales and body scales arranged in a wasplike pattern feeds while sitting conspicuously upon vegetation. (Courtesy of C. W. Retten-meyer.)

undoubtedly enhances the efficiency of the warning by limiting the amount of information which a predator must learn for accurate feeding judgments.

## Mimicry: "I Am Someone Else"

Cerambycid beetles resembling ants, bees, and wasps are common (Linsley, 1959). A number of insects effectively imitate social wasps (Fig. 8-8). Many distasteful lycid beetles are red and black (Linsley et al., 1961); so are a wide variety of other palatable insects, including members of nearly every insect order.

Insects do not imitate only environmental objects such as twigs, leaves, and thorns. They also copy each other. By imitating another individual which is unpalatable or dangerous (a **model**), individuals can gain an increased protection from predators which confuse them with the aposematic individual. **Mimicry** is the resemblance of one organism (the **mimic**), usually in color, pattern, form or behavior, to another organism (the model). It is worth noting that the resemblance of one organism to another can only be considered mimicry if both are found together. When similar forms are found in different areas, it is more likely to be an example of convergence.

Mimicry may be olfactory, tactile, or auditory, but visual mimicry has received the most attention and study. Appropriate behavior is necessary for an effectively functioning system, with mimics acting in ways that enhance their deception. Ant mimics, for example, must (and do) run alertly over the ground in the style of true ants.

There are several types of mimicry. When a *palatable* (unprotected) species evolves an appearance similar to that of an unpalatable, venomous, or otherwise protected one, the situation is known as **Batesian mimicry**, after H. W. Bates (1862) who first observed this phenomenon in the Amazon. Batesian mimicry is quite common, especially in butterflies (Fig. 8-9). Proving that one is actually dealing with a Batesian mimicry situation is often difficult, however. One good example is provided by the work of the Browers in Trinidad.

### CASE STUDY: ARTIFICIAL BATESIAN MIMICRY WITH *Hyalophora* MOTHS

In attempting to verify Batesian mimicry, a powerful argument would be to show that mimics are actually preyed upon less in natural situations than are otherwise identical non-mimetic prey. In the 1960s, Brower and colleagues (Brower et al., 1964, 1967) attempted to do just that.

DANAUS PLEXIPPUS                               LIMENITIS ARCHIPPUS

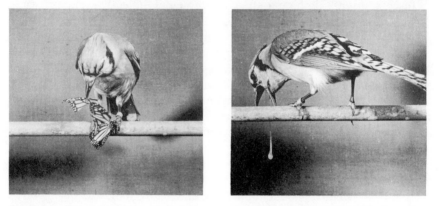

**Figure 8-9** Batesian mimicry: the bright orange monarch butterfly, *Danaus plexippus* (*top left*) and its mimic, the viceroy, *Limenitus archippus* (*top right*), which belongs to another family and differs in being slightly smaller and possessing an extra black line in the wings. When presented to a naive blue jay, the toxic monarch will be readily eaten (*bottom left*). However, ingestion is closely followed by vomiting (*bottom right*), and thereafter the blue jay will refuse both monarchs and viceroys. (Upper photographs by T. Eisner; lower photographs by L. P. Brower, from Brower, 1969.)

As artificial mimics, the Browers chose the diurnal silk moth, *Hyalophora promethea* (referred to in Chapter 5). This palatable moth did not normally occur in Trinidad, where their study was conducted, but was approximately the same size and shape as two native species of butterflies known to be unpalatable. By painting the wings, the Browers were able to produce moths with the same color pattern as the native species; using just black paint (which matched the natural background color) on others served as a control for the painting process.

After painting, the moths were released. Recapture for counting was easily accomplished by holding virgin female *Hyalophora* as attractant lures. However, the experimental results were surprising, for when averaged over several days, no difference was found in the percentage of mimic and control males returning. More mimics than controls returned the first few days, but later, fewer mimics than controls returned. Was mimicry advantageous or not?

Taking a different tack, the Browers painted other *Hyalophora* to be very conspicuous, unlike any other moth or butterfly in Trinidad. When compared with the controls, a significantly smaller percentage of these were recaptured, suggesting that the conspicuous experimental moths were more readily found by predators. Next, they repeated their first experiments and released large numbers of mimics and controls over several days in a single area; they also did a second experiment in which small numbers were released in many single areas. Recaptures of the first group paralleled the early results, with the altered "mimic" moths returning less frequently than controls. In the second group, however, the experimentally produced mimicry conferred a substantial advantage (Table 8-2).

Thus, the Browers' experimental results indicate that mimicry works only when the mimics are rare. A palatable, mimic tends to lose its protective advantage if it becomes too numerous, for predators discover the deception easily. Once this occurs, the predators soon learn to discriminate between the models and mimics. Then, because the brightly colored pseudoaposematic insects are much more easily spotted than the control individuals, they are at a net disadvantage.

More recently, the Browers' results have been reinterpreted, although their basic technique has been judged sound. Waldbauer and Sternburg (1975) contend that the control moths, though edible, were actually close mimics of another highly toxic native Trinidad butterfly, *Battus*. Furthermore, the conspicuously painted moths, rather than being unique, resembled several palatable Trinidad butterflies. Hence, results of the Brower group would be expected, for the controls would also have been protected by their similarity to the *Battus* butterflies. That is, the Browers actually compared artificial mimics of one model with mimics of another model.

Sternburg and colleagues (Sternburg et al., 1977) subsequently used *Hyalophora*, now called *Callosamia*, to repeat the experiment in Illinois. Here they painted their control moths with yellow stripes to resemble the edible, nonmimetic yellow form of the tiger swallowtail. The experimental group was painted black to resemble another toxic swallowtail, *Battus philenor*, a model for a widespread Batesian mimicry complex. They recaptured a significantly smaller proportion of the yellow-painted control moths, and those captured exhibited more wing

**Table 8-2**    Summary of the results of experiments with artificial mimicry in a natural environment[a]

|  | Control | Experimental |
| --- | --- | --- |
| **EXPERIMENT 1.** *Hylaphora* altered to mimic *Parides*. | | |
| Release | 51 | 52 |
| Recapture | 18 (35%) | 16 (31%) |
| **EXPERIMENT 2.** *Hylaphora* altered to mimic *Heliconius*. | | |
| Release | 162 | 156 |
| Recapture | 44 (27%) | 45 (29%) |
| **EXPERIMENT 3.** *Hylaphora* altered to be conspicuous and unique. | | |
| Release | 42 | 43 |
| Control | 16 (38%) | 8 (19%) |
| **EXPERIMENT 4-A.** *Hylaphora* altered to mimic *Parides* and released in high numbers at one locality. | | |
| Release | 414 | 414 |
| Recapture | 129 (31%) | 95 (23%) |
| **EXPERIMENT 4-B.** *Hylaphora* altered to mimic *Parides* and released at different times in many localities. | | |
| Release | 194 | 199 |
| Recapture | 37 (19%) | 53 (27%) |

[a] In experiments 1 and 2, *Hylaphora* was altered to mimic *Parides* and *Heliconius* butterflies, respectively, and no significant advantage was conferred to the mimics in either situation. In experiment 3, *H. promethea* moths were painted to be conspicuous and unique and the results indicated that such markings were disadvantageous to a palatable butterfly. In experiment 4, both groups were altered to resemble *Parides*; group A was released at high numbers at one locality with the same results as in experiments 1 and 2, whereas group B was released at different times in many localities. The results of the latter are consistent with the hypothesis that advantage is conferred only when mimics are less common than models. Based on Brower et al. (1964 and 1967).

injury attributable to greater predation by birds. Thus, their results demonstrate the selective advantage of one color pattern over another and support the Batesian mimicry hypothesis.

In natural situations Batesian mimic species are as a rule far less common than their models. However, occasionally a mimic *species* is quite numerous, and then an interesting phenomenon occurs. Consider, for example, the African swallowtail butterfly, *Papilio dardanus* (Fig. 8-10), the female of which exists in half a dozen different color forms. In

**Figure 8-10** Batesian mimicry in *Papilio dardanus*, showing the typical black and yellow nonmimetic form (*a*) and two of the approximately half a dozen mimic-model pairs which exist in varying proportions throughout the African swallowtail's range. The form *planemoides* (*b*) mimics the distasteful danaine model, *Amauris albimaculata* (*b*[1]); In Uganda this form comprises 20% of the females in the complex, but it is unknown in South Africa. The form *cenea* (*c*) mimics the distasteful danaine model, *Bematistes poggei* (*c*[1]); This form comprises 85% of the *P. dardanus* females in South Africa, but only 7% of the females in the Uganda complex. (Photography by P. M. Sheppard, from Wigglesworth, 1964; see also Ford, 1964.)

different regions of its range, different color assortments appear, and the balance of the percentage of the forms varies as well. In addition, other nonrelated butterflies in some of these areas produce mimics with the same coloration. The establishment of such **mimetic polymorphisms** is relatively easy to explain on theoretical grounds. Because palatable mimics must be less common than their models, a mimetic form can increase in abundance only up to a certain point before losing its advantage. Strong pressure would exist toward development, within part of the population, of mutants resembling other distasteful species. These mimics, too, would enjoy an advantage only in population numbers related to the local abundance of their model. In this way, a complex "balanced polymorphism" would be established.

A high population of Batesian mimics is not the only mechanism which may lead to polymorphism in a prey species. When a food type is encountered very commonly, many predators, particularly birds, quickly develop a strong search image; creatures fitting this image are fed upon disproportionately. This phenomenon by predators results in what has been termed **apostatic selection**, a situation in which the relative rate of predation rises more rapidly than does the relative rate of encounter. Such frequency-dependent selection can help in the maintenance of several morphological forms ("morphs") in a single species.

Many aposematically colored insects look superficially quite alike. Consider the various bees and wasps with their bands of yellow and black or the so-called tiger stripe butterfly complex of the Neotropics (Fig. 8-11). Such a case, in which a group of relatively unrelated insects, *all* distasteful or otherwise protected, come to mimic one another, has been termed **Müllerian mimicry** after Müller (1879) who first noted its occurrence, also in Brazil. The advantages of such a system are obvious. A predator may need to make several trials before he learns to recognize which color patterns are to be avoided. If the number of such patterns is very large, the possibilities for confusion and error are compounded. If a group of aposematic species converge in appearance, all of them will benefit, sharing the number of attacks needed in the learning process and reducing the confusion and learning time of predators. Müllerian mimicry, therefore, runs counter to Batesian mimicry on an important point. While polymorphism is favored in Batesian mimicry, it runs contrary to the advantages of the Müllerian system.

Batesian and Müllerian mimicry are probably labels for extreme cases in what is essentially a mimicry continuum. Within a mimicry complex, for example, some members may be more palatable than others (see Fig. 8-11). Whether these less protected individuals should be considered Müllerian mimics for which selection for bad taste has been relaxed, or Batesian mimics which have subsequently evolved (or are in the process

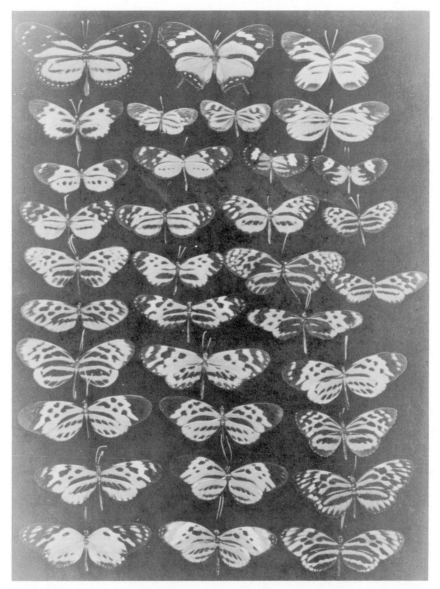

**Figure 8-11** The tiger stripe butterfly complex of the Neotropics is characterized by yellow, brown, black and orange-striped butterflies which generally fly between 2 and 7 meters above the forest floor in the same habitat. Most members of the complex belong to the family Ithomiidae, but members of the families Danaidae (first specimen, *top left*), Pieridae (specimens 2, 3, and 7), and Nymphalidae (fifth specimen) are also involved as well as a day-flying moth (seventh specimen) and several species of *Heliconius* (specimens 30–34). These species have converged in appearance to form a Müllerian mimetic complex in which there is a spectrum of unpalatability, with some species considerably more palatable than others. (From Papageorgis, 1975. Reprinted by permission, *American Scientist*, Journal of Sigma Xi, The Scientific Research Society of North America.)

331

of evolving) a bad taste, is a nearly impossible question to resolve. As another example, there are certain mimetic cerambycid beetles which often feed upon the distasteful lycid beetles that serve as their models. Such mimics might be alternatively Batesian or Müllerian, depending upon how recently they had fed upon a lycid (Eisner et al., 1962).

Other mimicry variations also occur. For example, monarchs (see Fig. 8-9) have been mentioned as a classic Batesian model. However, not all monarchs are equally disagreeable. Various food plants are utilized in different parts of the butterfly's range, and palatability depends on the amount of cardenolide supplied to the butterflies by the particular species of milkweed they ate as larvae. In a sense, then, Brower has argued, those individuals who are nonprotected chemically from predation are acting like mimics of their fellow species mates who are distasteful. Thus, since one species can include both mimics and models, recognition of a new category of mimicry becomes necessary. **Automimicry**, based on a palatability dimorphism, is a special case of Batesian mimicry within a species (Brower, 1969).

For many insects (and other arthropods as well), changes in shape, size, and behavior during growth and metamorphosis are common. Different instars may imitate entirely different models, in what may be regarded as the "I Led Three Lives" syndrome, or **transformational mimicry** (Fig. 8-12).

Beetle and other guests of ants (Chapter 4 and Figs. 4-16, 4-17) sometimes come to greatly resemble their hosts; mimetic resemblance which facilitates cohabitation with a mimic's host, its model, has been termed **Wasmannian mimicry**, (Rettenmeyer, 1970). Wickler (1968) and Rettenmeyer (1970) cover insect mimicry in depth. Aggressive mimicry is discussed in Chapters 4 and 6.

### Passive and Systemic Chemical Defenses: 'I Am Not Tasty"

Many interactions of insects with potential predators involve chemical agents, called allomones (see Chapter 5). The study of these interactions and the chemicals forming their basis is the subject of the rapidly expanding field of chemical ecology. Allelochemic defenses are widespread, but it is among the invertebrates, especially the arthropods, that chemical defense seems to have reached its peak of diversification.

Chemical defenses require an energy expenditure and therefore tend to be conservative in nature. The energetically least expensive strategy would seem to be to "borrow" deterrents present in one's host plants and utilize them against predators. Monarch butterflies normally sequester cardiac glycosides from their hosts which are potent emetic agents to

**Figure 8-12** Transformational mimicry in larvae of the eastern tiger swallowtail butterfly, *Papilio glaucus*. First instar larvae resemble amorphous bird droppings (*left*) and rest in full view at the center of a leaf. Notice the small nibble missing from the tip of the leaf; these larave feed only at night, when their diurnally active predators have ceased hunting. Older larvae occupy a similar position but have grown too large to effectively resemble bird droppings; colored bright green and possessing eyespots, they perform a striking snakelike display (*right*) when disturbed. A third line of defense is shown in Fig. 8-23. (Photographs by the authors; for a similar example in spiders, see Reiskind, 1970.)

birds (see Brower, 1969, and Fig. 8-9). However, such dependence on a plant's toxic principle may be an evolutionary constraint in its own right. For example, when Brower succeeded in raising monarchs on cabbage instead of milkweed, they lacked chemical protection.

An alternative is to synthesize one's own antipredator chemicals. This allows more evolutionary freedom. For example, aquatic Hemiptera differ from their terrestrial relatives in producing aromatic instead of straight-chain hydrocarbon defensive compounds. The apparent significance of this difference is that the aromatic compounds are much more repellent to fish (Hepburn et al., 1973).

Behaviorally, the use of systemic chemical defenses covers a range that overlaps at the active end with actual forms of attack. Like plant chemical defenses, some insect defenses are strictly passive: parts of the insect must be eaten before an effect is felt. However, an insect normally can ill afford to allow body parts to be ingested before a predator learns of its mistake. An individual facing predation would have an obvious advantage were it immediately sufficiently bad tasting or bad feeling for the predator to reject it relatively unharmed. Some distasteful insects have cuticular outgrowths such as hollow, brittle spines which are easily broken by predators. Many caterpillars have developed a coat of long hairs that render them unacceptable to most predatory birds, but some species gain still further protection by the inclusion of **urticating hairs**, delicate hollow organs each containing a minute quantity of irritating venom secreted by a gland at their base. When brushed against, these fragile hairs are broken and act like tiny poisoned barbs.

Another way to maximize the chance of escaping predation is to provide relatively nonharmful ways for predators to taste one's toxic blood. Distasteful lycid beetles are particularly prone to bleed from the wings, which possess swollen, easily ruptured veins. Other insects actually have active control over the release of their blood when under attack (Fig. 8-13). Such **reflex bleeding**, often occurring from leg joints, is highly developed in the Mexican bean beetle (Fig. 8-14), which may even rotate its leg around to bring the blood-laden knee in closest contact to the point of stimulation.

**Figure 8-13** A day-flying moth (a member of a large Müllerian mimicry complex of Neotropical ithomiine butterflies) bleeds reflexively and foams the blood with air from its spiracles when gripped with forceps. (Courtesy of R. E. Silberglied.)

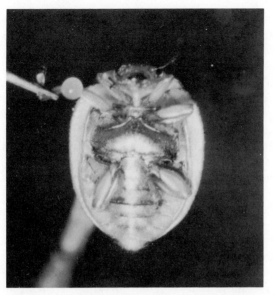

**Figure 8-14** The Mexican bean beetle, *Epilachna varivestis*, emitting a droplet of blood from the tibiofemoral joint of a leg that is being pinched in forceps; notice that a localized stimulus elicts a response only from the nearest leg. The blood, in addition to being distasteful to predators, clots so quickly that it gums up the mouthparts and appendages of predators. Attacking ants may become stuck together in groups if they contact one another after being contaminated with the bean beetle blood. (Courtesy of T. Eisner; from Eisner, 1970.)

When handled or disturbed, a wide variety of arthropods respond by regurgitating or defecating (see Fig. 5-9). Often very effective as predator repellents, such **enteric discharges** might almost be considered a form of short-range attack. The familiar frothy "tobacco juice" regurgitate of some grasshoppers, for example, is quite toxic to mammals. It is a topical irritant to eyes, may induce vomiting when swallowed, and may cause severe symptoms and even death when injected. When tethered grasshoppers placed beside ant colonies are induced to regurgitate, the fluid causes their assailants instantaneously to disperse and begin intense cleaning movements. When pieces of cut-up grasshopper, some treated with regurgitate, are placed along ant trails, only the untreated pieces are carried away by foragers (Eisner, 1970).

**Startle: "I Am Not What You Thought!"**

Disguise is a useful device against enemies only to a point. Once an insect is discovered, a second line of defense is clearly advantageous.

Systemic chemical defenses provide an essentially passive second line. In this section and the next we consider some more active approaches.

One common method of achieving a startle effect is through **flash coloration**, the sudden exposure of previously hidden colors and/or patterns when threatened. **Eye spots** (Fig. 8-15), commonly found on the wings of Lepidoptera and occasionally among many other insects, are a classic example. The sudden bright display of eyelike markings by a previously cryptic individual is often combined with other behaviors that heighten the startle effect (Fig. 8-16; see also Fig. 8-18 and Blest, 1957a). The more plastic the eye markings appear (by arrangement of colors in concentric rings and/or by rhythmic shaking of the wings), the greater the startling effect upon birds. That eye spots are mimicking owl eyes has been suggested on the grounds that insectivorous birds may have an innate fear of these stimuli, owls being *their* predators.

Mimicry of eyes of other predators may also occur. Three species of *Caligo* (the so-called "owl butterfly") in Trinidad have enormous eye spots (15–20 mm in diameter) on their hindwing undersides; *Eryphanis* has smaller (6–7 mm), more irregular markings. At rest, all four species hold their wings vertically so that only one side is visible to an observer at a time, but the eye spots are continuously displayed. Stradling (1976) has suggested that *Caligo* are mimicking *Hyla* tree frogs (Fig. 8-17), while *Eryphanis* are mimicking *Anolis* lizards, two vertebrate predators common in the resting microhabitat of the butterflies. To birds viewing

**Figure 8-15** Diagram illustrating the inherent conspicuousness of an eye spot, which attracts attention to itself in preference to a variety of other, and even larger, objects in the visual field (see also Fig. 8-18). (From Cott, 1940.)

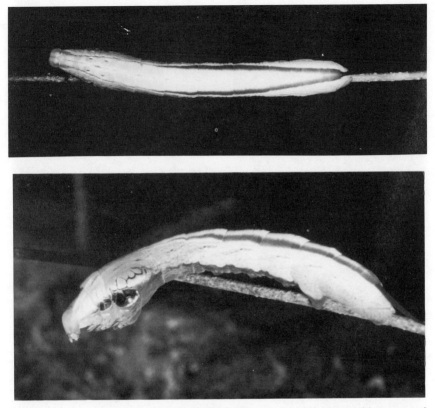

**Figure 8-16** A Neotropical sphinx moth caterpillar in normal resting position on a twig (*above*). When disturbed, it suddenly swells the anterior end of the body revealing previously hidden eye spots (*below*) which cause it to resemble a small snake. Its habit of repeatedly "striking" at the intruder behaviorally complements the startle effect. Note how the thoracic legs are held well forward to further enhance the reptilian appearance. (Courtesy of C. W. Rettenmeyer.)

the butterflies from a distance, the eye spots do not interfere unduly with the overall cryptic pattern. However, the lizards, which have limited binocular vision, maintain territories based on body size; seeing *Eryphanis* at close range as larger rivals of their own species, the lizards avoid them. At the same time, the larger *Caligo* resemble full-grown tree frogs which are quite capable of eating lizards, and thus the lizards avoid them as well.

Both fright-inducing flashed eye spots and continuously displayed mimetic ones may also have another important function, namely, the

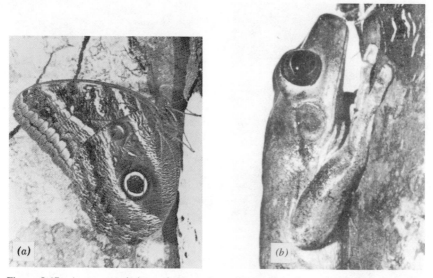

**Figure 8-17** Apparent mimicry of arboreal carnivores by a large Neotropical butterfly. The butterfly, *Caligo* (*a*), with a wing span of 12–15 cm, rests by day on tree trunks. Its mimetic pattern on the undersides of the wings includes not only eyes but head profile, pectoral region, and amphibian tympanum. The resemblence to the profile of the tree frog, *Hyla crepitans* (*b*), a widespread predator that also spends the day resting on tree trunks is striking. (From Stradling, 1976.)

detraction of a predator's attention from more vulnerable body parts. Attacking the eye spot, a predator gains only a piece of wing for its efforts. Many of the small lycaenid butterflies commonly known as "blues," for example, have eye spots on the underside of the hind wings and, close to them, long wing "tails" which somewhat resemble antennae. Kept in motion by the resting butterfly, it has been suggested that they help draw attack away from the true head, where a wound would be fatal. Correspondingly, any scheme of camouflage must also give the conspicuousness of eyes special consideration. Masking or modifying the appearance of the eyes has taken many forms, perhaps the most widespread of which are disruptive eye masks, which operate on the principle that if an eye can be made to appear in another shape, then it will cease to resemble an eye (see Fig. 8-3). A number of other insects such as some fulgorid bugs (Fig. 8-18) have complete false heads, which may serve to misdirect a predator's attack.

Flash coloration is not limited to eye spots. The sluggish Australian katydid *Acripeza reticulata* lives in leaf litter and probably escapes

**Figure 8-18** A *Lanternaria* bug (Fulgoridae) from Panama has a bizarre hollow sham "head" with alligatorlike markings which cause it to resemble a small carnivore (*top*). If this first line of defense fails, it opens its wings to reveal vivid eye spots on the hind wings (*bottom*). (Courtesy of R. E. Silberglied.)

predation because of its grayish, bumpy exterior. When disturbed, females raise their wings to reveal a vivid abdominal color pattern of reds, blues, and black. Sometimes, flash coloration works its effect in reverse, serving to make normal camouflage more effective through sudden contrasts. Thus, for example, many grasshoppers, cryptically

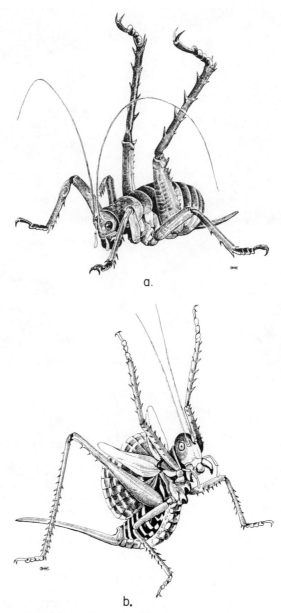

a.

b.

**Figure 8-19** Secondary defense display in the orthopterans (*a*) *Deinacrida heteracantha* from New Zealand and (*b*) *Neobarettia spinosa* from Mexico in response to attacks initiated by predators which have discovered these normally cryptic species. (Courtesy of D. Otte; from Otte, 1977.)

colored to match the stony or sandy soil when they settle, have hindwings of brilliant red, blue, or yellow. Highly conspicuous in flight, they conceal the hindwings so abruptly upon landing that they appear literally to vanish.

A wide variety of other startling display behaviors have also evolved in otherwise cryptic insects. Certain katydids perform stereotyped defensive displays when discovered (Fig. 8-19). Some moths react like whirling dervishes, dramatically flapping their wings about and rocking from side to side, thus warming up their flight muscles while temporarily startling their predator. In this way they can often escape before the predator is able to react.

The strategy of startling a potential predator includes still another widespread behavior. As a fleeing escapee might slip free from a jacket grasped by his assailant, so many insects elude potential predators by leaving behind various dispensable body coverings. Perhaps the most familiar of these are the easily shed scales which cover the wings of butterflies and moths. The loose hairs of adult caddisflies, the waxy powder on aleurodid "white flies," and the scales of Thysanura and Collembola may be similar protections against entrapment. Legs of crane flies, which readily break off, a phenomenon known as **autotomy** (see Fig. 8-29J), represent similar dispensable body parts. An unusual defensive use of detachable barbed setae is employed by certain carpet beetle larvae (Dermestidae). These insects can hopelessly entangle small predators such as ants with these interlocking hairs borne in prominent tufts on their abdomens (Nutting and Spangler, 1969).

Detachable coverings need not be strictly morphological. Some insects utilize artificial shields for protection. Gathering on their back a variety of debris, sometimes even including the sucked-out remains of prey, these larvae construct trash packet shields (Fig. 8-20). Other species are able to maneuver their shield with considerable agility against predators (Fig. 8-21). Certain chrysomelid larvae carry a case into which they retreat when disturbed (Wallace, 1970).

### Attack: "Turning The Tables"

Many adaptations which appear to have originally evolved for feeding have come to form part of the insect defense arsenal—powerful mouthparts, leg spines, raptorial forelegs, and a whole host of other such morphological features. These are often used highly effectively. But perhaps nowhere has the combination of feeding and defense been more highly refined than in the development and use of insect poisons. The best example occurs among the aculeate Hymenoptera, where glands

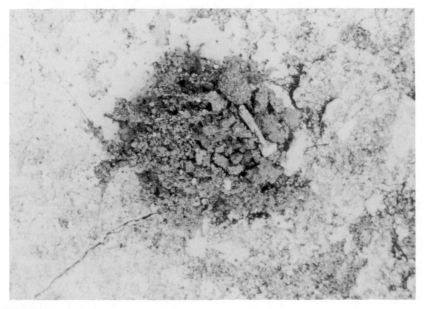

**Figure 8-20**  A predator on other insects, this African assassin bug nymph disguises and/or defends itself from its own enemies by carrying a packet of trash over its body leaving only the antennae visible. Similar trash packets are carried by larvae of lacewings (Chrysopidae). (Courtesy of R. E. Silberglied.)

associated with the egg-laying apparatus secrete venom and the ovipositor itself has been transformed into a sting. Males are incapable of stinging, although some have a well-developed spine (pseudosting) at the end of the abdomen which is realistically thrust at molesters and may sometimes even draw blood (Fig. 8-22). Solitary wasps (Pompilidae, Sphecidae, and Eumenidae) primarily use their venom to paralyze prey, but among the social bees, wasps, and ants the sting and venom have come to be retained solely for defense. Occasionally such stings can cause violent and severe reactions in man, an example of **anaphylaxis**, that is, a generalized reaction to foreign proteins to which the body has become sensitized.

Swallowtail and parnassian butterfly caterpillars suddenly evert a two-pronged **osmeterium** (Fig. 8-23) secreting intensely odorous butyric acid derivatives. Hemiptera such as *Rhopalus* collect secretions in depressions on the body wall; when bothered, they dip their legs into the accumulated chemical, then wipe them on the enemy target. Most nonstinging arthropods discharge their chemical defenses only in response to direct contact stimulation, but exceptions occur.

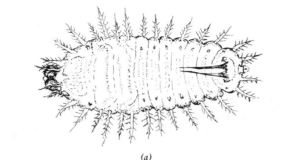

(a)

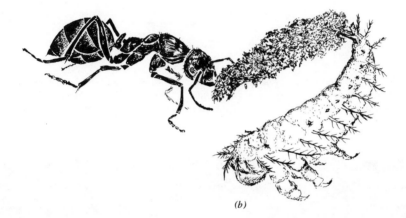

(b)

**Figure 8-21** The chrysomelid beetle larva, *Cassida rubiginosa,* forms its shield of cast skins and dried feces on a fork projecting from its hind end and held over its back (*a*). As fresh, wet faces are added near the base by means of telescoping anal abdominal segments, they add to the shields effectiveness. Through branched spine "sensors" around the periphery of its body, the larva ably responds to a predator's probing by interposing the shield between itself and its enemy (*b*), often in such a way that fecal material is smeared upon the offender. Ants, among the larva's chief enemies, are quickly repelled. Even those casually encountering the pasty material immediately flee and clean themselves. (Drawings by J. W. Krispyn from photographs by the authors; see also Eisner et al., 1967.)

Spraying is perhaps the most common method of applying noninjected secretions, and it is often accomplished with a high degree of accuracy and sometimes over a considerable distance. One of the most striking is *Anisomorpha buprestoides*, the large two-striped walking stick of the southeastern United States (Fig. 8-24). The predaceous reduviid bug, *Platymeris rhadamantus*, sprays its saliva directionally several feet in response to predator attack. The toxic fluid, ordinarily used to kill prey, is similar to cobra venom (Edwards, 1961). Some of the commonest ant

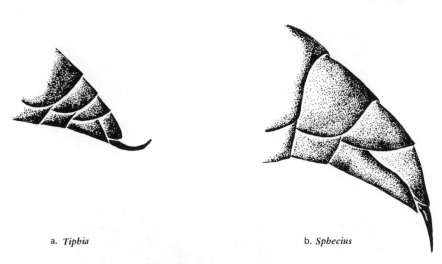

a. *Tiphia*                    b. *Sphecius*

**Figure 8-22** Pseudostings projecting from the abdomens of two unrelated male wasps, *Tiphia* sp. (Tiphiidae) (*a*), and the cicada killer, *Sphecius speciosus* (Sphecidae) (*b*). In both species the stinglike spine is formed by the last abdominal sternite. (Drawing by J. W. Krispyn.)

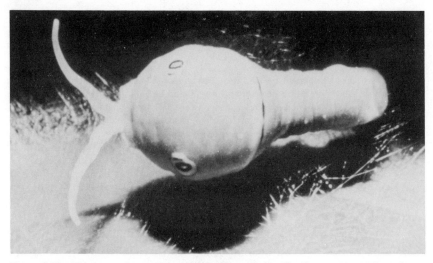

**Figure 8-23** The mature larva of the tiger swallowtail, *Papilio glaucus*, responds to danger by abruptly everting its forked defensive gland, the osmeterium, from beneath its neck integument. The side nearest the danger tends to be extruded furthest; in some cases, the caterpillar may even arch its body and wipe the horns directly upon the offender. The osmeterium, which contains butyric acid, is commonly colored bright yellow or red. Thus, the defense is probably a combination of a visual threat and a repellent odor. (Photograph by the authors; see also Fig. 8-12.)

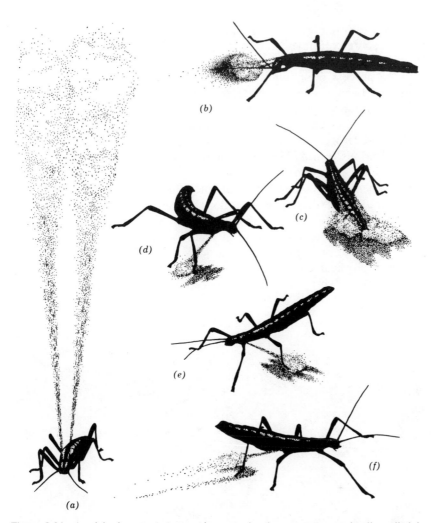

**Figure 8-24** A stick, insect, *Anisomorpha,* spraying in response to stimuli applied in various ways. Bilateral discharges are elicited by tapping the dorsal thorax (*a*), touching both antennae with a heated probe (*b*), or pinching rear of abdomen (*c*). Unilateral discharges are induced by pinching the right foreleg (*d*), the left middle leg (*e*), or the right hind leg (*f*). This insect is exceptional in that it will spray approaching birds from a distance, whereas most nonstinging insects require direct contact stimulation before discharging their chemical defenses. (Courtesy of T. Eisner, from Eisner, 1965. Copyright © 1965 by the American Association for the Advancement of Science.)

species use formic acid as a poison spray and thus have been used as a natural source of this acid from Roman times until quite recently; stereotyped postures control direction (Fig. 8-25).

In the process of spraying an arthropod inevitably contaminates itself with a least some of the chemical; as a result, after each discharge it may enjoy a period of partial or complete invulnerability to further attack. Such protection can be extremely important, particularly in defense against predators such as ants which may appear *en masse* (Eisner, 1970). Various morphological adaptations prolong the effectiveness of such residual secretions (Fig. 8-26).

Like pheromones (see Chapter 5), allomones also tend to be blends of compounds. The defensive secretions of some pentatomid bugs, for example, include as many as 18 components. The functional significance of such mixtures appears to lie in their increased chance of repelling diverse sorts of predators, combined with the ability of one chemical to alter the physical–chemical properties of another. For example, the whipscorpion's spray is 84% acetic acid, (hence the animal's common

**Figure 8-25** A carpenter ant, *Formica integer,* responding to an alien stands on its "tiptoes," bends its abdomen under its body, and sprays formic acid at the intruder. Formic acid is a potent irritant and effectively repels a variety of potential ant predators. It also serves as an alarm pheromone, alerting nestmates to the source of disturbance. (Courtesy of A. Hefetz.)

**Figure 8-26** The pupal stage of a *Chrysomela* leaf beetle which hangs fully exposed to predators on the undersides of the host plant. Persistent odor of the larval defensive secretion impregnates the shed skin, which remains attached (arrow) and acts as a potent repellent to foraging ants. A second larva preparing to pupate is also present. (Photograph by the authors; see also Wallace and Blum, 1969.)

name, vinagaroon); another 5% is caprylic acid, a penetrant that greatly enhances the effectiveness of the acetic acid (Eisner, 1970).

The active principles of most defensive secretions are highly volatile substances of low molecular weight, usually strongly odorous and irritating and in some cases even painful upon inhalation. Often, they are present in very high concentrations which apparently cannot be tolerated systemically even by the insects that produce them. How do these arthropods withstand their own discharges? Externally, immunity may be gained through possession of an especially impervious integument. The cuticle of Hemiptera with defensive glands, for example, is gener-

**Figure 8-27**  Bombardier beetle discharging its hot quinone secretion which early collec-

ally impermeable to hemipteran secretion unless abrasion takes place. The only Hemiptera with permeable cuticle found so far are species that apparently lack defensive glands or freshly molted individuals with epicuticles not yet fully formed (Remold, 1962, cited in Eisner, 1970).

Being membranous saclike invaginations of the body wall, the storage glands are lined internally with a cuticular membrane. Thus, the secretions do not come in contact with living tissue during storage. But how can the glands *produce* the poisons without poisoning themselves? One hypothesis is that many arthropod gland cells possess certain more or less elaborate cuticular chambers and ducts; synthesis of toxicants occurs within the lumen of these cuticular organelles, not in the cytoplasm of the living gland cells associated with them. Such a situation has been shown in certain tenebrionid beetles. However, other arthropods have glands constructed in such a way that chemical precursors of the secretion are mixed only at the moment of discharge. Such **reactor glands** allow the production of extremely toxic substances. The millipede *Apheloria corrugata,* for example, can generate about 0.6 mg hydrogen cyanide, several times the lethal dose for a mouse (Eisner et al., 1967). Likewise, bombardier beetles mix hydroquinones and hydrogen peroxide from one glandular compartment with appropriate enzymes from another to generate benzoquinone-containing secretions, which the accompanying release of free oxygen blasts out of the glands at temperatures up to 100°C (Fig. 8-27).

Not all defensive secretions are purely chemical in effect. Many arthropods manufacture sticky or slimy materials which mechanically hinder predators. Lithobiid centipedes, for example, discharge sticky threads from their posterior legs, entangling ants, lycosid spiders, and other potential enemies. From their pointed cephalic nozzles *Nasutitermes* soldiers (Fig. 8-28) eject a sticky, resinuous terpene-containing secretion which dries quickly in the air. Mechanically incapacitating or even killing insect predators, it also acts as an alarm substance inducing other termites to converge upon the site (Moore, 1968).

---

tors had described as "burning the flesh to such a degree that only a few could be captured with the naked hand." The abdominal tip can be accurately aimed to spray an attacker, here mimicked by grasping the left foreleg with forceps. (*b*) Dorsal view of the abdomen with the elytra partially cut away to reveal the pair of glands. Each consists of a glandular tissue (G) drained by a highly coiled duct (unraveled on one side) whose secretion (an aqueous solution of hydroquinones and hydrogen peroxide) is stored in a large cuticle-lined reservoir (R). Upon disturbance, the muscle (M) at the mouth of the reservoir contracts allowing some of the secretion to flow into the vestibule (V) which contains enzymes which trigger the almost instantaneous and explosive reaction. (Photo by T. Eisner and D. Aneshansley; drawing by D. Alsop and T. Eisner. See Aneshansley et al., 1969.)

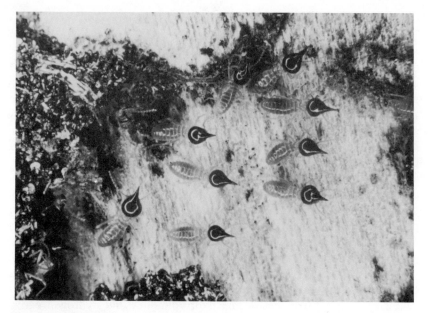

**Figure 8-28**  Soldiers of *Nasutitermes* sp. termites orienting to a predator. They possess a hollow "squirt gun" nozzle on the front of the head from which they shoot threads of a sticky secretion produced in their frontal glands. Their mandibles are degenerate and nonfunctional. Attacking arthropods find the tacky thread almost impossible to remove; it is also a potent irritant. (Courtesy of R. E. Silberglied; see Moore, 1968, and Eisner et al., 1976.)

## GROUP ACTIONS FOR DEFENSE

Some nasute termites make daylight forays above ground in true military fashion; columns of workers are guarded against arthropod enemy trespass by flanking rows of soldiers oriented with their spraying snouts pointed outward. If its nest is tapped, the mound-building ant, *Formica rufa*, rushes out and hundreds of workers eject acid spray toward the sound of the disturbance (see Fig. 8-25). Some male and female walking sticks pool their defensive resources even when not mating; pair formation may occur as early as the nymphal stages, the male riding atop the female from then on.

Not surprisingly, defense through group action is particularly well developed in gregarious and social insects (see Figs. 5-9, 6-11, 8-28, and 10-14). The use of chemical communicating substances for warning purposes (alarm pheromones and/or allomones) is widespread (see Chapter 5) and especially well developed among ants. When such a substance is of sufficient quantity, it triggers rapid dispersal of nearby individuals. Highly volatile, it fades rapidly unless reinforced, thereby minimizing overreaction.

For some species which regularly aggregate for various purposes, group "panic" may be a good defense strategy. Humphries and Driver (1967, 1970) term the unpredictable random flight of fleeing animals **protean displays** and suggest that such unsystematic behavior may more effectively stymy a predator than would an orderly retreat. Many predators must fixate upon one individual to successfully attack. When erratic movements are undertaken by a number of closely spaced individuals, such fixation is rendered quite difficult. Thus, for example, when disturbed, swarms of a wide variety of insects will literally seem to "explode" into a mass of erratically swirling individuals (see Fig. 6-10). At the same time, such a defensive behavior probably also has elements of a mass startle effect.

**Mobbing**, a type of group defense in which a predator is harrassed by a number of maneuverable prey individuals, has long been known among colonially nesting birds. A variety of social and/or communally-living insects have also developed this effective behavior (see Fig. 10-25). *Bembix* sand wasps, for example, nest in aggregations; when a potential predator approaches, a score of male wasps may fly up at it, buzzing ominously and loudly but harmlessly. When mobbing, the prey animals may approach, threaten, or occasionally physically attack a predator. Often, though the predator is not actually injured, it moves off. This is probably not only due to the harassment but also to the fact that its probability of successfully capturing a prey is very low amidst all the commotion.

## THE PREDATOR–PREY RELATIONSHIP: SOME THEORETICAL CONSIDERATIONS ON DEFENSE

Insects are often their own worst enemies. Aphids are preyed upon by a wide variety of insect enemies from adult pemphredonine wasps to coccinellid beetle larvae, syrphid flies, and lacewings. Yet these small, soft-bodied creatures are far from helpless (see Fig. 5-12). Solitary wasps and their prey play games of cat and mouse, and the prey does not always lose (Fig. 8-29). A formicine ant stumbling into a foreign colony is quickly hit with venom spray; the victim responds with much grooming and rubbing of its mouthparts on the ground. The nests of certain tropical social wasps are suspended on a single stalk, which is continually treated with secretions from the wasps' abdominal glands

**Figure 8-29** Interactions of the solitary sphecid wasp *Liris nigra* with its cricket prey. Broad arrows indicate possible order of phases of behavior. Movements of wasp and prey are indicated by narrow dashed and solid arrows, respectively. Initially, the wasp tracks its prey by scent (*a*) or sights it visually (*f*). The wasp attacks (*b*), but the cricket jumps away or walks away (*g*). The wasp seizes one cricket leg and attempts to sting, but the cricket escapes by dropping its leg (autotomy, *j*) or by kicking the wasp away (*i*). When pursuit is temporarily disrupted (*d*), the wasp stops and assumes the alert "head up" position from which she scans the immediate environment. At (*h*) the wasp attacks prey in its burrow. (From Steiner, 1968.)

(Fig. 8-30). Ants are repelled from walking upon stalks so treated but otherwise enter and destroy developing brood within the nest.

Protective adaptations in insects are intimately related to the behavior and physiology of their predators, a basic concept which provides insights into a wide range of phenomena. Relatively a great deal of

**Figure 8-30** The nest of a tropical social wasp, *Polistes* sp., is attached to the end of a long pedicel the opposite end of which forms the only point of direct access to the nest from the substrate. The pedicel appears highly vanished due to the wasp's continual application of sternal gland secretion to it. Of almost universal occurrence in social Vespidae, the function of the sternal glands has been experimentally demonstrated only in a few species. (Courtesy of R. L. Jeanne; see Jeanne 1970.)

attention in the past has focused upon insects as prey of vertebrates, but defense against invertebrates has often tended to be overlooked or its importance unappreciated by investigators. Likewise, few have remembered that insects are also continuously under seige by less conspicuous but nonetheless significant microorganisms, such as bacteria, fungi, viruses, nematodes, and parasites. In these contexts, allomones may have long-term antagonistic effects. Such antibiotic effects generally have received little research attention.

At the same time, certain aspects unique to vertebrate predation have had an important role in shaping arthropod defenses. For example, vertebrate predators form search images of specific prey. Vertebrates are also noted for their ability to be conditioned to avoid unpleasant stimuli. Furthermore, most tend to avoid stimuli similar to, but not identical with, the original conditioning stimulus, a learning phenomenon called **stimulus generalization**. Experimental studies have revealed that under certain conditions, stimulus generalization provides advantage to a wide gamut of prey even in cases where they may appear only vaguely alike to our perception. It is precisely these aspects of vertebrate prey recognition which are commonly exploited in insect mimicry (see Table 8-1). Batesian and Müllerian mimicry alike depend upon stimulus generalization. On the other side of the coin, mimetic polymorphism provides a method of circumventing the establishment of a specific search image.

The potential predator as an evolutionary force has important implications in every form of defense. For example, in considering color crypsis, the optic system of the predator must be considered. We have already mentioned the very different spectral and focal world of the insect and the vertebrate (Chapter 6). In auditory communication (Chapter 7) we considered that many protest sounds might actually be vibrational signals. Consider also for a moment various systemic chemical defenses the evolutionary and ecological implications of which have been examined by Eisner (1970). For example, the blood and tissues of meloid beetles contain cantharadin, long known as Spanish Fly and traditionally, though incorrectly, reported to have aphrodisiac properties; actually, the chemical is capable of inducing severe systemic effects when ingested. It is present in such quantities in some species that swallowing a single beetle could cause acute toxicity in some vertebrates (Carrel and Eisner, 1974).

How, asks Eisner, can a systemic poison be adaptive? There appears little chance for the predator to associate the ill effects with the causative act if noxious effects are delayed, as they often are. It appears of little warning value, for the carrier must actually be ingested before the poison exerts its effect. In addition, since such toxins require energy in their production or incorporation, their development would seem to be disadvantageous.

Often the question cannot be answered. Again, lethal dosages depend on the predator involved, a fact often unknown. Theoretically, however, it is also plausible that through time with the use of lethal poisons a predator might be transformed into a species from which the tendency to capture the lethal prey had been selected out entirely. (The extent to which such evolutionary pressures might have influenced the development of restricted feeding habits is unknown, but this hypothesis may provide some interesting new insights into that subject.) In still other cases, even if predatory organisms neither die nor discriminate against the causative agent, the substance may impair vigor or fecundity; any substance which does this to the predators of an area is advantageous to the species that produces the poison.

How can the sacrifice of ingested individuals be justified in adaptive terms? The answer may lie within the concept of altruism (see Chapter 10), in which the surviving insects are such close kin to those sacrificed that their survival is nearly equivalent genetically to the survival of the sacrificed individual itself. Support may be found in the observation that many poisonous arthropods form dense, highly localized, and often quite static aggregations, conditions that would be expected to favor a high degree of inbreeding. Behaviors such as egg placement in one or a few concentrated masses (a common oviposition pattern in butterflies) also increase the likelihood of genetic relatedness of adjacent individuals.

Crypsis and its opposite, aposematism, are considerably different defensive strategies. For saturniid moths, striking life history adaptations exist which correlate to the apparent defensive strategy. For example, unpalatable aposematic species tend to have a longer life span following reproduction than do cryptic, presumably palatable species. The apparent explanation is that even after reproduction distasteful moths can confer some protection on their siblings by serving as a predator's learning experience. In contrast, the greatest benefit a cryptic species could confer to its relatives would be to remove itself from the population soon after breeding so as not to chance that a predator might develop a search image for its species (Blest, 1963).

## SUMMARY

For organisms as small as most insects, discretion is the better part of valor, and avoidance of actual conflict is the preferred strategy. Not surprisingly, passive modes of protection comprise the greater part of the insect defense arsenal. However, a second line of defense is clearly advantageous, and usually occurs as well.

Insect defenses may be classified according to the prey-to-predator message implicit in a particular behavior or structural adaptation. Crypsis, or camouflage, is the imitation of certain environmental features. Melanism in the peppered moth illustrates the degree of protection which crypticity can afford and the speed with which its evolution can occur.

Aposematic coloration or structure conspicuously warns of special capacities for defense. Combinations of red, yellow, or orange on black are most common. Vertebrates, especially birds, appear to be the primary targets of the warning.

Mimicry is the adaptive resemblance of one organism (the mimic) to another organism (the model). The traits of the model that confer some benefit to the mimic may include unpalatability, efficient physical and chemical defenses, and morphology (including color patterns). Mimicry occurs in several types. Aggressive mimicry involves a resemblance which allows the mimic to approach its prey, the model. Wasmannian mimicry involves resemblances which facilitate cohabitation with a mimic's host, its model. Batesian mimicry results from the resemblance of a normally unprotected mimic to a protected model. Müllerian mimicry results when both model and mimic are unpalatable, so that their conspicuous coloration and pattern serve mutually as a learning stimulus to predators. Automimicry is a special type of Batesian mimicry which occurs within a single species when some members are unpalatable while others are not. Batesian and Müllerian mimicry, while distinct, are not mutually exclusive. However, they differ in that for a predator the existence of Batesian mimicry complexes is detrimental, because they result in avoidance of "good" food. Müllerian mimicry complexes are advantageous to the predator, since they result in pursuit of fewer injurious or unpalatable items.

Passive and systemic chemical defenses are widespread and diverse. Some insects sequester their allelochemic substances directly from their host, while others synthesize their own antipredator chemicals. Urticating hairs, reflex bleeding, and enteric discharges are common methods of providing predators with a systemic chemical "sample"; behaviorally, the use of allomones covers a range that overlaps at the active end with actual forms of attack.

Startle includes such methods as flash coloration, eye spots, sudden erratic movements and the shedding of various detachable coverings and/or body parts; all are based on the strategy of surprising a predator sufficiently to allow time for escape.

Attack involves becoming sufficiently aggressive to turn the tables on a potential predator. Many of the adaptations used appear to have evolved originally for feeding or other purposes. Blends of chemical compounds are common weapons, usually either injected or applied by spraying or daubing.

Defense through group action is particularly well developed in gregarious and social insects. Commonly, it involves alarm pheromones and/or allomones. Protean display, or group "panic," is another useful strategy, as is mobbing.

The protective adaptations of insects are intimately related to the behavior and physiology of their enemies, which include not only vertebrates and other invertebrates but also a host of microorganisms. The concept of altruism has applications to the adaptive justification of systemic poisons and an explanation of differential life spans in cryptic and aposematic insects.

For additional information about insect defensive behaviors, the following are particularly useful: books by Blum (1978), Cott (1940), Edmunds (1974), Ford (1964), Portmann (1959), and Wickler (1968); and reviews by Blum (1971), Brower (1971), Duffey (1977), Eisner (1970, 1972), Eisner and Meinwald (1966), Rettenmeyer (1970), Robinson (1969a), Roth and Eisner (1962), and Weatherston and Percy (1970). Brower's (1968) film provides a better appreciation for the intricacies of antipredator behaviors than words can convey.

## SELECTED REFERENCES

Aneshansley, D., T. Eisner, J. M. Widom, and B. Widom. 1969. Biochemistry at 100°C: the explosive discharge of bombardier beetles (*Brachinus*). *Science* **165**: 61–63.

Bates, H. W. 1862. Contributions to the insect fauna of Amazon Valley. *Trans. Linn. Soc.* (London) **23**: 495–566.

Bishop, J. A. and L. M. Cook. 1975. Moths, melanism and clean air. *Sci. Amer.* **232**: 90–99 (January).

Blest, A. D. 1957. The function of eye spot patterns in Lepidoptera. *Behaviour* **11**: 209–226.

Blest, A. D. 1963. Longevity, palatability and natural selection in five species of New World saturniid moth. *Nature* **197**: 1183–1186.

Blum, M. S. 1971. Arthropod defensive secretions. In *Chemical Releasers in Insects. Pesticide Chemistry*, Vol. 3, A. S. Tahori (ed.). Gordon and Breach, New York, pp. 163–176.

Blum, M. S. 1978. *Chemical Defenses of Arthropods*. Academic Press, New York, in press.

Brower, L. P. 1968. *Patterns for Survival: A Study of Mimicry and Protective Coloration in Tropical Insects*. Amherst College, Amherst, Mass., 16-mm color film, 26.5 minutes.

Brower, L. P. 1969. Ecological Chemistry. *Sci. Amer.* **220**: 22–29 (February).

Brower, L. P. 1971. Prey coloration and predator behavior. In *Topics in the Study of Life: The BIO Source Book*, Section 6, *Animal Behavior*, Harper and Row, New York, pp. 66–76.

Brower, L. P., J. V. Z. Brower, F. G. Stiles, H. J. Croze, and A. S. Hower. 1964. Mimicry: differential advantage of color patterns in the natural environment. *Science* **144**: 183–185.

Brower, L. P., L. M. Cook, and H. J. Croze. 1967. Predator responses to artificial Batesian mimics released in a Neotropical environment. *Evolution* **21**: 11–23.

Carrel, J. E. and T. Eisner. 1974. Cantharidin: potent feeding deterrent to insects. *Science* **183**: 755–757.

Coppinger, R. P. 1970. The effect of experience and novelty on avian feeding behavior with reference to the evolution of warning coloration in butterflies. II. Reactions of naive birds to novel insects. *Amer. Nat.* **104:** 323–335.

Cott, H. B. 1940. *Adaptive Coloration in Animals.* Methuen, London, 508 pp.

Duffey, S. S. 1977. Arthropod allomones: chemical effronteries and antagonists. Proc. XV Internat. Congr. Entomol., Washington, D.C., pp. 323–394.

Edmunds, M. 1974. *Defense in Animals. A Survey of Anti-Predator Defenses.* Longman, New York, 358 pp.

Edwards, J. S. 1961. The action and composition of the saliva of an assassin bug, *Platymeris rhadamanthus* Gaerst. (Hemiptera, Reduviidae). *J. Exp. Biol.* **38:** 61–77.

Eisner, T. 1965. Defensive spray of a phasmid insect. *Science* **148:** 966–968.

Eisner, H. E., D. W. Alsop, and T. Eisner. 1967. Defense mechanisms of arthropods. XX. Quantitative assessment of hydrogen cyanide production in two species of millipedes. *Psyche* **74:** 107–117.

Eisner, T. 1970. Chemical defense against predation in arthropods. In *Chemical Ecology,* E. Sondheimer and J. B. Simeone (eds.). Academic Press, New York, pp. 157–215.

Eisner, T. 1972. Chemical ecology: on arthropods and how they live as chemists. *Verhandl. Deutsch. Zool. Gesellsch.* **65:** 123–137.

Eisner, T. and J. Meinwald. 1966. Defensive secretions of arthropods. *Science* **153:** 1341–1350.

Eisner, T., F. C. Kafatos, and E. G. Linsley. 1962. Lycid predation by mimetic adult Cerambycidae (Coleoptera). *Evolution* **16:** 316–324.

Eisner, T., E. van Tassell, and J. E. Carrel. 1967. Defensive use of a "fecal shield" by a beetle larvae. *Science* **158:** 1471–1473.

Eisner, T., I. Kriston, and D. J. Aneshansley. 1976. Defensive behavior of a termite *(Nasutitermes exitiosus). Behav. Ecol. Sociobiol.* **1:** 83–125.

Ford, E. B. 1964. *Ecological Genetics.* 2nd ed., Methuen, London, 355 pp.

Hepburn, H. R., M. J. Berman, H. J. Jacobson, and L. P. Fatti. 1973. Trends in arthropod defensive secretions and aquatic predator assay. *Oecologia* **12:** 373–382.

Humphries, D. A. and P. M. Driver. 1967. Erratic display as a device against predators. *Science* **156:** 1767–1768.

Humphries, D. A. and P. M. Driver. 1970. Protean defence by prey animals. *Oecologia* **5:** 285–302.

Jeanne, R. L. 1970. Chemical defense of brood by a social wasp. *Science* **168:** 1465–1466.

Kettlewell, H. B. D. 1959. Darwin's missing evidence. *Sci. Amer.* **200:** 48–53 (March).

Kettlewell, H. B. D. 1961. The phenomenon of industrial melanism in Lepidoptera. *Annu. Rev. Entomol.* **6:** 245–262.

Kettlewell, H. B. D. 1973. *The Evolution of Melanism. The Study of a Recurring Necessity. With Special Reference to Industrial Melanism in the Lepidoptera.* Clarendon, New York, 424 pp.

Linsley, E. G. 1959. Mimetic form and coloration in the Cerambycidae (Coleoptera). *Ann. Entomol. Soc. Amer.* **52:** 125–131.

Linsley, E. G., T. Eisner, and A. B. Klots. 1961. Mimetic assemblages of sibling species of lycid beetles. *Evolution* **15:** 15–29.

Moore, B. P. 1968. Studies on the chemical composition and function of the cephalic gland secretion in Australian termites. *J. Insect Physiol.* **14:** 33–39.

Müller, J. F. T. 1879. *Ituna* and *Thyridia*: a remarkable case of mimicry in butterflies. *Trans. Entomol. Soc., (Proc.)* **1879**: 20–28.

Nutting, W. L. and H. G. Spangler. 1969. The hastate setae of certain dermestid larvae: an entangling defense mechanism. *Ann. Entomol. Soc. Amer.* **62**: 763–769.

Otte, D. 1977. Acoustical communication in Orthoptera. In *How Animals Communicate,* Sebeok, T. A. (ed.). Indiana Univ. Press, Bloomington, 1344 pp.

Papageorgis, C. 1975. Mimicry in Neotropical butterflies. *Amer. Sci.* **63**: 522–532.

Portmann, A. 1959. *Animal Camouflage.* Ann Arbor Science Library, Univ. Michigan Press, Ann Arbor, 111 pp.

Reiskind, J. 1970. Multiple mimetic forms in an ant-mimicking clubionid spider. *Science* **169**: 587–588.

Rettenmeyer, C. W. 1970. Insect mimicry. *Annu. Rev. Entomol.* **15**: 43–74.

Robinson, M. H. 1969a. Defenses against visually hunting predators. In *Evolutionary Biology,* Vol. 3, T. Dobzhansky, M. K. Hecht, and W. C. Steere (eds.). pp. 225–259.

Robinson, M. H. 1969b. The defensive behaviour of some orthopteroid insects from Panama. *Trans. Roy. Entomol. Soc. Lond.* **121**: 281–303.

Roth, L. M. and T. Eisner. 1962. Chemical defenses of arthropods. *Annu. Rev. Entomol.* **7**: 107–136.

Rothschild, M. 1972. Colour and poisons in insect protection. *New Scientist:* 318–320 (May 11).

Ruiter, L. de. 1955. Countershading in caterpillars. *Arch. Neerl. Zool.* **11**: 1–57.

Silberglied, R. E. and A. Aiello. 1977. Defensive adaptations of some Neotropical long-horned beetles (Coleoptera: Cerambycidae): antennal spines, tergiversation and double mimicry. *Psyche* **83**: 256–262.

Steiner, A. L. 1968. Behavioral interactions between *Liris nigra* Van Der Linden (Hymenoptera: Sphecidae) and *Gryllus domesticus* L. (Orthoptera: Gryllidae). *Psyche* **75**: 256–273.

Sternberg, J. G., G. P. Waldbauer, and M. R. Jeffords. 1977. Batesian mimicry: selective advantage of color pattern. *Science* **195**: 681–683.

Stradling, D. J. 1976. The nature of the mimetic patterns of the brassolid genera, *Caligo* and *Eryphanis. Ecol. Entomol.* **1**: 135–138.

Waldbauer, G. P. and J. G. Sternberg. 1975. Saturniid moths as mimics: an alternative interpretation of attempts to demonstrate mimetic advantage in nature. *Evolution* **29**: 650–658.

Wallace, J. B. 1970. The defensive function of a case on a chrysomelid larva. *J. Ga. Entomol. Soc.* **5**: 19–24.

Wallace, J. B. and M. S. Blum. 1969. Refined defensive mechanisms in *Chrysomela scripta. Ann. Entomol. Soc. Amer.* **62**: 503–506.

Weatherston, J. and J. E. Percy. 1970. Arthropod defensive secretions. In *Chemicals Controlling Insect Behavior,* M. Beroza (ed.). Academic Press, New York, pp. 95–144.

Wickler, W. 1968. *Mimicry in Plants and Animals.* World University Library, McGraw-Hill, New York, 255 pp.

Wigglesworth, V. B. 1964. *The Life of Insects.* World, Cleveland, 360 pp.

# 9

# Reproductive Behavior

"Lovebugs Menace Florida's Tourists" a recent newspaper headline declared. How could such a poetically named insect constitute a "menace"? In the first place, the lovebug is not a true bug but a fly. Unlike its more notorious relatives, it does not bite or transmit diseases, nor is it a pest at picnic outings. Simply stated, the problem is sheer numbers of individuals with a propensity for prolonged aerial copulation—hence the name, lovebug (Fig. 9-1). Their peculiar attraction to odors associated with automobiles further promotes their image (Callahan and Denmark, 1973). To drive through a swarm of lovebugs is an unforgettable experience, as any north Florida native or tourist in May or September will attest. The state legislature has become sufficiently alarmed by the semiannual lovebug invasion that entomologists have been hired whose specific assignment is to study the reproductive behavior of the little flies.

Elsewhere we have mentioned several examples of sexual behavior. Among night-flying moths (Chapter 5) we examined sexual signaling through long-distance olfactory cues. In the dragonflies and fireflies (Fig. 9-4 and Chapter 6) we viewed two primarily visually mediated systems. The use of sequential cues from a number of sensory modalities, characteristic of the majority of courtship systems, was illustrated by the Douglas fir bark beetle (Chapter 5) with its interplay of pheromonal and acoutic signals and by butterflies (Chapter 6) whose courtship typically combines tactile, visual, and olfactory stimuli.

These examples left several questions unanswered, however. Why are the courtship rituals of some species so intricate, even bizarre, while many others mate with apparently only minimal preliminaries? What are the factors which influence the choice of a mate? Why are some species territorial? Why do individuals of some species mate but once and others promiscuously? To answer, let us briefly explore the spectrum of arthropod reproductive behavior and then focus on the sorts of selective pressures which may result in a particular insect courtship pattern. A vast literature covers various aspects of insect courtship; a good general review is provided by Engelmann (1970).

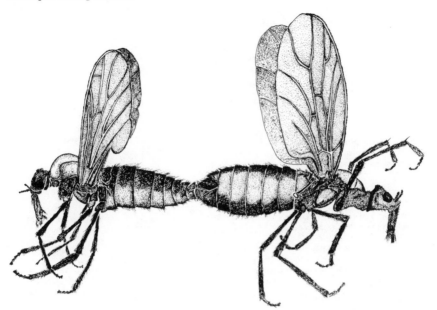

**Figure 9-1** Lovebugs, *Plecia nearctica* (Diptera: Bibionidae), in typical copulatory position (female on the right). In the laboratory, such pairs copulate for an average of 56 hours! (Drawing by J. W. Krispyn; see Thornhill, 1976c.)

## ARTHROPOD MATING BEHAVIOR

Mating behavior includes all those events surrounding fertilization or insemination (the meeting of sperm and eggs outside or inside the female's body, respectively). Among the arthropods as a whole, direct coupling, or copulation, is a comparatively recent development (Fig. 9-2); primitive aquatic groups undoubtedly practiced external fertilization during oviposition, in a manner analogous to some fish, and all primitively aquatic arthropod groups still include some members with this mating type. All terrestrial and secondarily aquatic arthropods, however, practice internal fertilization, although even after emergence to land many of the more primitive practiced only indirect transfer of sperm. In many cases, males left **spermatophores** (sperm transfer containers) attached to the female or to the substrate where females were attracted to them. In other cases, however, the spermatophores were introduced into the female's gonopore by various methods. In general, the greater the population density of a species, the less intimate the contact between sex partners needs be. Sometimes, in species living in

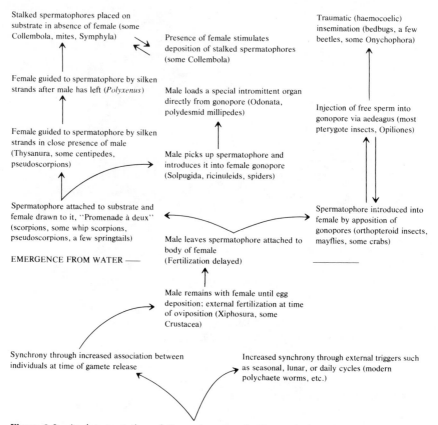

**Figure 9-2** An interpretation of the major steps in the evolution of mating behavior of arthropods. (Compiled from various sources, especially Alexander, 1964.)

very humid habitats, an almost complete dissociation of individual males and females occurs, each sex reacting to the spermatophore itself rather than to each other. Collembola provide good examples of this syndrome of primitive mating habits. No doubt, spermatophores evolved independently many times, for their form and mechanism of dehiscence varies widely (Fig. 9-3); most terrestrial arthropods have some sort of spermatophore or traces of its existence even today. For a review of indirect sperm transfer in soil arthropods, see Schaller (1971).

Nearly all of the higher insects (Pterygota) copulate in the true sense of the word. Variation in the manner in which this is accomplished, however, resembles an insectan Kama Sutra; more positions are known here than in all other Arthropoda combined (Alexander, 1964). Appar-

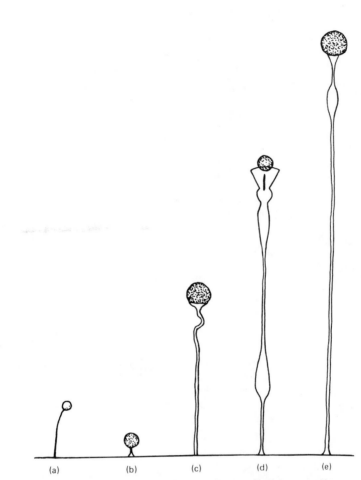

**Figure 9-3** Some examples of arthropod spermatophores, all drawn to the same scale: (*a*) Collembola (*Orchesella* sp.); (*b*) Diplura (*Campodea* sp.); (*c*) Oribatida (*Belba* sp.); (d, e). Pseudoscorpionida (*Chthonius* sp., *Obisium* sp.). In each, males first stick a rapidly hardening secretion to the substrate with their genital opening, then taper the secretion into a stalk by raising the body; the sperm droplet placed at the top becomes spherical because of its viscosity. Hooks, levers, and other adornments on the spermatophores of some, especially scorpions, apparently facilitate recognition, orientation, and uptake by females. Within the female, the sperm packet swells by absorption of fluid and bursts to liberate the activated sperm. (From Schaller, 1971. Reproduced, with permission, from the Annual Review of Entomology, volume 16. Copyright 1971 by Annual Reviews Inc. All rights reserved.)

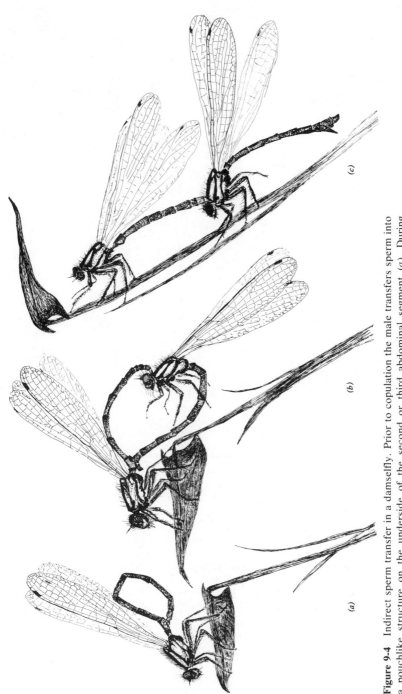

(c)

(b)

(a)

**Figure 9-4** Indirect sperm transfer in a damselfly. Prior to copulation the male transfers sperm into a pouchlike structure on the underside of the second or third abdominal segment (a). During copulation, the female bends the tip of her abdomen forward to contact the male's accessory genitalia. The male (dorsal) uses claspers at his abdominal tip to grasp the female's prothorax, thus completing the "wheel position" characteristic of this group (b). Pairs of damselflies or dragonflies often fly about for hours in tandem, the male holding the female by his anal claspers (c); in some species the female oviposits during tandem flight. (Drawing by J. W. Krispyn.)

364

ently, however, most can be derived from a primitive female-above position.

A modified type of indirect sperm transfer is found in the Odonata (Fig. 9-4) and in spiders. Males of more advanced insects such as grasshoppers and butterflies place their spermatophores directly into the female genital opening. Direct insemination involving a penis occurs in some members of the Hemiptera and in most holometabolous insects except Lepidoptera. Bedbugs have a rather bizarre form of copulation, in which sperm is injected subcutaneously into the female's blood-filled body cavity, a behavior popularly termed "traumatic insemination."

## Function and Complexity

Two closely related sympatric species of *Melittobia* wasps (see Fig. 9-21) differ strikingly in the occurrence and sequence of different elements of their courtship. In *M. chalybii,* males pump the female's antennae slowly but continuously and only periodically lift their middle legs; complete courtship requires an average of 10 minutes. Males of morphologically similar species "A" pump the female's antennae in rapid alternation with an up-and-down movement of the *hind* pair of legs; courtship duration for this species is slightly less than 2 minutes. The two ecological homologues may parasitize the same host in a single back yard, but the different courtship patterns will not permit them to cross-mate (Evans and Matthews, 1976).

An attractive analogy states that insect courtship operates like a ratchet regulated by innate releasing mechanisms or physiological filters; at each stage, some signal on the part of one participant elicits a response in the other, which in turn elicits a new action by the initiator, and so on. Thus, the pair clicks along toward the inevitable consummatory act while the ratchet mechanism precludes inappropriate matings or steps being performed out of sequence. Reproductive isolation is generally assumed to be a principal function of such reaction chains. But does courtship actually fill this function? To answer such a question, one must turn to analytical analyses of individuals reared in isolation in order to exclude learning or imprinting. One investigation which has elucidated such reaction chains in particularly fine detail is that of Stich (1963) on the crane fly, *Tipula oleracea.*

CASE STUDY: COURTSHIP IN THE CRANE FLY, *Tipula oleracea*

Crane flies, which look rather like delicate, fragile, oversized mosquitoes, include a great number of species, generally similar in appearance and tending to occupy the same sorts of habitats. While the males mate repeatedly, females typically mate but once. Under such conditions, a complex series of reciprocal signals would be predicted; in what specific ways might this be accomplished?

Stich raised crane flies in the laboratory, keeping them isolated from one another as adults. When the adults were five days old, he began mating trials. They mated readily, in a very quick series of actions and reactions (Fig. 9-5) taking as little as 15 seconds and usually lasting but 90 seconds. Though the female often began the courtship by touching a male's long threadlike legs, it was apparent that the male took the most active role in courtship as a whole. True to the ratchet analogy, if Stich interrupted the courtship pattern by removing the female, reintroducing her never allowed the pattern to resume. Each time, the males began courtship anew from the beginning.

The first detectable step of the crane fly courtship occurs when two individuals happen to contact each other's legs; a male's response is to grab the leg with his own. Stich began amputating legs before introducing live females to males; six, five, four, three, two, or one leg—it made no difference to sexually active males. Only when a female had no legs at

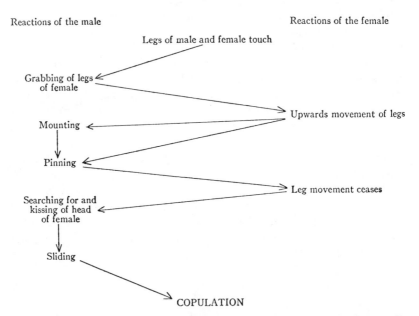

**Figure 9-5**  Stages of a normal courtship sequence in the crane fly, *Tipula oleracea* (From Stich, 1963.)

all did males fail to respond. Stich picked up amputated legs in his tweezers and offered them. Males immediately grabbed them. Was any discrimination taking place at this step at all? Apparently not in relation to sex, because dead males or their legs evoked the same grabbing as females did. In fact, any leglike tactile stimulus seemed sufficient; males even readily grabbed paraffin-stiffened threads dangled near them. But when stiffened threads of different thicknesses were presented, those of the "wrong" diameter were immediately released. Stich repeated the experiment again, this time using legs of different diameters from other species of *Tipula*. Male *T. oleracea* grabbed at them all but immediately let them go, a discrimination that may help to ensure that the potential mate belongs to the proper species.

Receptive females respond to the male's grasp by raising one or more of their legs upward, whereupon the male assumes a mounting position above the female's body. What is the functional significance of this? Stich found that males refused to mount a dead female with artificially stiffened legs but would mount a variety of movable models, even a simple 1 mm long wire if it were fashioned into a pair of movable legs. When he caged two males together, they grabbed each other whenever their legs touched; in over 100 trials, however, neither ever responded by upward leg movement nor did they ever exhibit a mounting reaction.

Next, a male crane fly tries to pin down the female's raised legs. Sometimes, this is accomplished in less than 3 seconds; at other times, the female may resist, and the ensuing struggle over leg position can extend over a period of 3 minutes. On the hunch that this great variability might be associated with sexual receptivity, Stich presented males with previously mated females; these readily elicited the males' grabbing and mounting reactions, but all their attempts at pinning down the females' legs failed. Sexually unreceptive females kept at least one leg continually raised, eventually causing the male to leave. Thus, previously mated or otherwise unreceptive females effectively terminate courtship attempts at this point.

After successfully pinning a female so that she remains motionless, a male begins to "search" for her head, by licking her with his mouthparts while he moves progressively forward over her body. When he reaches her head, he "kisses" the back of it and stops his search. To determine the stimuli for this particular element in the courtship sequence, Stich exposed males to variously altered females (Fig. 9-6). Male reactions to these models clearly indicated that both the female's body and head finally were necessary as a stimulus at this point in the courtship.

The final step in the courtship sequence occurs when the male slowly steps backward, sliding his abdomen back over the female's, and bends

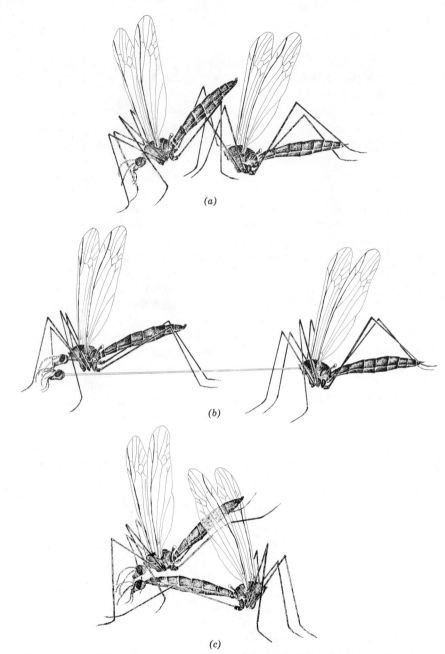

**Figure 9-6**  Searching reactions of a male crane fly toward altered females. (*a*) Decapitated female; searching extends beyond the body until male contacts substrate. (*b*) Female head glued onto paper strip; male uses strip as guide to reach head; percent completion declines as strip length increases. (*c*) Female head attached to end of abdomen, model presented so that male and female heads point in same direction; male orients toward head, attempts copulation through neck cavity. (Drawings by J. W. Krispyn after Stich, 1963.)

his abdomen in such a way that the male genitalia come into contact with the female's. Does the female head, through the kissing reaction, serve as the releaser for this behavior, or does the abdomen function as the sign stimulus? Male responses to reversed-head females (Fig. 9-6c) showed clearly that the head, not the abdomen, was the sign stimulus.

Thus, crane fly courtship facilitates decision making with respect to sex, species, and physiological condition of the participants. At each stage of the courtship series, different behaviors and different body parts are employed as specific signaling devices in a touch-oriented chain ideally suited for such weak-flying, drab denizens of the woods, and courtship proceeds like a rigid sequence of filters excluding all but conspecific sexually active males and receptive unfertilized females. (For another detailed analysis of a courtship reaction chain, see Fig. 6-12.)

Insect courtship displays have at least three common functions: to stimulate and maneuver females into copulation, to facilitate the meeting of solitary individuals, and to facilitate species and sex identification (but see also p. 394). Mating among normally predatory species entails an additional problem, namely, potential mates may be mistaken for food. For insects such as preying mantises, robber flies, and scorpion-flies, for example, a principal function of courtship is **appeasement**, that is, inhibition of the normal predatory instincts of the participants, especially the female. In spiders, where the danger of becoming a copulatory snack is accentuated because males tend to be much smaller than their mates, appeasement has been accomplished in a great variety of fashions (Fig. 9-7). Among mantids, courtship by the male has sometimes been described as a "sneak attack"; often the female devours her mate while copulating (Fig. 9-8).

**Courtship** (or **nuptial**) **feeding**, where the male provides a food "gift" to his "bride" (Fig. 9-9), has evolved independently in diverse predatory groups and appears to be another mechanism for appeasement. However, other explanations for courtship feeding are also possible. One is that males with prey are more conspicuous to females (see Fig. 9-15). Females probably also enjoy a selective advantage by consuming prey at a point where they need extra protein and calories to convert into eggs; male fitness is also increased by providing such prey, since the success of his genetic contribution to the next generation is intimately tied to the survival and reproductive success of each of his mates (Thornhill, 1976a).

Observations of nuptial feeding have so far been largely descriptive rather than the experimental studies that would shed real light on this phenomenon. However, it is worth noting that a nonpredatory bug,

*(a)*

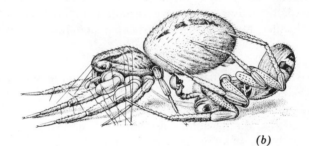

*(b)*

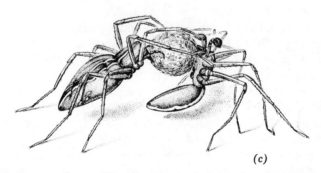

*(c)*

**Figure 9-7**  Aspects of courtship and mating in three species of spiders: (*a*) Male wolf spiders, *Lycosa amentata*, illustrating two positions of the strikingly adorned pedipalps which are waved in a species-characteristic "dance" as they slowly approach the female in the prelude to a mating attempt. Before approaching more closely, the female must respond with a signal indicating her receptivity. (*b*) The male of a crab spider, *Xysticus cristatus*, quickly binds his mate with strands of silk attached to the substrate prior to attempting to inserting his sperm-containing palp into her genital aperture. (*c*) Males of the pisaurid spider, *Pisaura mirabilis*, present a silk-wrapped prey to potential mates. Males then copulate safely while the female consumes the gift. (From Alcock, 1975a.)

370

**Figure 9-8** A pair of copulating mantids, photographed in the field in Costa Rica. The female has eaten the head, prothorax, and prothoracic legs of the male, while the genitalia remain coupled. (Courtesy of R. E. Silberglied; see also Roeder, 1935.)

**Figure 9-9** Nuptial feeding in a scorpionfly, *Bittacus apicalis*. Through this habit, widespread in the Bittacidae, females gain increased reproductive success as measured by increased oviposition. (Courtesy of R. Thornhill; see Thornhill, 1976b.)

*Stilbocoris natalensis,* has been found in which the male offers the female a seed as a nuptial gift. In an elaborate ritual, he transfers the seed after injecting it with saliva which apparently partially predigests it. A male without a seed may court, but he is unable to induce the female to copulate (Carayon, 1964).

In addition to male cannibalism and food gifts as sources of nutriment, females may also receive nourishment from glandular secretions of males before, during, or after mating. Bastock (1967) provides an overview of courtship functions in animals.

**Pollination and Reproductive Behavior**

Approximately 65% of all flowering plant species are known to be insect pollinated. Such plants are sometimes termed *entomophilous;* correspondingly, the insects which serve as pollen vectors are termed *anthophilous.* As a group, entomophilous plants generally have relatively showy and conspicuous flowers (which may however appear quite different to an insect's eye than to our own, see Fig. 6-6). They produce pollen over a period of time and characteristically produce nectar as well. In contrast, wind-pollinated plants have rather small and inconspicuous flowers and generally lack nectar. Such *anemophilous* plants produce copious amounts of pollen in concentrated flushes. Corn is a familiar example; most grasses and their relatives are anemophilous, a not surprising fact since grassland habitats tend to be relatively windy and open.*

Pollination is often closely interwoven with other aspects of insect behavior (Proctor and Yeo, 1972). Coevolved adaptations are common. One particularly striking case, the fig wasp, was examined in Chapter 4. A few other examples now will illustrate some of the variety that interrelationships between pollination and reproductive behavior may exhibit.

One of the better studied yet still incompletely understood pollination systems involves the coevolution of certain orchids and male euglossine bees in the Neotropics (Dodson, 1970, 1975). Relatives of bumble bees, the Euglossini comprise over 200 species all males of which share two peculiar morphological traits: dense hair "brushes" on the front feet and

---

* Certain bambusoid grasses which grow deep inside tropical forests where wind circulation is nil have secondarily evolved insect pollination systems (see Soderstrom and Calderon, 1971), an example which serves to reemphasize the importance of ecology and especially population and community structure in understanding insect–plant coevolution.

greatly swollen but hollow hind tibiae. Orchid pollen grains are contained in specialized packets called **pollinia**, which adhere to pollinating insects in a variety of locations characteristic of each given orchid species (Fig. 9-10). The approximately 30 genera of orchids which contain species attractive to these bees all have characteristic strong fragrances but lack nectar, the usual reward for flower visitation. A male bee attracted by the odor approaches a blossom from downwind and upon landing alternately brushes the surface of the basal flower petal (labellum), with his front tarsal brushes, then hovers in front of the flower while scrubbing his legs together as if to transfer some substance to the cavity in his hind tibiae (Evoy and Jones, 1971). Individual bees may stay with a given blossom for up to 90 minutes, with repeated bouts of brushing and hovering–transferring. Toward the end of prolonged visits the male becomes much less wary and his overall behavior suggests nothing less than increasing intoxication!

The exact reason behind fragrance collection by the males continues

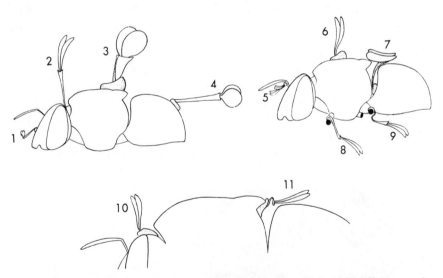

**Figure 9-10** Outlines of male euglossine bees showing the pollinia of 11 different species of orchid each deposited in precise locations on the bee's body. Upon entering a flower, the bee either squeezes down a narrow pathway and brushes against the sticky "handle" of the pollinia or trips a "trigger" mechanism which causes the pollinia to be ejected (sometimes with rather great force) onto the insect's body. On a visit to another conspecific flower, the same sequence occurs, except that as the pollinator leaves, the attached pollinia are forced into the sticky stigmatic cavity on the flower and are removed before the insect receives its new pollinia. (From Dressler, 1968, *Evolution* 22:202–210.)

to elude scientists, although several hypotheses have been advanced. Most favored is that the males utilize components of the floral fragrance to form leks (see p. 382) where mating occurs. In extensive chemical investigations of these fragrances Dodson et al., (1969) have isolated approximately 60 distinct chemicals and shown that each orchid species has a characteristic fragrance spectrum. One of the commonest compounds is cineole (Fig. 9-11). When tested alone under field conditions, cineole attracted 35 of 57 different species of male euglossine bees native to the area. But when one or more of the other compounds was combined with the cineole, the number of attracted species declined dramatically.

Another insect–plant symbiosis revolving about reproduction is that of a little moth, *Tegeticula (Pronuba) yuccasella*, and the spanish dagger plant, *Yucca*. During the day, yucca moths rest with folded wings within the half-closed flowers of their hosts, becoming active only at dusk as the flowers open fully. As evening begins, the female moth starts collecting a huge load of pollen, using her modified mouthparts to gather it from the anthers (Fig. 9-12) while swinging her head like a caterpillar during feeding. Shaping the pollen into a pellet which may be three times as large as her head, she holds her load firmly against her neck and front trochanters and flies off to another plant. The moth rests momentarily upon a flower, then deposits her eggs into the floral ovary, piercing where the walls are thinnest so that the ovipositor enters without touching the ovule within. Then, climbing up the style, she packs the pollen mass into a deep depression in the upper part of the stigma. For oviposition, she always chooses newly opened flowers; never has the female been seen to oviposit in older flowers. Oviposition stimulates the

**Figure 9-11** A male euglossine bee attracted to cineole daubed on the label of this commercial chemical. (Courtesy of R. E. Silberglied.)

**Figure 9-12** The yucca moth, *Tegeticula (Pronuba) yuccasella*, gathering pollen from an anther, a task to which she is structurally well suited. Spinous prehensile maxillae help hold the large loads of pollen required for yucca fertilization. (Courtesy of C. W. Rettenmeyer.)

yucca ovule to enlarge, in a manner similar to the swelling of plant tissue following gall midge attack. The moth eggs hatch in about a week, and larvae develop in another week. Falling to the ground, they pupate in silken cocoons intermixed with soil to await emergence a few days before the yucca blooms again. The yucca, in turn, continues to develop its seeds, for the number destroyed by the moth's activities has been very small—rarely more than a dozen and usually less (Riley, 1892).

*Ophrys* is a European genus of orchids that does not secrete nectar, nor is its pollen available to most insects. Female aculeate Hymenoptera are never seen visiting its flowers, but males of certain species of wasps and bees approach the flowers, their mating instinct stimulated by chemicals exuded from the orchid's basal petals. As they shift about, attempting but never successfully completing copulation (Fig. 9-13), the wasps loosen the pollinia, which then stick to their bodies in positions specific to the orchid species involved (Kullenberg, 1961; Kullenberg and Bergström, 1973, 1976). A similar phenomenon occurs in Australia, where quite unrelated orchids are assaulted by sexually stimulated male ichneumonid or thynnid wasps (Stoutamire, 1974). By coming to possess the necessary releasers for male sexual behavior, the orchids have evolved a system of "pseudocopulation" for pollination.

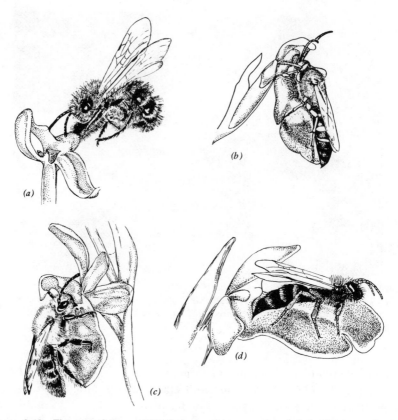

**Figure 9-13** Flowers of the orchid, *Ophrys,* trigger copulatory behavior by males in a number of unrelated genera of aculeate wasps (a,b) and bees (c,d). After a chemically mediated attraction, vital tactile stimulation is provided by the form and construction of the orchid labellum and by characteristics of the hairs upon it, including their direction, length and grouping. Each of the approximately 30 *Ophrys* species depends upon different Hymenoptera species for pollination accomplished by entering the flower frontwards or backwards, depending on the insect species: (a) *Campsoscolia ciliata* on *O. speculum,* (b) *Argogorytes mystaceus* on *O. insectifera,* (c) *Colletes cunicularius* on *O. "arachnitiformis-sphecodes"* of *sphecodes* type, (d) *Andrena* sp. on *O. fusca.* (From Kullenberg and Bergström, 1976.)

## COURTSHIP AND CONFLICT

About 10 minutes before an adult female crab hole mosquito leaves her pupa, males begin to congregate about the emergence site. When the pupal cuticle begins to split, a frantic and violent scuffle breaks out between the males as each attempts to establish possession. The victor monopolizes the emerging female by standing over her and holding her in his legs; mating is established by the time she is free from the cuticle (Downes, 1966). Similar male congregations occur at the emergence of female *Megarhyssa* wasps (Fig. 9-14).

Among some desert Orthoptera, silent satellite males become courtship parasites on calling males by intercepting females attracted to the acoustic signalling of the latter (Otte and Joern, 1975). Such "cheating" may not be uncommon in other groups. In swarm-mating species such as empidid flies (Fig. 9-15), where females mate more than once, swarms consist not only of males bearing prey but also include individuals without prey or with empty prey skins. By such "cheating," these latter males may obtain extra copulations over the male who must capture and prepare a new prey for each mating while depending on these other males to give their mates enough nutrient to promote the development of eggs (Alcock, 1973).

Males and females of a wide variety of organisms are greatly different

**Figure 9-14** An aggregation of males of three species of the ichneumonid wasp, *Megarhyssa*, assembled over female about to emerge through the bark; all three species parasitize horntail larvae in the stump. The stimuli that alert males to females' emergence are thought to be auditory but not species specific; once the female emerges, males of the other species disperse and conspecific males compete for the opportunity to mate. (Photographs by the authors; see Heatwole, et al., 1964.)

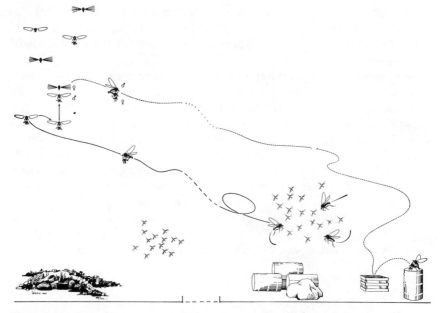

**Figure 9-15** Courtship in an empidid fly, *Rhamphomyia nigripes*. Males first capture a small insect, such as a midge or mosquito, then fly to a mating swarm to which females are attracted. Females choose males by hovering above them; coupling and prey transfer then occur, and the two fly out of the swarm and alight on vegetation to complete copulation. (From Downes, 1970, see also Downes, 1969. Courtesy Entomological Society of Canada.)

in appearance; insects are no exception. Often the sexes look like two different species. In some aculeate wasps, in fact, matching of species has been possible only when males and females have been discovered *in copula* (see Fig. 3-9). Some of these cases may be sex-linked mimicry in which each sex belongs to a different mimicry complex (Evans, 1968). It would be a mistaken oversimplification to explain all sexual dimorphism in terms of courtship and mating. In birds, for example, there is a strong relationship between sexual dimorphism and unstable environments; ecological partitioning of resources sometimes results. A number of cases of sexual dimorphism in both lizards and birds have been clearly shown to be based on a difference in food preferences between the sexes. In other cases, however, the answer seems to hinge directly upon competition and conflict centered about courtship and mating, particularly as explained through the concept of **sexual selection.** First developed by Darwin in 1871, this concept seeks to explain the reason for sexual divergence in terms of the differential ability of individuals of different genetic types to acquire mates. Competition for mates among the members of one sex, Darwin reasoned, was responsible for the evolution of those traits peculiar to that sex, including all the various anatomical, physiological, and behavioral mechanisms involved in mate selection. Richards (1927) was one of the first to focus on the subject of sexual selection in insects.

In certain instances, strong selective pressure must exist for a potential mate to be "convinced" that a suitor is fully viable. In theory, females should be especially particular and discriminating about potential mates, and mates should respond by performing increasingly elaborate courtship displays or otherwise satisfying females of their superior contribution to her fitness. Furthermore, males should compete with one another as each attempts to maximize the number of offspring he leaves; intimidating or threatening other males and/or successfully inseminating more females before others should be favored in this situation.

Sexual selection actually has a number of subcomponents (Fig. 9-16), but two major types of competition appear. One depends upon choices between the sexes ("the power to charm," in Darwin's words); this has been termed **epigamic** (or **intersexual**) **selection.** The other is based on interactions between males or, less commonly, between females; Darwin's "power to conquer other males in battle" is now called **intrasexual selection.** Pure epigamic display has been likened to a contest between salesmanship and sales resistence. As so well expressed by Wilson (1975, p. 320),

"The sex that courts, ordinarily the male, plans to invest less reproductive effort in the offspring. What it offers to the female is chiefly evidence that it is fully normal and physiologically fit. But this warranty consists of only a brief performance, so that strong selective pressures exist for less fit individuals to present a false image. The courted sex, usually the female, will therefore find it strongly advantageous to distinguish the really fit from the pretended fit. Consequently, there will be a strong tendency for the courted sex to develop coyness. That is, its responses will be hesitant and cautious in a way that evokes still more displays and makes correct discrimination easier."

Epigamic selection in insects has been well demonstrated in *Drosophila* (Spieth, 1968, 1974; Petit and Ehrman, 1969) where the females not only require intense stimulation from a given male but usually must be courted several times before indicating receptivity. Such switching on of receptivity may have a hormonal basis.

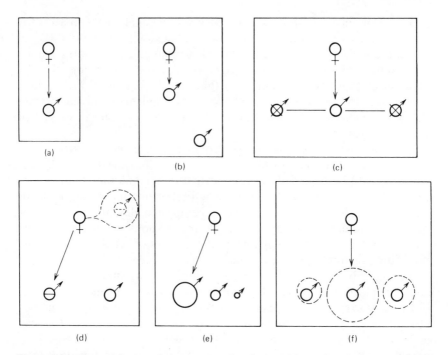

**Figure 9-16** Types of mate choice open to females: (*a*) No choice, female accepts any available conspecific mate; (*b*) female accepts most available mate; (*c*) female accepts any victor of a competition among available males; (*d*) female chooses on basis of comparison against some absolute standard; (*e*) female selects an extreme phenotype, simultaneously rejecting other equally available but less flamboyant males; (*f*) female chooses among territories or position in a lek. (From Otte, 1974; see also Faugeres et al., 1971. Reproduced, with permission, from the Annual Review of Ecology and Systematics, Volume 5, Copyright © 1974 by Annual Reviews, Inc. All rights reserved.)

## Male Competition

Intrasexual selection is based on aggressive exclusion among members of the courting sex. Just as among vertebrates, this sometimes takes the very direct form of intense fighting. Hercules beetles, for example, clash in spectacular battles (Fig. 9-17) which both attract females and intimidate other males (Beebe, 1947). The horns of male rhinoceros beetles and mandibles of male stag beetles are also apparently used in this way.

Intrasexual competition is not limited to behaviors before insemination. A great number of ingenious postcopulatory devices also exist among animals as a whole. By far the greatest diversity of these occur in the insects, probably because female insects, unlike vertebrates, are able

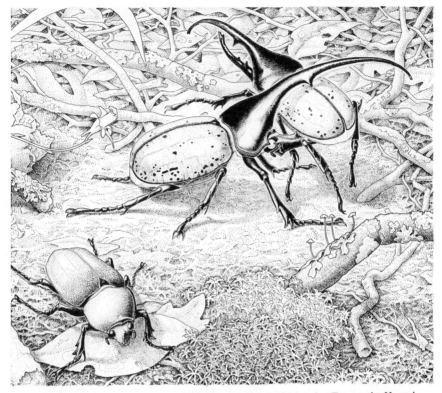

**Figure 9-17** Intrasexual competition in a Tropical scarabid beetle. Two male Hercules beetles (*Dynastes*) clashing over dominance and access to a nearby female. Grappling, pinching, jerking, and lifting one another with their huge horns, they ram at one another repeatedly, then rear up vertically together, after which the temporary victim is slammed down or carried some distance and banged to earth. Sometimes, uninjured, the fallen beetle will renew the battle; more commonly, it makes its escape. Throughout, the female remains passive, even though the victor may occasionally pick her up and carry her aimlessly about. There is apparently no precopulatory display in this species. (Drawing by Sarah Landry. Reprinted by permission of the publishers from *Sociobiology: The New Synthesis* by Edward O. Wilson, Cambridge, Mass.: The Belknap Press of Harvard University Press, Copyright © 1975 by the President and Fellows of Harvard College.)

to store viable sperm from a single insemination for use during the remainder of their reproductive lives (several years in the case of some social species). The female's sperm storage organ, or **spermatheca,** is a pouchlike structure whose opening to the genital tract is controlled by a sphinctor muscle through which sperm are parceled out as eggs are laid. In cases where sperm are deposited in packets and maintain their cohesion during storage, they presumably would be used in the reverse order from which they were obtained. That is, the sperm of the last male to mate would be the first to be used in egg fertilization (Brower, 1975). Thus, a late-arriving male might still profit by attempting to mate with an inseminated female.

Obviously, it is not to a male's advantage to have his own sperm subsequently replaced by another's. How can he ensure that his sperm will be the most likely to fertilize the female's eggs? One way is simply to monopolize the female so that other males do not have access to her. Such a situation occurs in certain dung flies (Fig. 9-18). Continuing to maintain a territory has a similar effect, as does merely staying in physical contact with one's partner without actual genital contact; the tandem flight of many Odonata is a familiar example (see Fig. 9-4c). So does prolonged copulation, which occurs widely among insects such as the lovebug and the queen butterfly. Another widespread device is the mating plug, formed by coagulation of male accessory gland secretions in the female's genital tract; such plugs prevent sperm leakage and/or physically prevent subsequent matings with other males. Parker (1970a, 1974, 1978) further discusses the evolutionary consequences of sperm competition in insects.

After members of the courting sex have aggressively excluded a portion of the competition, classical intrasexual selection theory states that the more passive sex simply chooses a potential mate from among the elite group of winners or the single winner. In the process of acquiring a vigorous partner, she often also acquires something else of great value—a share in his set of resources. How males partition these resources among themselves brings us to the twin topics of territoriality and dominance.

### Territoriality and Dominance

Male carpenter bees vigorously chase a variety of small objects tossed near them. Male dragonflies drive off other conspecific males (see Fig. 6-9). Males of Hawaiian *Drosophila,* euglossine bees, Asian fireflies, and certain dragonflies all gather at specific sites, called **leks,** where they

**Figure 9-18** Fighting between males of the yellow dung fly, *Scatophaga stercoraria*. The female at the bottom is copulating with the male above. At the same time, the male is attempting to repel a rival male (top, partly out of view) with thrusts by his middle legs. Competition among males for mates is exceedingly intense and continues even after copulation, with the inseminating male remaining mounted and continuing to attempt to dislodge rivals. (Courtesy of W. A. Foster; see Parker, 1970b.)

display and compete for the attention of females. Parasitic wasp males (*Nasonia*) defend host fly puparia from which females are starting to emerge (King et al., 1969).

The phenomenon of **territoriality**—broadly, any space-associated intolerance of others and, more narrowly, an intolerance based on real estate holdings—has long been well known in vertebrates, often in association with aggressive behavior. Among insects it has received less emphasis, although it is a surprisingly common and widespread occurrence. Table 16-1 in Price (1975) provides a sampling of the diversity of territoriality in insects. The papers by Alcock (1975b), Campanella and Wolf (1974), Campanella (1975), Otte and Joern (1975), Velthuis and Camargo (1975a, b), and Waage (1974) are pertinent recent additions.

Territoriality is almost always, but not exclusively, associated with competition for mates or food. (Other examples include social insects that defend their nests and sometimes foraging trails from competitors and intruders (see Hölldobler, 1976).) Its function is to partition priority of access to resources which are limited and tied to a particular area. Territoriality is not always associated with overt aggression, for some types of space exclusion can be mediated solely through display. Nor does it even always require the owner's presence; some territorial insects, such as female apple maggot flies (Prokopy, 1972) and certain parasitic wasps (Fig. 9-19); see also Fig. 5-13), depend upon persistent chemical marks which have a repellent effect on conspecifics.

Since territorial activities incur an expense of energy, selection for territorial behaviors tends to operate only under a limited set of conditions (Otte and Joern, 1975). Among these are that the desired but limited resource should be relatively localized in a readily defensible situation and that animals should stand to obtain more of the resource by defending the area against competing individuals than by searching for new resources to exploit. For example, males of the cicada killer wasp, *Sphecious speciosus* (Fig. 9-20), establish themselves on perches which

**Figure 9-19** A female *Trissolcus basalis* (Hymenoptera: Scelionidae) marking an egg in which it has just oviposited. Such chemical marking, produced by drawing the ovipositor tip across the cap of the host egg, serves to repel other conspecific females and minimizes the likelihood of multiple ovipositions on a single host. (Courtesy of C.S.I.R.O., Division of Entomology, Canberra, Australia; see also Wilson, 1961.)

**Figure 9-20**  A male cicada killer wasp, *Sphecious speciosus* on its territorial perch overlooking a nesting area. Note the characteristic erect antennae and alert posture. The same individual returns day after day to occupy the same perch. If artificially removed, another individual will generally take over the perch within a day. (Courtesy H. E. Evans; see Lin, 1963.)

overlook highly clumped nesting areas containing emerging females. Investment in defense of such perches may be rewarded by extra copulation opportunities for the territory owner.

In most vertebrates fights among conspecifics are rarely fatal, the combatants being inhibited or breaking off as a result of various submissive postures. In insects, in contrast, murder and even cannibalism may be a rather normal phenomenon. One of the more striking examples of aggression occurs among male *Melittobia*, a genus of tiny parasitic wasps (Fig. 9-21). Predisposing their behavior in this direction is an unusual breeding structure—males are incapable of dispersal and thus their only mating opportunities are with sisters emerging from the same host. Upward of 150 progeny from a single mother develop gregariously and emerge from a single host, with an overwhelming preponderance (95%) being female. Inside the host's skin or pupal case, the few brothers aggressively compete among themselves for the opportunity to copulate with their monandrous sisters. In mild forms, such

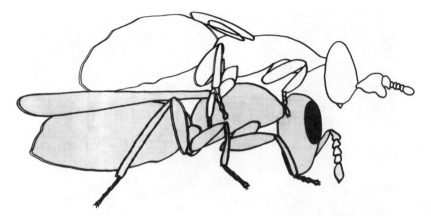

**Figure 9-21** A pair of *Melittobia* wasps (Eulophidae) engaged in precopulatory courtship. A highly dimorphic species, the male (*above*) differs from the female in being blind and flightless (although possessing wing stubs) and in having highly modified antennal scapes which at first appear to be the insect equivalent of deer antlers, but in actuality turn out to be claspers used to grasp the female's antennae during courtship. (From Evans and Matthews, 1976.)

aggression consists of intensive grappling bouts in which contestants may lose an appendage; at its extreme, the first males to emerge selectively decapitate their unemerged brothers. Females themselves do not participate in the fray; once mated they disperse to search for new hosts (Matthews, 1975).

Aggression may also be resolved through the establishment of **dominance,** a ranking of individuals on the basis of real or apparent authority, strength, influence, etc. Examples of a dominance hierarchy include *Polistes* paper wasps, dragonflies, crickets, and passalid beetles. The dominance concept implies an ability to recognize individuals and remember relationships and would be expected to evolve in species with long-lived adults.

In some species of *Polistes*, groups of females join to begin their nests together, whereupon each fertile potential queen then competes in laying eggs and relatively rigid hierarchies are established. Dominant females maintain their reproductive superiority by three means: laying the greatest number of eggs, physically removing and eating the eggs of subordinates who occasionally succeed in oviposition, and demanding and receiving the greatest share of food. Interactions between dominant and subordinate individuals are often matters of posture, the dominant individual rising on her legs above the subordinate who crouches and lowers her antennae (Fig. 9-22). A conspicuous side-to-side vibration of

**Figure 9-22** Dominance and subordinance behavior in *Polistes fuscatus* (Vespidae). The female on the left shows the relatively elevated posture characteristic of a dominant individual. The subordinate (*right*) has been seized by the hind leg and crouches with antennae lowered. (Courtesy of M. J. West Eberhard; from West Eberhard, 1969.)

the abdomen (tail wagging) also is performed more frequently by the dominant female (West Eberhard, 1969).

Male crickets confined to a limited space will also establish a rank order which is stable for some time (see Chapter 7). Age, size, possession of a territory, opportunity to copulate, and the results of previous fights determine the rank order, and acoustic signals (rival song) of a dominant individual inhibit the aggressive tendency of a subordinate one. If the rival male does not retreat, he may respond with his own rival song and, at closer range, with antennal lashing. Until the hierarchy is established, aggression includes singing, jumping at each other, and biting with the mandibles (Alexander, 1961).

Blum and Blum (1978) include several detailed studies on mate competition.

## THE PHYSIOLOGICAL CONTROL OF MATING BEHAVIOR

Until now, the impression has perhaps been given that courtship is an invariable, rigidly controlled interaction between the sexes which once initiated progresses inevitably to the copulatory conclusion. In reality, however, things are rarely that simple. On one occasion, a male encountering a female may court her intensely; another time, he may simply ignore her. Numerous quantitative studies of courtship, such as that diagrammed in Figure 9-23, make it apparent that factors such as age, experience, and physiological condition can alter courtship patterns. For example, advancing age of male *Nasonia vitripennis* wasps has been positively associated with a general slowing of four out of five courtship parameters (Barrass, 1960).

Among the vertebrates in particular, such variations in responsiveness have historically been attributed to changes in "motivation"; however, such changes are probably due to no one single underlying process but rather to a great number of separate physiological effects. The same is to

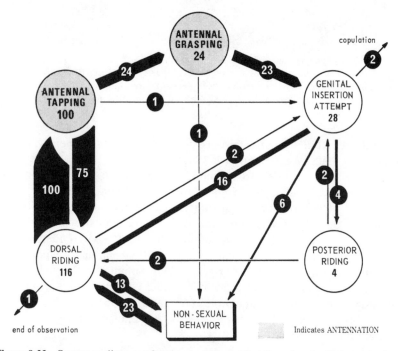

**Figure 9-23** Sequence diagram of male courtship in the blister beetle, *Meloe strigulosus*, representing 23 courtship bouts in five pairs of beetles during an observation period of 175 minutes. Large circles represent courtship elements. Variability in the sequence of components is indicated by the arrows pointing away from a particular circle, the sum of which equals the frequency of this component given by the numeral inside the large circle. (From Pinto, 1972.)

be expected for the insects. The existence of neuroendocrine involvement in insect mating behavior appears to be broadly correlated with type of reproductive cycle (Barth, 1965; Barth and Lester, 1973). Neuroendocrine control tends to occur in those insects which have a long adult life (e.g., cockroaches) and in which the life span includes time periods when mating is inappropriate (perhaps even impossible) alternated with repeated reproductive periods. For other insects which are short-lived as adults (e.g., mayflies), the female must quickly attract a male and lay eggs since she will die within a few days. With such a premium on efficiency, mating is more efficiently built into the developmental process to appear automatically with adulthood and thus can be divorced from extensive endocrine controls. This relationship applies not only to neuroendocrine control of sexual signaling but also to other aspects of reproduction such as sexual receptivity and perhaps oviposition.

Most insects, of course, fall between these extremes. A more common reproductive pattern includes a single adult period of feeding and oocyte maturation, followed by mating and then by a period (or periods interrupted by feeding) of oviposition. For most insects which feed as adults and require neuroendocrine factors for maturation of oocytes, some type of neuroendocrine sexual regulation is the rule. In insects that do not feed as adults, such as many Lepidoptera, females from which the juvenile hormone-producing apparatus has been removed can still lay fertile eggs; in at least some, however, neuroendocrine factors are necessary for proper environmentally cued release of sex pheromones.

## Signaling of Sexual Readiness and Female Receptivity

In insects chemical sex attractants predominate (see Chapter 5), although other sensory modalities may also be used. For indicating the location of a receptive potential sexual partner, the specificity and long-range effectiveness of pheromones are unexcelled. These may function as attractants, excitants, arrestants, identifiers of species and/or sex, or releasers or have several of such effects serially or concurrently. Either sex may produce them—sometimes both sexes have their own, each with a distinctive role—and the neuroendocrine system participates in this communication.

In the polyphemus moth, *Antheraea polyphemus,* for example, female calling behavior (see Fig. 5-7) is initiated only after exposure to a volatile chemical, *trans*-2-hexenal, present in the leaves of the host plant, and then only at a specific time during the photoperiodic cycle (see Fig. 5-18). The result is that mating occurs on or near the larval food. Riddiford

(1974) showed that removal of the corpora cardiaca–corpora allata complex in female pupae abolishes calling behavior in the adults, whereas allatectomy alone does not affect normal calling behavior nor the ability to mate. Reimplantation of the CC–CA complexes at the outset of adult development did not restore this behavior. However, when virgin females which had not been exposed to *trans*-2-hexenal were injected with blood from calling females, 80% began calling within $2\frac{1}{2}$ hours. Thus a "calling" hormone can be recovered from the blood and presumably acts on the female's nervous system to trigger calling behavior. Figure 9-24 summarizes the hormonal control of reproductive behavior in this moth.

In ovoviviparous cockroaches which are characterized by cyclical ovipositions and pregnancies alternating with mating bouts, juvenile hormone is implicated in controlling the release and probably the production of female sex pheromones. Allatectomized females of *Byrsotria fumigata* fail to produce sex pheromone and do not attract males. When these same females are artificially coated with the sex pheromone, normal attraction and mating occur. If active corpora allata are implanted, pheromone production ability is restored; the same effect results from injections of synthetic juvenile hormone analogs. Thus, hormones regulate pheromone production and release so that mating will occur at the appropriate stage in the reproductive cycle of the emitter and/or in the proper environmental context. In theory, neuroendocrine control of sexual signaling should apply equally to nonchemical communicative modes.

Female insects usually pass through three distinct behavioral states: young virginity, unreceptive to courtship; mature virginity, responsive to male advances; and mated, typically unresponsive toward males but showing intense oviposition behavior. The switch from one state to another is often abrupt and often hormonally induced. Most work on receptivity development has been done with cockroaches and Diptera (especially mosquitoes). In fact, the first evidence that hormones mediated the onset of insect sexual behavior was provided by cockroaches, in a series of seemingly contradictory experiments that underscored the dangers of generalizing from a single strain of laboratory animal.

In 1960, Engelmann reported that when he allatectomized newly emerged females of *Leucophaea maderae*, 70% of them never displayed sexual receptivity. When he reimplanted corpora allata, however, normal receptivity ensued. Roth and Barth attempted to duplicate Engelmann's results, but to their surprise, their allatectomized females became receptive and mated normally. To resolve the conflicting results, Engelmann and Barth repeated their experiments, this time both using

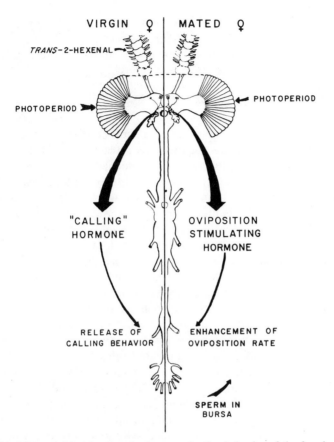

**Figure 9-24**   Hormonal involvement in the reproductive behavior of the female silkmoth, *Antheraea polyphemus*. In the virgin female (*left side*), "calling" hormone released from the corpora cardiaca in response to host plant odors activates release of the attractant pheromone and assumption of a characteristic calling position. After the female has mated, the presence of sperm apparently triggers the bursa copulatrix to secrete a humoral "factor," and the female switches to the mated behavior mode (*right side*). Later in the photoperiodic cycle, the corpora cardiaca release a hormone greatly increasing oviposition rates and flight activity. (Courtesy of L. M. Riddiford; from Riddiford, 1974.)

Engelmann's cockroaches. Nearly all their control roaches mated, usually within a few minutes after the introduction of males. Allatectomy, on the other hand, once again led to a high percentage of nonmated females, and those that did mate required over an hour before males could induce them to copulate. The different experimental results were due to strain differences, which emphasizes again the importance of the role of heredity in behavioral study (see Chapter 1). The interested reader may wish to refer to Barth and Lester (1973) or Engelmann (1970) for additional details as well as references to the original *Leucophaea* studies.

## Refractoriness in Mated Females

Most female insects will no longer accept courtship advances after successful mating. For some species, this **refractoriness** is temporary; in others it lasts for life. A great variety of mechanical or chemical stimuli may be responsible. Female grasshoppers, for example, show secondary defensive behavior rather than receptivity toward males once a spermatophore is placed in the genital tract. In some cockroaches which also respond to such mechanical stimuli, postmating inhibition can be artificially stimulated by inserting glass beads into the bursae of virgin females. In others insects whose effective mating stimuli appear to be primarily chemical, juvenile hormone withdrawal or the release of chemicals from the maturing ovaries are apparently involved. Often, secondary reproductive structures such as male accessory glands or female bursae release agents that effect the switch-over.

A multiple-mechanism control is found in some Diptera, where mating often results in a rapid, neurally mediated refractory state, followed by another more persistent, hormonally induced one. For example, as the bursa copulatrix of *A. aegypti* fills with seminal fluid, an immediate (probably neural) switch-off of receptivity occurs. For the rest of their lives, however, females remain behaviorally unreceptive, because a chemical (matrone) from the male accessory glands passes through the bursa copulatrix walls into the haemolymph and acts on the nervous system. In fact, injecting matrone into the haemocoel is an equally effective producer of such refractoriness (Craig, 1967).

The implicit assumption that "mated" equals "inseminated" should be carefully evaluated in each case. For example, freshly emerged flying female *A. aegypti* mosquitoes appear to mate readily but actually are unable to accept the transfer of sperm for at least 24 hours (Lea, 1968). Certain tiny parasitic wasp females will mate only once; thereafter they rebuff the overtures of the males who nevertheless persistently attempt to court them. However, an occasional virgin female, having been intensely stimulated by male courtship, reacts as though inseminated and thereafter behaves like a mated female even though her actual status is still virgin (Matthews, 1975). In behavioral studies, insemination should be the criterion for a successful mating; conclusive demonstration requires either dissection or rearing progeny.

Because the stimulation of oviposition and the onset of refractoriness are very closely related, changes in oviposition rate have often been taken as a quantitative measure of the extent of a female's switch from a "virgin mode" to a "mated mode." However, mating also stimulates oogenesis, leading to an increased egg maturation rate and therefore a higher oviposition rate. Making such a distinction between changes in oogenesis rate and changes in oviposition rate is difficult, and only a few workers have done it.

### The Role of Hormones in Male Sexual Behavior

Usually, male sexual behavior matures along with the metamorphosis of the nervous system; following ecdysis, expression of full copulatory behavior awaits only suitable stimuli. In most such cases, the corpora allata are not involved, although sometimes juvenile hormone lowers the male's behavioral threshold to the female. There are, however, many exceptions to this, particularly among longer-lived insects which have a restricted breeding season. Some Orthoptera are a case in point. The endocrine control of male sexual behavior has been best studied in the grasshoppers and locusts, where some species require the corpora allata for the maturation of male sexual behavior. For example, allatectomy of newly ecdysed males of the desert locust, *Schistocerca gregaria*, and the red locust, *Nomadacris septemfasciata*, completely prevents the onset of male sexual behavior; if, however, several pairs of active corpora allata are transplanted into an allatectomized male, sexual behavior reappears. Interaction between the corpora allata, the brain's neurosecretory system, and environmental parameters provides an environmentally responsive control over the onset of sexual behavior (Pener, 1972). In some locusts and grasshoppers, for example, male diapause is photoperiodically controlled. Under long-day conditions, males show little or no sexual behavior, apparently due to cessation of neurosecretory activity in the brain. Sexual behavior resumes, however, when active corpora allata are implanted from locusts reared under short day conditions. Barth and Lester (1973) and Engelmann (1968) provide reviews of hormonal involvement in insect reproductive behavior.

### THE EVOLUTION OF COURTSHIP

Sexual reproduction, nearly all biologists agree, has evolved because of the much greater speed with which new genotypes are assembled under this system. Such diversity, they argue, is highly adaptive. Sexually reproducing populations are more likely than asexual ones to adjust to changed environmental conditions through the creation of new genetic combinations. Thus, sexual reproduction has increasingly become the mode.

Sex would seem to be the ultimate social act. Yet, as Wilson (1975) has pointed out, in many ways sex can quite properly be considered an antisocial force in evolution. A survey of the animal kingdom indicates an inverse relationship between sexual reproduction and degree of social evolution. The "ideal" society would lack conflict and would have a high degree of selfless behaviors and coordination, a situation most likely to evolve where all members of the group are genetically identical. Within the invertebrates, the highest forms of sociality are found in groups that create new colony members by budding—the sponges, tunicates, etc. In the insects, increased sociality is strongly dependent on a modification of the "shearing force of sexuality", namely, the haplodiploidy found among the Hymenoptera (see Chapter 10). Sexual reproduction dilutes genetic relationships, causing a one-half reduction in genes shared among parents and offspring. Thus, a conflict of interest inevitably arises as each mate, parent, or offspring strives to increase its personal genetic fitness at the expense of others. In this view, courtship and sexual bonding are means by which organisms override the "antagonism and tension" resulting from genetic differences induced by sexual reproduction.

### Mating Systems and Parental Investment

Those aspects of a species organization which determine ways in which the sexes come together for breeding are called its mating system. Three principal types occur in insects. **Monogamy** exists where each breeding adult is mated to only one member of the opposite sex. Life-long monogamy is rare among insects; however, its existence is notable in termites, where a king and queen form a pair bond for life (Fig. 9-25). **Polygamy** is the state in which an individual has two or more mates, none of which mates with other individuals. It is of two forms. One is **polygyny,** where a single male mates with several females; this is the most common system in insects. The reverse case, **polyandry,** which occurs when a single female mates with several males, is relatively rare in insects, as in animals generally. (Thus, the terms polygamy and polygyny are often used synonymously.) Honey bees, however, have this type of mating system, for the queen mates with several males in succession during her nuptial flight; the act of copulation is fatal to drones (Gary and Marston, 1971).

**Figure 9-25**  Male and female termites running in tandem while searching for a crevice in which to begin a nest. Both have broken off their wings moments earlier upon alighting and will remain paired for life. Copulation occurs some hours or even days later after the pair have excavated a nuptial chamber. (Photograph by E. S. Ross.)

**Parental investment** may be defined as behaviors (investments of time and/or energy) that increase the probability of some offspring surviving to reproduce at the cost of the parent's ability to generate additional offspring (Trivers, 1972; Wilson, 1975). Different strategies of parental investment exist. For example, one species may invest highly in a few offspring, while another may produce many but invest very little in any one of its young. Within a given species, the sexes generally differ in their degree of parental investment (Fig. 9-26). Males and females make unequal physiological investments in the production of gametes. The clutch of eggs a female lays may more than equal her total body weight, whereas even over an entire season sperm produced usually represent a mere fraction of this. Moreover, the commitment of time and energy required in such aspects as egg maturation, birth, or oviposition are borne solely by the female.

In most insects, a male's only reproductive role is to provide a set of genes. One result of the general imbalance between the parental investments of the sexes is that a reproductive mistake has much greater consequences for a female. The ''coyness'' mentioned earlier is one

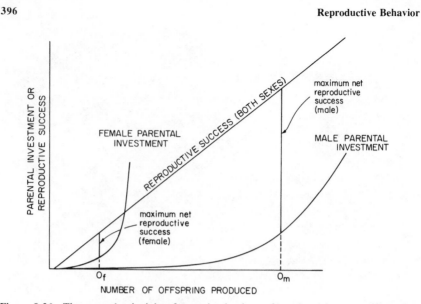

**Figure 9-26** The central principle of sexual selection reformulated in terms of parental investment. In the common situation illustrated here, a conflict of the sexes arises because the optimum number of offspring ($O_f$ for the female, $O_m$ for the male) differs between them. Because the female must expend a greater effort to create offspring, her greatest net production comes at a lower number than in the case of the male. Under these conditions the male is likely to turn to polygamy. (Reprinted by permission of the publishers from *Sociobiology: The New Synthesis* by Edward O. Wilson, Cambridge, Mass.: The Belknap Press of Harvard University Press, Copyright © 1975 by the President and Fellows of Harvard College. Modified slightly from R. L. Trivers, 1972.)

strategic result; another is that in almost every species, it is the female that chooses a mate, not the reverse. Females are under selective pressure to pick a male with a superior set of genetic instructions; and to the extent that such discrimination is possible, the outcome tends to be that a select group of males with superior traits mate with a large proportion of the female population. It should be to a female's advantage to mate with a male that signals genetic quality, no matter how many times he may have already mated. Therefore, one should expect insect species, like most other animals, to be fundamentally polygamous (or more properly, polygynous).

Monogamy, where it occurs, is generally an evolutionarily derived condition. Fidelity tends to evolve only under certain ecological circumstances, usually when the advantages of cooperation in rearing offspring outweigh the personal advantages to both partners of seeking extra mates. Under most conditions, parental care—adding to the inequality of parental investment—will merely reinforce the polygynous system. In some few species, however, the male's contribution to the fitness of his mate and offspring is extensive. Territories and male parental care provide two striking examples. Although the basic theory of parental investment is built on parameters difficult to test in practice, such exceptional cases where males have taken on the burden of parental care provide decisive tests in an indirect way.

Among the Hymenoptera, where so many cases of "typical," that is, female, parental care exist, male participation in parenting occurs only in two unrelated sphecid wasp genera; there is, however, a genetic explanation (see Chapter 10). Outside the Hymenoptera, it has been documented in two other orders—the Coleoptera and the Hemiptera.

Belostomatid bugs are moderate-sized to large aquatic predaceous Hemiptera. Two aspects of their reproductive behavior are unique. One is that females invariably attach their eggs to the backs of males, where the eggs remain until they hatch (Fig. 9-27). The other is that the females go their own way shortly after mating and oviposition rather than sharing in parental care as some other insect pairs do (Smith, 1974). Prior to the turn of the century, reflecting a prevailing chauvinistic attitude as to the "proper" sex roles, it was generally believed that the egg bearers were females carrying their own eggs. Some authors even described the oviposition process, attributing the feat to a long protrusile ovipositor which could be extended over the female's body. In 1899, F. W. Slater (a woman) set the record straight in a delightful account (cited in Smith, 1975) not totally emancipated from chauvinistic overtones:

"That the male chafes under the burden is unmistakable; in fact, my suspicions as to the sex of the egg-carrier were first aroused by watching one in an aquarium which was trying to free itself from its load of eggs, an exhibition of a lack of maternal interest not expected in a female carrying her own eggs."

A few years later another entomologist, while mentioning the "indignity" of such male servitude, noted the "peculiar fact" that copulation took place in connection with oviposition in the species he observed.

CASE STUDY: REVERSED SEX ROLES IN THE GIANT WATER BUG, *Abedus herberti*

Detailed studies of male belostomatid egg brooding have been conducted by Smith (1975, 1976b) in an attempt to answer several questions about such reversed sex roles. For example, how necessary is male parental care? Upon an average male's back, about 97% of the eggs can be

**Figure 9-27**   An egg-encumbered male of the belostomatid bug, *Abedus herberti,* with the female in the act of laying eggs which are individually glued to her mate's wings. Male brooding behavior also includes a feeding inhibition toward early instar nymphs which is not shared by the nymph's mothers or unencumbered males. (Courtesy R. L. Smith, see Smith, 1976b.)

expected to hatch. Does this high egg survival rate depend upon eggs being carried about upon the back of a live male water bug?

Removing egg pads from six males, Smith placed each egg pad in a separate fingerbowl in the laboratory; three were covered with water and three were allowed to dry. None of the eggs survived; those kept in water developed a fungus after about four days, and those allowed to dry became desiccated. Repeating the experiment but first washing the eggs with distilled water, Smith got the same result. Next, he killed egg-bearing males and left them with eggs intact in their containers. These eggs also failed to hatch. Attachment to a healthy male appeared essential for eggs to develop normally and hatch. But why?

Brooding males, Smith noticed, tended to prefer shallow water and seemed to continually be adjusting their position to keep their egg pad at or above the water's surface. Perhaps regular exposure to atmospheric air was a requisite for normal development. To test this, Smith placed an egg pad in a finger bowl and added water to a depth of less than 3 mm. Evaporative losses tended to expose the unattached ends of the eggs in the center of the pad, so he added water daily. After two weeks, he was excited to find that some of the centermost eggs began hatching. And although only 14 of the 72 eggs in the pad finally hatched, the success was a notable improvement over previous results. Nevertheless, it still appeared that the live brooding male was providing further services beyond simply exposing the eggs regularly to atmospheric air.

Observing brooding males of a related species of belostomatid in the field, Smith (1976a) noticed that submerged egg-bearing males frequently pulled their hind legs vigorously and repeatedly over their egg pad from front to rear, while unencumbered males only rarely stroked their backs, and then apparently only as part of normal grooming behavior. Perhaps, he postulated, such stroking also was necessary. A. herberti seemed to show analogous behavior of a slightly different form. Egg-carrying males spent nearly one third of their submerged time in "brood pumping"; resting on submerged vegetation or stream bottom substrate, they typically rocked their bodies repeatedly forward and backward, pivoting on their middle legs while remaining in a fixed position. Prior to this, Smith also noted, males of both species patted or touched the eggs with their hind legs. Was this behavior an essential precursor to the pumping or stroking?

Smith quickly peeled off egg pads after a pumping bout had been initiated but while the bug was still under water. The males all continued to pump. However, when they "felt" for the eggs later and found them gone, they no longer would pump. Thus, Smith concluded that it was through patting and stroking that the bug was informed of the status of his egg pad.

Do males ever care for eggs that they have not fertilized themselves? Smith's observations on Abedus courtship revealed a behavioral system apparently designed to minimize this possibility. In contrast to most "normal" species where the female determines when and whether mating and oviposition will occur, courtship and oviposition in Abedus prove to be under overwhelming male domination. The male maintains absolute control over the female in a mating sequence as long as he is

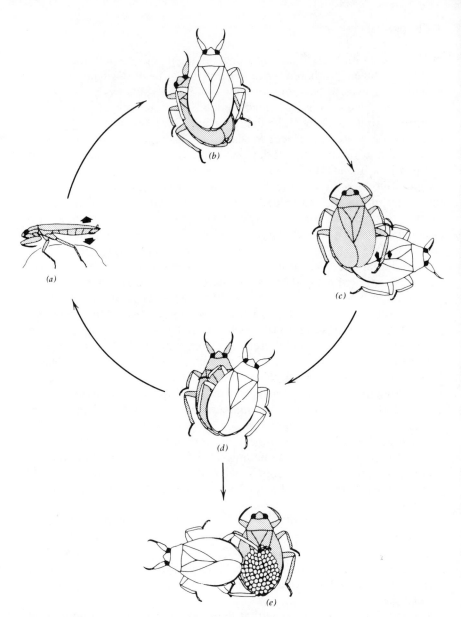

**Figure 9-28** The copulation–oviposition cycle of the giant water bug, *Abedus herberti*. After preliminary sparring between the sexes, receptive males perform display pumping (*a*) in the manner of brood pumping but at a much more vigorous pace. Females respond by climbing on the male's back (*b*) as if to oviposit. With one hind leg, the male manipulates the female off into a copulatory position (*c*). About 1 minute after intromission, the male abruptly "scrubs" one hind leg on the female's hemelytra. Immediately the female repositions herself (*d*) and begins to lay eggs, starting at the apex of the male's hemelytra. After about 5 minutes, although the female has laid only three eggs or less, the male's

receptive. He commands her to copulate, guides her into the copulatory position, signals his desire to uncouple, and determines the length of time she is given to oviposit (Fig. 9-28). Moreover, he will allow no oviposition to take place at all until after the female mates with him at least once. Thus, male giant water bugs have developed a whole set of courtship and mating behaviors which maximize individual male fitness and preclude altruistic brooding. By "insisting" on copulation prior to oviposition, the male ensures that the first eggs he receives will have been fertilized by him. Limiting the period of oviposition tends to minimize the chance of "error," which would be defined as acceptance of eggs fertilized by another male.

What mating system occurs? The answer is somewhat unclear. Apparently under natural conditions males typically carry the eggs of only one female. However, Smith was able to induce successful polygamous matings in the laboratory. The latter system might well be advantageous to both sexes, since brooding extra eggs probably entails little added risk for a male and fitness would be enhanced for a female by ovipositing on the back of as many willing males as possible.

Smith did not attempt to experimentally assess the direct caloric cost to males incurred by brooding. However, it seems evident that the brooding male is exposed to extra risks. The additional weight of developing eggs would impair his swimming and perhaps reduce his ability to escape predators. (Smith did measure swimming speeds of egg-carrying and unencumbered males and found that the latter could swim considerably faster.) Additionally, the extra egg weight may impair their feeding efficiency. The behaviors involved in the father's parental care undoubtedly also increase risk of predation. For example, because water bugs normally rest motionless, protected by their cryptic coloration, pumping or stroking behaviors probably make them more vulnerable by betraying their location. It may be partly to balance such risks that water bugs have developed a tremendous bite.

## OVIPOSITION BEHAVIOR

A female locust or grasshopper begins tapping the soil with the tip of her abdomen and probing with her ovipositor. Rejecting surfaces where the ground is hard, she comes to a soft and sandy place where she begins to

temporary quiescence ends. Forcing the female out of position with his hind leg, the male begins a new bout of vigorous display pumping. This cycle continues until the female's total egg clutch is deposited on the male's back (e). (Adapted after Smith, 1975.)

dig. Raising her body on her first two pairs of legs, she arches the tip of her abdomen downward and scrapes particles of soil sideways and upward with valves at its tip. As a hole forms and deepens, her abdomen lengthens considerably. (If she is a migratory locust, *Schistocerca,* she may dig to a depth of 14 cm.) To maintain necessary body pressure, she expands her airsacs and swallows air into her crop and midgut caecae. Once the hole is dug, if the soil is saline or dry, she rejects it. If the soil is moist, however, she begins to lay groups of eggs (Fig. 9-29). To maintain haemolymph pressure, she continues to pump air into her tracheal system by vigorous ventilatory movements of her head. About 20 minutes later, her airsacs occupy the whole cavity of the first five abdominal segments, and the volume of her tracheal system has more than doubled from the start of oviposition. Finally, when all her eggs have been laid, she forms a frothy plug in the mouth of the hole, then

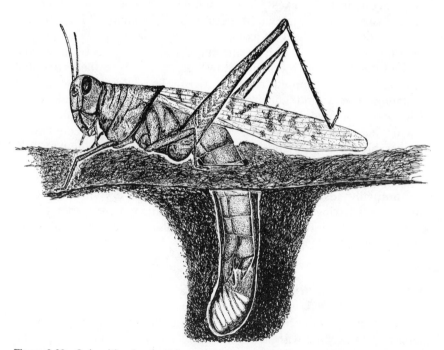

**Figure 9-29** Oviposition by the migratory locust, *Schistocerca.* Digging is accomplished primarily by opening and closing the dorsal and ventral ovipositor valves. After deposition, the eggs absorb water via a special pore in one end and swell to twice their initial size. Successful breeding occurs only where there is rainfall (see Fig. 3-11), and final choice of site is positively influenced by the presence of other ovipositing locusts. (Drawing by J. W. Krispyn.)

withdraws her abdomen completely and scrapes dirt over the top of the hole with her hind tibiae. All together, the oviposition process has taken her about 2 hours (Uvarov, 1966).

Insects deposit their eggs in an almost infinite number of ways. Sometimes both parents participate, but more commonly gravid females have sole responsibility for the fate of their eggs. In general, the behavioral chain of events leading to oviposition closely parallels that used in food location (see Chapter 4). Because oviposition mistakes may severely lower reproductive success and hence individual fitness, the stimulus–response sequences tend to be quite complex, minimizing chances for error.

## Site Location and Selection

Mosquitoes are attracted to water, influenced by the presence of vegetation and amount of light reflected from the surface. They do not, however, always oviposit once they have reached it. When they land upon the water's surface, some judge salt content and others, pH of the water through their tarsal sensilla to determine whether or not to lay eggs. Likewise, female cabbage butterflies, *Pieris brassicae,* are usually attracted to blue or yellow; when ready to oviposit, however, they become attracted to green surfaces. Landing upon a plant, the female drums upon it with her forelegs, sensing through her tarsal receptors whether the plant contains mustard oil; if it does, she hangs upside down and oviposits her eggs on the underside of its leaves. Buprestid beetles of the genus *Melanophila* are attracted to burned trees over distances of several kilometers through an oriented response to infrared radiation emanating from forest fires.

A great diversity of stimuli influence the behavior of egg-laden female insects. Also included are characteristics of the host which negatively modify the behavior of the ovipositing female. Thus, plant-derived chemical factors inhibit oviposition by boll weevils. One species of cotton was found which reduced boll weevil oviposition by 40% (see Kogan, 1975). (Needless to say, efforts are under way to breed this factor into commercial cotton species!) In other cases, the stimuli may be only indirectly related to actual larval feeding preferences, a situation which may have bizarre consequences (Fig. 9-30).

Site selection generally has two phases. First, there is a general reaction to the environment; this is followed by a final selection mediated by quite specific responses. At different stages of site selection the stimuli which are operating usually vary as well. This is particularly true among phytophagous insects. For example, oviposition by the

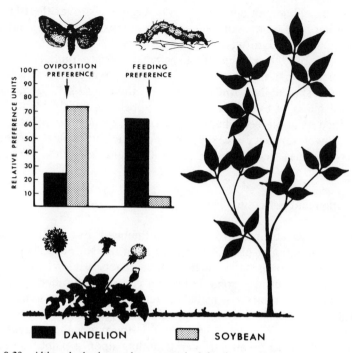

**Figure 9-30** Although the larvae have a marked feeding preference for dandelion, adult females of the noctuid moth *Autographa precationis* (Guenee) clearly prefer soybeans over dandelions for oviposition, apparently because the shape of the dandelion leaves is a less effective oviposition stimulus. (From Kogan, 1975.)

tobacco hornworm moth has two principal phases: approach and landing. Approach is largely mediated by visual cues and is not particularly discriminatory. Landing, on the other hand, is quite discriminatory and is based on olfactory responses to smells emanating from host plants. After landing, contact chemostimulation elicits egg deposition.

Among species that do not practice any form of parental care following egg deposition, proper egg placement is particularly crucial. In the final stages of site selection, considerable time and energy may be spent on fine discriminations regarding a wide variety of factors related to food availability, food suitability, and predator pressures. One such discrimination relates to the host's suitability from the standpoint of egg or larval load; many insects have evolved the ability to assess this and modify oviposition accordingly. Female bean weevils, *Callosobruchus maculatus,* do not deposit their eggs in mung beans at random but assess each potential host for its ability to sustain another offspring by simple comparisons between the present bean and the previous one. A bean that is larger or bearing fewer eggs than the last one encountered is most likely to receive an egg. The resultant nonrandom pattern of oviposition increases larval survival by 70% over what would occur if eggs were laid randomly (Mitchell, 1975).

Likewise, some species of hover flies (Syrphidae) whose larvae feed on aphids are sensitive to aphid density and lay their eggs in numbers proportional to the abundance of the aphid population. In a similar vein, females of the parasitic wasp *Tiphia popilliavora* control the sex of eggs they lay according to the size of their beetle host, *Popillia*. In third instar larval hosts, the parasites lay fertilized female eggs, but in smaller second instar hosts they lay smaller nonfertilized male eggs (Brunson, 1937).

At the same time, there is strong selective pressure for discrimination of potential danger from parasites and/or predators. Among cannibalistic species such as preying mantises, even one's own young may be a threat to siblings. For example, the young of female heliconiine butterflies in the Neotropics will eat eggs and one another. Ovipositing females often spend considerable time inspecting the host plant prior to oviposition; it has been suggested that this is a visual search for other *Heliconius* eggs or larvae. Probably the search is for other egg predators and parasites as well; one study estimated that over 90% of the eggs were killed by parasites (Gilbert, 1975). Such pressures may have led to uniformity in egg dispersion becoming a widespread trait among many parasitic and phytophagous insects (see also Fig. 9-19).

## Reproductive Rates and Parental Care

The survival of adult female carabid beetles from the end of one breeding season to the start of the next has been found to be inversely proportional to the amount of reproduction done in that first breeding season (Murdoch, 1966). Similar relationships probably hold for many intermittently breeding species.

**Reproductive effort** includes not only the caloric content of eggs and sperm but the whole variety of phenomena involved in the production of reproductively successful offspring: the *energies* expended in seeking mates, searching for appropriate oviposition sites, building nests, guarding eggs or young, feeding young, etc., and the *risks* resulting from the performance of these behaviors. How much effort and/or exposure to danger will be optimal for maximizing the number of viable offspring, and how should this be apportioned among those produced? Phylogeny offers little help in answering this, for within a particular taxon fecundity often varies enormously. Instead, one must turn to ecological factors and consider theories whose applicability ranges beyond insects (see Emlen, 1973, Brown, 1975, or Wilson, 1975, for an expanded account).

In an ideal environment, one with no predators and no intraspecific competition for resources, a population would increase at its maximum rate, $r$, a value simply obtained by substracting the population's death rate due to old age from its birth rate. The whole reproductive thrust of every species is directed toward maintaining $r$ as close to maximum as possible. In the nonideal real world, two very different strategies for doing this are possible, depending on the amount of competition and rigor of the environment. One may increase birth rate or decrease death rate. Uncrowded and nearly ideal environments tend to favor increases in birth rate, raising $r$ in the process. For this reason, selection that acts to raise the maximum rate of population increase has been called **r-selection** (more properly, "$r_{max}$ selection"). Species that are $r$-selected are opportunists, able to quickly discover new habitats. However, they are rarely able to persist successfully for long in such habitats, since they tend to be poor competitors. They often have what may be termed "big bang" reproduction, good colonizing ability, and short adult life spans. They can be termed **r-strategists.**

Other species live in approximate equilibrium with each other; their densities do not fluctuate much. Their population levels persist at just about carrying capacity, $K$, the number of individuals that the environment can support. There are no uncrowded habitats to find and exploit; however, being able to survive where one is, particularly for a long period, takes on new importance. There is little advantage to producing large numbers of young, but behaviors which increase the survival of one's young become important. These species can be termed **K-strategists.**

It must be appreciated that $r$ and $K$ are in reality the endpoints of a continuum. Related to all of this is the matter of energy allocation.

## Energy Allocation and the Evolution of Reproductive Rates

While some cockroaches deposit their oothecae rather indiscriminately in their environment, others incubate them internally, eventually giving birth to live young. Certain earwig and dipluran females brood their egg clutches prior to hatching. Females of the weevil *Byctiscus populi* construct rolls from young poplar leaves; within these cylinders eggs are deposited and larvae develop (Fig. 9-31). Other insects appear considerably less choosy about the fates of their offspring. Often they literally broadcast large numbers of eggs indiscriminately about the environment

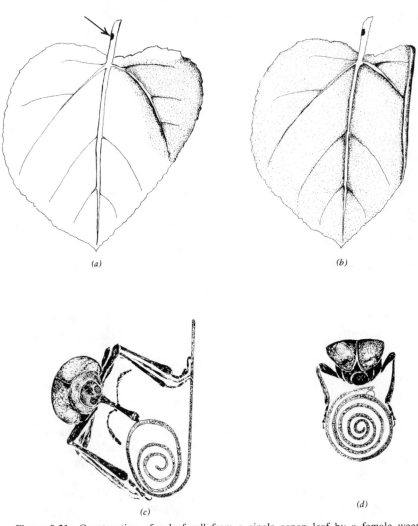

(a)

(b)

(c)

(d)

**Figure 9-31** Construction of a leaf roll from a single aspen leaf by a female weevil, *Byctiscus populi:* (*a*) After the leaf petiole is excavated (*arrow*), the beetle walks about on part of the leaf blade half (stippled area), perforating the epidermis with its tarsal claws and sometimes its jaws; resultant loss of turgor pressure makes that part of the leaf more flexible. (*b, c*) The beetle positions itself parallel to the leaf edge, sinks in its claws, and draws its legs up toward each other, using the snout to help guide the developing roll. Then, creeping inside the roll, the weevil chews a longitudinal slit through one layer; reversing position, she deposits a single egg (not shown) (*d*) Leaf rolling continues to completion, at which point the last edge is glued to the roll with anal secretion. (Drawing by J. W. Krispyn after Daanje, 1975.)

and then leave. For example, bombyliid flies spray their miniscule eggs in almost any depression in the soil by hovering briefly while making flicking movements with their abdomens (see Fig. 10-9). Walking sticks continuously drop single eggs apparently at random as they move through the forest foliage. Parasitic wasps of the family Trigonalidae leave thousands of eggs on foliage likely to be consumed by the caterpillars which serve as intermediate hosts. What adaptive bases might there be for such diverse patterns?

Any organism has only a limited lifetime with a limited amount of time and energy to partition among the various activities of its life. Clearly, this must be divided between efforts expended for reproductive activities and those expended for maintenance activities. One attempt to grasp the underlying logic behind the ways in which various organisms have come to partition their time, energy, and body resources is through the **principle of allocation** (Wilson, 1975). Each animal must divide its time and energy among three major requirements, which are (in their usual order of descending importance) food, defense against predators, and reproduction. For each species this total time/energy will be divided in a different manner depending upon environmental and evolutionary constraints. However, just as in any other budget, to the extent that one priority is easily satisfied, more expenditure can be devoted to activities of the other priorities.

The principle of allocation is well illustrated among that group of ichneumonid wasps which parasitize the Swaine jack pine sawfly, *Neodiprion swainei*. Since each species in the group is attacking the same host, each has approximately the same amount of food to utilize and to allocate to different functions. However, wide differences in such allocation are evident. These differences may be found in almost any parameter one wishes to choose, but a particularly revealing aspect is egg production.

Should a parasitic wasp produce many eggs or few? At first glance, large egg production seems to be a better strategy. However, the fecundity of the different wasp species in this jack pine sawfly complex varies from as low as 30 eggs to as many as 1000 in a lifetime (Fig. 9-32). Price's (1973, 1974) studies on this complex confirm that fecundity is related to a number of ecological factors, especially ease in finding hosts to serve as food for the progeny. Those wasps attacking the abundant, readily discovered early stages of the sawfly larvae which occur in the tree canopy tend to have high fecundity. However, the number of their offspring to survive is held in check by a relatively high mortality among the host larvae from predation and hyperparasitism.

Because of such mortality, later stages of the host are less abundant;

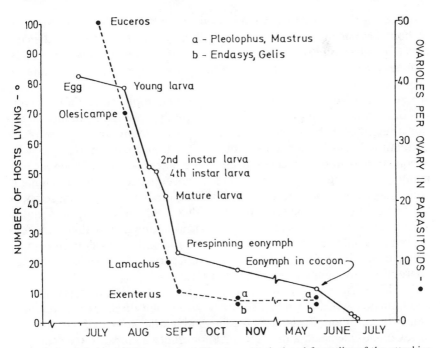

**Figure 9-32** Relationship between host life stage attacked and fecundity of the attacking parasite, as illustrated by the Swaine jack pine sawfly, *Neodiprion swainei,* and some of its ichneumonid wasp parasites. Solid line represents the survivorship curve of the sawfly host with the stages present at different times of the year indicated. The number of ovarioles per ovary in selected parasites (dashed line) is plotted in synchrony with the time each species most commonly attacks the host. (From Price, 1974, in *Evolution* **28:** 76–84.)

they are also better concealed, having moved down into forest floor litter. Ichneumonids attacking these stages have lower fecundity, there being little advantage to producing many more eggs than one can find hosts to leave them in. However, their offspring suffer less mortality in these better protected hosts.

### Egg Production and Fecundity

At first glance, the reproductive capacity of a given species would seem to be easily determined. In truth, however, records of egg laying or total egg output rarely give a true picture of the reproductive potential of any species. A variety of factors, both internal and external, influence total egg production.

Nutrition is probably the single most important factor. Many insect species, particularly those having a short adult life span, can lay eggs without having ingested any proteinaceous food. The short-lived mayflies (Ephemeroptera) are a prime example of this phenomenon, which is termed **autogeny.** Some autogenous insects never eat at all as adults or ingest only water or carbohydrates such as nectar. Most autogenous species, however, lay relatively few eggs under such conditions; given the opportunity to ingest proteinaceous foods, they will produce additional egg batches. In at least some cases there appears to be a correlation between autogeny and larval nutrition, those individuals with adequate nutritional reserves being more likely to be autogenous. In many cases autogeny also appears as a consequence of adverse climatic conditions. For example, autogeny in mosquitoes is common in temperate and cold regions, where hosts and adult food sources are sometimes scarce. High arctic mosquitoes are voracious biters, but there are few mammals available upon which to feed. Confined to feeding on nectar, they can still lay a few eggs to assure propagation of the species until an appropriate host appears (Spielman, 1971).

A wide variety of environmental factors such as light, temperature, and humidity wield an influence not only by their direct effects but also indirectly through effects upon feeding and mating activities. Day length, for example, has such a dual effect. Reproductive diapause (see Chapter 3) is common, as is hibernation, which may be thought of as a type of reproductive winter diapause. Both appear to be photoperiodically induced.

Finally, any attempt to ascertain an insect's reproductive capacity must take note of the fact that mature eggs may simply not be deposited. For example, some unmated females may mature a full complement of eggs but retain some or all of them until after mating. In other cases, where the proper substrate or host is not available, even fertile eggs may be retained and later reabsorbed.

In summary, it is doubtful whether data from laboratory colonies can ever realistically be equated with egg laying in natural environments. Insects reared in the laboratory generally live under better conditions, especially with respect to temperature and the availability of food, but natural ecological relationships are difficult to sustain.

## Modes of Insect Reproduction

In Chapter 1 we considered oviparity, ovoviviparity, and viviparity among cockroaches, and it seemed quite straightforward. By far the numerical majority, oviparous insects lay eggs. Ovoviviparous insects do the same but retract their eggs back into a brood pouch to develop, extruding them into the world again once they become nymphs. Viviparous insects, going a step further, "feed" their developing young within the brood pouch with nutrients and water.

The picture is not quite this simple. First, this area of insect reproduction badly needs further study. Most published reports on the topic are simply descriptive and speculative rather than providing sorely needed information on reproductive physiology and adaptive features. Many viviparous insects, for example, are probably classified as ovoviviparous simply because they are poorly known. Since there is no information on whether they obtain nutrients from their mother, it is guessed that they do not. Likewise, a really precise definition of true viviparity would confine the term to cases when the embryos obtain essential nutrients from their mother, without which they could not complete development; this assumption has almost never been tested in full. Second, one is faced with a quite common semantic problem. Discrete terms cannot adequately describe a continuum, and insect reproductive modes clearly form a continuum. For example, some insects have an extremely short retention of larvae; several chrysomelid beetle species, for one, have embryos which develop in the ovary so quickly that they hatch a few minutes after deposition. Should they be considered oviparous or ovoviviparous? They do not differ essentially from close relatives in which the embryos hatch within the genitalia just a few minutes earlier. Finally, in addition to the variety of reproductive modes mentioned above, insects also show a wide variety of more unusual types of development. A brief look at some of these should help us to gain some perspective on just how varied insect reproduction actually is. Additional information may be found in Chapman (1969).

In a smattering of insects and quite regularly among endoparasitic ones, each egg develops into a number of larvae rather than just one. This process, called **polyembryony,** theoretically would have the effect of increasing the reproductive potential of the insect involved, but because polyembryonic forms tend to lay fewer eggs than related "normal" species the actual net effect is not always much greater although the metabolic "cost" is lower.

In other cases insect eggs develop without being fertilized, a situation known as **parthenogenesis.** It is probably widespread as an occasional occurrence when a female fails to find a mate, for it has been recorded for nearly every insect order. In a number of species, however, it has become the normal method of reproduction, with both advantages and disadvantages. When only females are produced, the reproductive potential is much greater than if half the population are males, and a parthenogenetically reproducing female can spend all her time and energy in feeding and reproduction, without having to find a mate and/or court. However, there is an important drawback—lack of the genetic recombination which normally occurs at mating means a significantly slower adaptation to environmental changes.

In some cases this problem has been solved through an evolutionary compromise, namely, a life history of alternating parthenogenetic and bisexual generations.

Let us look at one example of such **alternation of generations.** On the underside of oak leaves one can often find galls where small wasps, *Neuroterus lenticularis,* overwinter. In the spring, only females emerge from these galls. The eggs they lay in catkin galls are all unfertilized, for there are no males. However, some eggs undergo meiosis to become haploid; others do not and thus remain diploid. As is generally true among Hymenoptera (and occasionally among other insects as well), haploid eggs are male and diploid eggs are female. In this way, a bisexual generation emerges in early summer. After mating, the females of this generation lay all fertilized eggs, which produce the females to emerge from oak leaf galls the following spring.

There are a few cases in which immature insects are able to reproduce. Apparently as a result of a hormone imbalance, development of the offspring usually begins while the parent is still a larva itself. Most insects that show such **paedogenesis** are also parthenogenetic and viviparous.

When all of these phenomena converge into a single life history, the situation can become incredibly complex, as aphids strikingly illustrate. In the Tropics, aphids are continuously parthenogenetic, but in temperate zones they have developed a complex alternation of generations. Throughout the summer, generations succeed one another so rapidly that while embryos are developing in a mother's egg tubes, they already have embryos developing in them. Thus, the female aphid is not only an expectant mother, but an expectant grandmother! In the fall males appear, and normal sexual and oviposition behavior are resumed. Eggs overwinter and, upon hatching the following spring, resume rapid-fire parthenogenetic reproduction once again. Overlaid upon this sytem there are usually also two other seasonal cycles: the occurrence of several aphid **morphs** (physiologically and morphologically distinct forms of the same species) and an alternation of host plants. For example, the first spring generation of bean aphids, *Aphis fabae,* emerge from overwintering eggs on spindle trees (a European species of *Euonymus*). This generation is entirely female and parthenogenetically produces either a winged female generation or several wingless generations of females which themselves give rise to winged forms. These migrate to bean plants, there to produce another series of wingless morphs. Ultimately, these produce both winged and wingless sexual forms. The winged ones return to spindle trees and produce females; the wingless ones go on to produce winged males which then join those females to mate and produce winter eggs.


## SUMMARY

Most arthropods are nonterritorial and nonsocial, have internal fertilization, and are polygamous. The mates do not stay together for parental care. When the group is restricted to this common majority, breeding habits are rather uniform. Among the more advanced arthropods, courtship is common. As with vertebrates, courtship initiation and direction rely upon special stimuli (releasers), usually coming from potential mates and often involving several sensory modalities. Usually, the displays are quite stereotyped and less variable than those of vertebrates.

Courtship is affected by life history constraints, ecological requirements, and the occurrence of related species. Common functions of insect courtship are to stimulate and maneuver females into copulation, to facilitate the meeting of solitary individuals, to facilitate identification, and to appease normally aggressive predatory mates. Sometimes species are effectively sorted prior to actual contact between the sexes; at other times, elaborate courtship rituals after contact serve the purpose of identifying mates as belonging to the proper sex and species. Coevolution of insects and plants has led to diverse examples of close relationships between plant pollination needs and insect reproductive behavior.

Sexual selection, a subcategory of Darwinian selection, helps to explain sexual divergence in behavior and morphology. Shaping the mechanisms that surround courtship and mating, it fosters a competition in which the outcome is production or nonproduction of offspring. Several modes of sexual selection exist, based on factors such as choices made among courting partners and various types of competition within the courting sex, including intraspecific aggression and territoriality.

The neuroendocrine system regulates much of reproductive behavior, particularly in species long-lived as adults and those with a well-developed reproductive diapause. In a "generalized" insect, juvenile hormones secreted by the corpora allata typically act as modifiers, setting the behavioral mode appropriate to the reproductive condition of the adult. Then, in the chain of specific events that lead to copulation, neurosecretory links appear, typically (but not necessarily exclusively) in the female, in association with specific phases of reproductive behavior and suitable environmental cues such as photoperiod. However, the exact behavioral role(s) of such hormones and other neurosecretions is still unclear for most insects.

Investment of the two sexes in reproductive activities usually tends to be unequal; males often invest no more than their sperm, which are small, numerous, and energetically cheap. Females, in contrast, not only produce much larger and more energetically expensive eggs but may also have the responsibility to search for suitable oviposition sites, build nests, guard young, etc., that is, activities requiring not only substantial additional energy allocation over that of males but considerably elevated risk levels.

Oviposition involves a chain of behavioral responses (not unlike those found in feeding behavior) resulting in selection of the oviposition site and assessment of its suitability. Reproductive rates as measured by egg production are closely intertwined with ecological factors. The evolution of various forms of postovipositional care represent "solutions" to the problem of reduction of parental fitness from offspring mortality due to parasites or environmental stress.

## SELECTED REFERENCES

Alcock, J. 1973. The mating behavior of *Empis barbatoides* Melander and *Empis poplitea* Loew (Diptera: Empididae). *J. Nat. Hist.* **7:** 411–420.

Alcock, J. 1975a. *Animal Behavior. An Evolutionary Approach.* Sinauer Associates, Sunderland, Mass., 547 pp.

Alcock, J. 1975b. Territorial behaviour by males of *Philanthus multimaculatus* (Hymenoptera: Sphecidae) with a review of territoriality in male sphecids. *Anim. Behav.* **23:** 889–895.

Alexander, R. D. 1961. Aggressiveness, territoriality, and sexual behavior in field crickets (Orthoptera: Gryllidae). *Behaviour* **17:** 130–223.

Alexander, R. D. 1964. The evolution of mating behaviour in arthropods. In *Insect Reproduction, Roy. Entomol. Soc. London Symp.* No. 2 K. C. Highnam (ed.), pp. 80–92.

Barrass, R. 1960. The effect of age on the performance of an innate behaviour pattern in *Mormoniella vitripennis* Walk. (Hymenoptera: Pteromalidae). *Behaviour* **15:** 210–218.

Barth, R. H. Jr. 1965. Insect mating behavior: endocrine control of a chemical communication system. *Science* **149:** 882–883.

Barth, R. H. and L. J. Lester. 1973. Neuro-hormonal control of sexual behavior in insects. *Annu. Rev. Entomol.* **18:** 445–472.

Bastock, M. 1967. *Courtship. An Ethological Study.* Aldine, Chicago, 220 pp. (especially pp. 51–72).

Beebe, W. 1947. Notes on the Hercules beetle, *Dynastes hercules* (Linn.) at Rancho Grande, Venezuela, with special reference to combat behavior. *Zoologica* **32:** 109–116.

Blum, M. S. and N. A. Blum (eds.). 1978. *Reproductive Competition and Selection in Insects.* Academic Press, New York (in press).

Brower, J. H. 1975. Sperm precedence in the Inidan meal moth, *Plodia interpunctella. Ann. Entomol. Soc. Amer.* **68:** 78–80.

Brown, J. L. 1975. *The Evolution of Behavior.* W. W. Norton, New York, 761 pp.

Brunson, M. H. 1937. The influence of the instars of host larvae on the sex of the progeny of *Tiphia popilliavora. Science* **86:** 197.

Callahan, P. S. and H. A. Denmark. 1973. Attraction of the "lovebug," *Plecia nearctica* (Diptera: Bibionidae) to UV irradiated automobile exhaust fumes. *Fla. Entomol.* **56:** 113–119.

Campanella, P. J. 1975. The evolution of mating systems in temperate zone dragonflies (Odonata: Anisoptera). II. *Libellula luctuosa* (Burmeister). *Behaviour* **54:** 278–309.

Campanella, P. J. and L. L. Wolf. 1974. Temporal leks as a mating system in a temperate zone dragonfly (Odonata: Anisoptera). I: *Plathemis lydia* (Drury). *Behaviour* **51:** 49–87.

Carayon, J. 1964. Un Cas d'offrande nuptiale chez les Hétéropteres. *C.R. Acad. Sci. Paris* **259:** 4815–4818.

Chapman, R. F. 1969. *The Insects. Structure and Function.* Elsevier, New York, 820 pp.

Craig, G. B. Jr. 1967. Mosquitoes: female monogamy induced by male accessory gland substance. *Science* **156:** 1499–1501.

Daanje, A. 1975. Some special features of the leaf-rolling technique of *Byctiscus populi* L. (Coleoptera: Rhynchitini). *Behaviour* **53:** 285–316.

Dodson, C. H. 1970. The role of chemical attractants in orchid pollination. In *Biochemical Coevolution,* K. L. Chambers, (ed.). Oregon State Univ. Press, Corvallis, pp. 83–107.

Dodson, C. H. 1975. Coevolution of orchids and bees. In *Coevolution of Animals and Plants* Gilbert, L. E. and P. H. Raven (eds.). Univ. of Texas Press, Austin, pp. 91–99.

Dodson, C. H., R. L. Dressler, H. G. Hills, R. M. Adams, and N. H. Williams. 1969. Biologically active compounds in orchid fragrances. *Science* **164:** 1243–1249.

Downes, J. A. 1966. Observations on the mating behaviour of the crab hole mosquito *Deinocerites cancer* (Diptera: Culicidae). *Can. Entomol.* **98:** 1169–1177.

Downes, J. A. 1969. The swarming and mating flight of Diptera. *Annu. Rev. Entomol.* **14:** 271–298.

Downes, J. A. 1970. The feeding and mating behaviour of the specialized Empidinae (Diptera); observations on four species of *Rhamphomyia* in the high Arctic and a general discussion. *Can. Entomol.* **102:** 769–791.

Dressler, R. L. 1968. Pollination by euglossine bees. *Evolution* **22:** 202–210.

Emlen, J. M. 1973. *Ecology: An Evolutionary Approach.* Addison-Wesley, Reading, Mass., 493 pp.

Engelmann, F. 1968. Endocrine control of reproduction in insects. *Annu. Rev. Entomol.* **13:** 1–26.

Engelmann, F. 1970. *The Physiology of Insect Reproduction.* Pergamon Press, New York, 307 pp. (Chapter 6 on Mating, pp. 57–106, is an excellent review.)

Evans, D. A. and R. W. Matthews. 1976. Comparative courtship behaviour in two species

of the parasitic chalcid wasp *Melittobia* (Hymenoptera: Eulophidae). *Anim. Behav.*
    **24:** 46–51.

Evans, H. E. 1968. Studies on Neotropical Pompilidae (Hymenoptera) IV. Examples of
    dual sex-limited mimicry in *Chirodamus*. *Psyche* **75:** 1–22.

Evoy, W. H. and B. P. Jones. 1971. Motor patterns of male euglossine bees evoked by
    floral fragrances. *Anim. Behav.* **19:** 583–588.

Faugeres, A., C. Petit, and E. Thibout. 1971. The components of sexual selection.
    *Evolution* **25:** 265–275.

Gary, N. E. and J. Marston. 1971. Mating behaviour of drone honey bees with queen
    models (*Apis mellifera* L.). *Anim. Behav.* **19:** 299–304.

Gilbert, L. E. 1975. Ecological consequences of a coevolved mutualism between butter-
    flies and plants. In *Coevolution of Animals and Plants,* L. E. Gilbert and P. Raven
    (eds.). Univ. Texas Press, Austin, pp. 210–240.

Heatwole, H., D. M. Davis, and A. M. Wenner. 1964. Detection of mates and hosts by
    parasitic insects of the genus *Megarhyssa* (Hymenoptera: Ichneumonidae). *Amer.
    Midl. Nat.* **71:** 374–381.

Hölldobler, B. 1976. Recruitment behavior, home range orientation and territoriality in
    harvester ants, *Pogonomyrmex*. *Behav. Ecol. Sociobiol.* **1:** 3–44.

King, P. E., R. R. Askew, and C. Sanger. 1969. The detection of parasitized hosts by
    males of *Nasonia vitripennis* (Walker) (Hymenoptera: Pteromalidae) and some possi-
    ble implications. *Proc. Roy. Entomol. Soc. London A*. **44:** 85–90.

Kogan, M. 1975. Plant resistance in pest management. In *Introduction to Insect Pest
    Management,* R. L. Metcalf, and W. H. Luckmann (eds.). Wiley, New York, pp.
    103–146.

Kullenberg, B. 1961. Studies in *Ophrys* L. pollination. *Zool. Bidr.* **34:** 1–340.

Kullenberg, B. and G. Bergström. 1973. The pollination of *Ophrys* orchids. *Nobel
    Symposium* (Stockholm) **25:** 253–258.

Kullenberg, B. and G. Bergström. 1976. Hymenoptera Aculeata males as pollinators of
    *Ophrys* orchids. *Zoo. Scripta* **5:** 13–23.

Lea, A. O. 1968. Mating without insemination in virgin *Aedes aegypti*. *J. Insect Physiol.*
    **14:** 305–308.

Lin, N. 1963. Territorial behaviour in the cicada killer wasp, *Sphecious spheciosus* (Drury)
    (Hymenoptera: Sphecidae), I. *Behaviour* **20:** 115–133.

Matthews, R. W. 1975. Courtship in parasitic wasps. In *Evolutionary Strategies of
    Parasitic Insects and Mites*. P. W. Price (ed.). Plenum Press, New York, pp. 66–86.

Mitchell, R. 1975. The evolution of oviposition tactics in the bean weevil, *Callosobruchus
    maculatus* (F.). *Ecology* **56:** 696–702.

Murdoch, W. W. 1966. Population stability and life-history phenomena. *Amer. Nat.* **100:**
    5–11.

Otte, D. 1974. Effects and functions in the evolution of signalling systems. *Annu. Rev.
    Ecol. Syst.* **5:** 385–417.

Otte, D. and A. Joern. 1975. Insect territoriality and its evolution: population studies of
    desert grasshoppers on creosote bushes. *J. Anim. Ecol.* **44:** 29–54.

Parker, G. A. 1970a. Sperm competition and its evolutionary consequences in the insects.
    *Bio. Rev., Cambridge Phil. Soc.* **45:** 525–568.

Parker, G. A. 1970b. The reproductive behaviour and the nature of sexual selection in

*Scatophaga stercoraria* L. (Diptera: Scatophagidae): IV. Epigamic competition and competition between males for the possession of females. *Behaviour* **37:** 113–139.

Parker, G. A. 1974. Courtship persistence and female guarding as male time investment strategies. *Behaviour* **48:** 157–184.

Parker, G. A. 1978. Evolution of competitive mate searching. *Annu. Rev. Entomol.* **23:** 173–196.

Pener, M. P. 1972. The corpus allatum in adult acridids: the interrelation of its functions and possible correlations with the life cycle. *Proc. Int. Study Conf. Current and Future Problems of Acridology, London, 1970,* pp. 135–147.

Petit, C. and L. Ehrman. 1969. Sexual selection in *Drosophila. Evol. Biol.* **2:** 157–191.

Pinto, J. D. 1972. The sexual behavior of *Meloe (Meloe) strigulosus* Mannerheim. *J. Kans. Entomol. Soc.* **45:** 128–135.

Price, P. W. 1973. Reproductive strategies in parasitoid wasps. *Amer. Nat.* **107:** 684–693.

Price, P. W. 1974. Strategies for egg production. *Evolution* **28:** 76–84.

Price, P. W. 1975. *Insect Ecology.* Wiley, New York, 514 pp.

Proctor, M. and P. Yeo. 1972. *The Pollination of Flowers.* Taplinger, New York, 418 pp.

Prokopy, R. J. 1972. Evidence for a marking pheromone deterring repeated oviposition in apple maggot flies. *Environ. Entomol.* **1:** 326–332.

Richards, O. W. 1927. Sexual selection and allied problems in the insects. *Biol. Rev., Cambridge Phil. Soc.* **2:** 298–364.

Riddiford, L. M. 1974. The role of hormones in the reproductive behaviour of female wild silk moths. In *Experimental Analysis of Insect Behaviour,* L. Barton-Browne, (ed.). Springer-Verlag, New York, pp. 278–285.

Riley, C. V. 1892. The yucca moth and yucca pollination. *Rep. Mo. Bot. Gard.* **3:** 99–158.

Roeder, K. D. 1935. An experimental analysis of the sexual behavior of the preying mantis *(Mantis religiosa). Biol. Bull.* **69:** 203–220.

Schaller, F. 1971. Indirect sperm transfer by soil arthropods. *Annu. Rev. Entomol.* **16:** 407–446.

Smith, R. L. 1974. Life history of *Abedus herberti* in central Arizona (Hemiptera: Belostomatidae). *Psyche* **81:** 272–283.

Smith, R. L. 1975. Bionomics and behavior of *Abedus herberti* with comparative observations on *Belostoma flumineum* and *Lethocerus medius* (Hemiptera: Belostomatidae). Ph.D. Dissertation, Arizona State University, Tempe, 171 pp.

Smith, R. L. 1976a. Brooding behavior of a male water bug, *Belostoma flumineum* (Hemiptera: Belostomatidae). *J. Kans. Entomol. Soc.* **49:** 333–343.

Smith, R. L. 1976b. Male brooding behavior of the water bug *Abedus herberti* (Hemiptera: Belostomatidae). *Ann. Entomol. Soc. Amer.* **69:** 740–747.

Soderstrom, T. R. and C. E. Calderon. 1971. Insect pollination in tropical rain forest grasses. *Biotropica* **3:** 1–16.

Spielman, A. 1971. Bionomics of autogenous mosquitoes. *Annu. Rev. Entomol.* **16:** 231–248.

Spieth, H. T. 1968. Evolutionary implications of sexual behavior in *Drosophila. Evol. Biol.* **2:** 157–193.

Spieth, H. T. 1974. Courtship behavior in *Drosophila. Annu. Rev. Entomol.* **19:** 385–406.

Stich, H. F. 1963. An experimental analysis of the courtship pattern of *Tipula oleracea* (Diptera). *Can. J. Zool.* **41:** 99–109.

Stoutamire, W. P. 1974. Australian terrestrial orchids, thynnid wasps, and pseudocopulation. *Amer. Orchid Soc. Bull.* **43:** 13–18.

Thornhill, R. 1976a. Sexual selection and parental investment in insects. *Amer. Nat.* **110:** 153–163.

Thornhill, R. 1976b. Sexual selection and nuptial feeding behavior in *Bittacus apicalis* (Insecta: Mecoptera). *Amer. Nat.* **110:** 529–548.

Thornhill, R. 1976c. Reproductive behavior of the lovebug, *Plecia nearctica* (Diptera: Bibionidae). *Ann. Entomol. Soc. Amer.* **69:** 843–847.

Trivers, R. L. 1972. Parental investment and sexual selection. In *Sexual Selection and the Descent of Man, 1871–1971,* B. Campbell (ed.). Aldine, Chicago, pp. 136–179.

Uvarov, B. P. 1966. *Grasshoppers and Locusts.* Cambridge Univ. Press, 481 pp.

Velthuis, H. H. W. and J. M. F. de Camargo. 1975a. Observations on male territories in a carpenter bee, *Xylocopa (Neoxylocopa) hirsutissima* Maidl. (Hymenoptera, Anthophoridae). *Z. Tierpsychol.* **38:** 409–418.

Velthius, H. H. W. and J. M. F. de Camargo. 1975b. Further observations on the function of male territories in the carpenter bee *Xylocopa (Neoxylocopa) hirsutissima* Maidl. (Anthophoridae, Hymenoptera). *Neth. J. Zool.* **25:** 516–528.

Waage, J. K. 1974. Reproductive behavior and its relation to territoriality in *Calopteryx maculata* (Beauvois) (Odonata: Calopterygidae). *Behaviour* **47:** 240–256.

West Eberhard, M. J. 1969. The social biology of polistine wasps. *Misc. Publ. Mus. Zool., Univ. of Michigan* No. 140, 101 pp.

Wilson, E. O. 1975. *Sociobiology. The New Synthesis.* Belknap Press of Harvard Univ. Press, Cambridge, Mass., 697 pp.

Wilson, F. 1961. Adult reproductive behavior in *Asolcus basalis* (Hymenoptera: Scelionidae). *Austr. J. Zool.* **9:** 739–751.

# 10
# Brood Care and Social Life

Webspinners (Embioptera) of all life stages live gregariously in a series of chambers constructed from silk spun from glands in the forelegs of adults. Male and female dung beetles in the genus *Geotrupes* dig a burrow in the soil together and store dung for their larva, which is tended for most of its life by the female. Earwig females (Dermaptera) lay their eggs in burrows and guard them until they hatch. Among the most highly developed social insects the brood form a core around which all activity is centered, and these brood are reared from egg to adult by adults.

Once a parent insect has fulfilled its responsibility for placing the egg (or larva in some instances) in an appropriate environmental situation, it may simply leave it. However, some continue to associate with the eggs and immature stages. Such behaviors are generally considered to be the precursors of the development of social behavior. For this reason, it is helpful to consider brood care and social life together.

## BROOD CARE

Certain female sawflies guard their eggs and young (Fig. 10-1). A pair of scarab beetles cuts out a chunk of freshly deposited manure, rolls it off, and then prepares an elaborate underground burial vault for it where eggs are eventually deposited. Young of the tingid *Gargaphia solani* orient to their mother and follow her from place to place. Brazilian pentatomid bugs, *Phloeophana longirostris,* apparently provide nourishment for their nymphs in addition to protecting them.

The initial stages of collaboration in the care of young probably center about oviposition, as for example, when a male damselfly accompanies his mate while she lays eggs (see Fig. 9-4). Since almost all insects are selective to one degree or another in their choice of oviposition sites, one might even say that the most fundamental rudiments of parental care are nearly universal, namely, the laying of eggs in the correct place at the correct time for proper development of young. Likewise, one might

**Figure 10-1** Maternal care in the Brazilian sawfly, *Themos olfersii*. Throughout the 20-some days required for incubation, the conspicuous orange and black female straddles her eggs. If disturbed, she displays one of a series of at least 10 different types of defensive or threatening reactions. Later, even when the fully sclerotized gregarious larvae migrate together to the basal portion of their host leaf and begin feeding, their mother continues to guard them, sometimes accompanied by other females that happen to be nearby. Apparently because of their distastefulness and warning coloration, the adults are only rarely subject to predation, and eggs and young also benefit from this protection. (Courtesy of B. Dias; from Dias; 1975. Copyright 1975 by Editora Vozes Ltda., Petrópolis, Rio de Janeiro.)

view internal incubation and the birth of live young as forms of parental care. Many nonsocial insects go far beyond these, however, by tailoring the environment to the needs of their young through various behaviors such as nest construction. When such an involvement of the adults in feeding and protection extends past the time of oviposition or birth, one may speak of **brood care**. As the involvement stretches ever further into the life of the offspring, various presocial and subsocial behaviors begin to appear. Among scarabaeid beetles, for example, one finds an evolutionary progression from species in which the adults simply amass manure provisions upon which eggs are laid to species such as *Copris* where the mother stays with her young until they reach adulthood. Finally, in groups such as the ants, adult offspring remain with the mother and assist her in the production of more young (Halffter and Matthews, 1966).

## The Ecology of Parental Care

During the past two decades, the combined efforts of a wide range of biologists have begun to result in a comprehensive theory of parental care. Four environmental "prime movers" appear to underly the evolution of parental care (Wilson, 1975a) not only among insects but among a wide variety of vertebrates as well.

The first of these is a stable, predictable environment, for the $K$-selection (see Chapter 9) which tends to prevail has certain demographic consequences: the animal tends to live longer, to grow larger, and to reproduce at intervals instead of all at once. If the habitat is structured, the animals will also tend to occupy a home range or territory or at least to return to regular feeding and/or refuge areas. These developments, in turn, all favor the evolution of parental care.

A second circumstance that may propel an animal toward parental care is the penetration of new, physically stressful environments which

**Figure 10-2** Maternal care in the membracid *Umbonia crassicornis*. The mother's presence and behavior has been shown to enhance the survival of the young. When a predator, such as a coccinellid beetle, comes into contact with the female she fans her wings and may also buzz, sometimes knocking the beetle to the ground. Similar behavior is recorded for over 30 species of membracids. (Courtesy of T. K. Wood, from Wood, 1976; see also Hinton, 1977.)

require some sort of protection of offspring, at least during the most vulnerable period of their development. When the environment is extremely harsh, many insect species guard their offspring. For example, the staphylinid beetle, *Bledius spectabilis,* occupies an unusually stressful habitat, the intertidal mud; females of this beetle both protect and feed their offspring.

Two other factors are also often involved. One is that sometimes it may be evolutionarily advantageous to have young present to help the parent find, exploit or guard certain specialized food sources. Finally, predator activity may favor a prolongation of parental investment which increases the probability that offspring will reach breeding age (Fig. 10-2).

All four of these can act singly or in combination to favor the evolution of parental care, and there is often considerable challenge in finding what factors actually lie behind an observed case of parental care. One instance in which this has been particularly well done involves a tropical stink bug, *Antiteuchus tripterus.*

CASE STUDY: PARENTAL CARE IN THE STINK BUG, *Antiteuchus tripterus*

*Antiteuchus tripterus* lays its barrel-shaped, unusually thin-shelled eggs, in compact masses almost always numbering 28, on the undersurface of leaves. For 15–16 days following oviposition, females remain with their clutch without feeding while the eggs develop and hatch and the nymphs grow to reach the second instar. Such parental behavior is not uncommon in the Pentatomidae, and representatives of several genera have been reported to have similar behavior. Why? Eberhard (1975) reasoned that the parent bug was defending her eggs against predators. His first experiment was simple. He removed brooding females from some batches of eggs while leaving others nearby untouched to serve as controls. Not one bug survived from 48 unguarded egg masses, most of which vanished, whereas 50% of the protected eggs produced bugs! Later observations revealed that several species of insects, especially foraging ants, fed on undefended eggs. Thus, the females' defense of their eggs against such "generalized" predators appeared to be highly effective, and necessary.

Nevertheless, a 50% mortality of the guarded eggs seemed surprisingly high, considering the time investment of the mothers. Eberhard presented artificial insect-sized models to brooding females. These elicited a repertoire of defensive behaviors, including waving antennae in the direction of the stimulus, tilting the body to form a shelf between the threat and the nymphs, general body "shuddering," scraping of the front legs along the periphery of the egg mass, and kicking backward with the middle and hind legs. Although like all other stink bugs *Antiteuchus* possessed glands which could spray a highly repellent chemical, this defense was never employed against the insect-sized stimulus. Eberhard assumed that such chemical defense was reserved for larger potential predators, for when the body was tilted for spraying, the bugs also vividly displayed aposematic orange bands along the sides of their abdomens.

Attempting to rear the eggs, Eberhard discovered that a number of them yielded, instead of bugs, tiny scelionid wasps of two different species. Why were they so successful despite constant maternal vigilance? Stereotypy in maternal orientation provided a clue. *Antiteuchus* females consistently aligned their bodies parallel with the axis of the leaf upon which their eggs were laid, generally facing toward the leaf tip. Observing them carefully when their overt behavior (kicking, scraping, etc.) indicated they were aware of the presence of a wasp attacking their eggs, Eberhard was unable to document a single instance when a mother bug turned and assumed a new orientation over its eggs. The consequence of this constant maternal orientation behavior (Fig. 10-3) was that, while females' defenses did effectively deter many wasp oviposition attempts in eggs situated on the front and sides of the egg mass, eggs to her rear were not effectively guarded. Analysis of the distribution of parasitized eggs within the mass confirmed this.

Would wasps attack more if eggs were not guarded at all? To exclude the foraging ants and other walking predators that had decimated unguarded eggs in his previous experiment, Eberhard ringed the trunk and branches of the host tree with sticky "tanglefoot." One of the scelionid wasps, *Trissolcus,* attacked the eggs at the same rate in the mother's absence. However, removing the mother caused significantly *reduced* parasitism levels for the other, *Phanuropsis*. In fact, parasitism was lessened by over half, from 64% on controls to 29% on unguarded eggs! These unexpected results led Eberhard to conclude that perhaps the presence of egg-guarding females served more or less as a visual "beacon" to this parasite, serving to identify location of potential hosts. Several observations seemed to support this idea. For one, *Phanuropsis* often approached or rested near quiescent individual bugs of both sexes. When a guarding female was removed several centimeters to one side of her eggs, female *Phanuropsis* waiting nearby typically ignored the undefended eggs and moved instead to realign themselves with the bug's

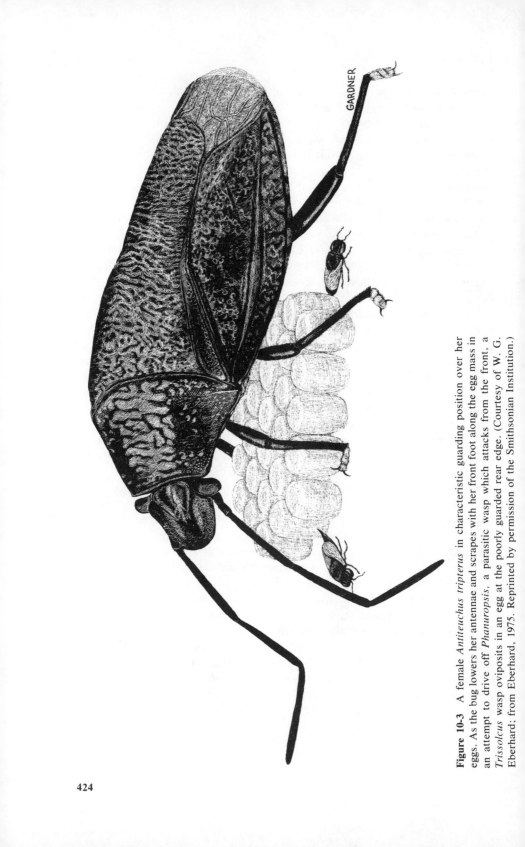

GARDNER

**Figure 10-3** A female *Antiteuchus tripterus* in characteristic guarding position over her eggs. As the bug lowers her antennae and scrapes with her front foot along the egg mass in an attempt to drive off *Phanuropsis*, a parasitic wasp which attacks from the front, a *Trissolcus* wasp oviposits in an egg at the poorly guarded rear edge. (Courtesy of W. G. Eberhard; from Eberhard, 1975. Reprinted by permission of the Smithsonian Institution.)

new position. That this response was visually mediated was supported by the additional observation that the wasps' searching activity was limited to daylight hours, peaking about midday.

Thus, even though the parental behavior of *Antiteuchus* seems well suited to combat activities of parasites and predators in general, she still suffers a net loss of eggs to at least one parasite species which, during the course of evolution, has turned her defensive habit to its own advantage for finding hosts.

For *Antiteuchus,* long adult life with survival enhanced by potent chemical defense and aposematism, coupled with the uncertainty of offspring survival due to predation and parasitism, has led to maternal care being exercised during the most vulnerable developmental stages of the young. Reproductive episodes are repeated two or three times during an individual's life rather than being put into one mass breeding attempt. Other aspects of the bug's ecology and natural history tend to fit the *K*-selected species criteria, notably the relatively high population densities per plant and the rather viscous population structure. Thus, parental behavior and ecology tend to be well correlated, as predicted, with risks of predation and parasitism acting as important selective factors in this instance. As we shall see, parasite pressure tends to be a recurring factor in parental care, particularly among nesting Hymenoptera.

## NESTING BEHAVIOR

Nest building, or **nidification,** is not an overwhelmingly common insect behavior. Yet it has reappeared with striking regularity among a few groups, particularly those closely associated with the soil. For example, nidification is found in certain earwigs, a few burrowing cricket genera (Fig. 10-4, see also Fig. 7-3), among a variety of beetles (Milne and Milne, 1976), and widely among both solitary and social Hymenoptera and Isoptera. Deferring the truly social species for later, let us focus upon Scarabaeinae beetles and the solitary wasps. In both, certain phylogenetic trends in nesting behavior are evident and correspond reasonably well with established taxonomic categories. Ecologically adaptative trends often displayed by various taxonomic groups in parallel are also evident.

### Nidification in Dung Beetles

When many species of the big green *Phanaeus* beetle in full flight detect an animal dropping, they immediately begin a circling descent, land on

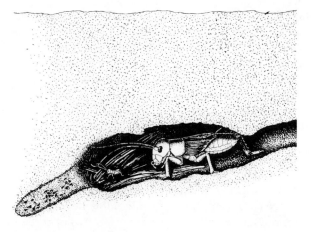

**Figure 10-4** A female burrowing cricket, *Anurogryllus muticus,* in her brood chamber. The chamber floor is lined with grass gathered by the female. To the right is the exit burrow; to the left, a filled defecation chamber. A single newly hatched nymph may be seen touching antennae with the mother, while an egg pile lies beneath her abdomen partly covered by soil. In this species, special miniature eggs are laid by the mother to serve as baby food; they are very attractive to the offspring, which crowd around them and fight for their possession. (Slightly modified from West and Alexander, 1963.)

this potential food, and then begin to dig a burrow near its edge. As the female pushes and rolls fragments of excrement toward and into this burrow, her activity attracts the male, who joins her and helps in the digging. At first the hole may function solely as a feeding burrow for the adults, but usually the female within soon begins to prepare one or several brood balls, that is, carefully rolled masses of excrement with a top egg chamber surrounded by a strong earth wall. As the larva develops, it feeds on the dung, which has been kept fresh and moist by its protective coating. Although the couple continue to live in the same burrow, they do not appear to give the brood ball further attention; any cracks in the outer wall are repaired by the larva itself, using its own excrement (Halffter et al., 1974).

The term "dung beetle" has commonly been applied to three subfamilies of the Scarabaeidae: the Scarabaeinae, the Aphodiinae, and the Geotrupinae, for all have dung-eating, or **coprophagous,** members. Only in the Scarabaeinae, however, are the vast majority of the species coprophagous, most feeding on the excrement of large mammals. No known adult Scarabaeinae have mouthparts capable of chewing or cutting solid food, hence when dung begins to desiccate it becomes unusable to them. Nest construction in the subfamily is nearly always the cooperative endeavor of a male–female pair. The association is a close one, and once a pair bond is formed it is usually not broken voluntarily by either individual until after the nest has been dug and provisioned in cooperation. Interestingly, there is no courtship of the ritualistic sort; apparently the prolonged cooperation in nidification in this subfamily serves the same purpose of species recognition by bringing the two members of the pair into repeated contact prior to copulation.

Halffter and Matthews (1966), responsible for the most comprehensive studies available on the natural history of dung beetles, recognize four major behavioral groups based on nidification (Fig. 10-5). Group I nesting behavior involves a brood burrow not differing in principle from

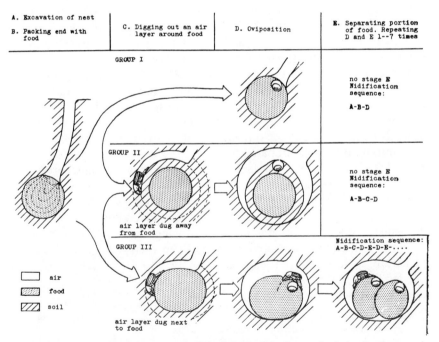

**Figure 10-5** A diagrammatic representation of three of the four major behavioral groups in the beetle subfamily Scarabaeinae, based on nidification. The primitive condition, in which larval food is simply packed into the tunnel end, has undergone various elaborations and modifications resulting in increased protection for the immature stages and the larval food. In group II the egg is isolated from direct contact with the food material, possibly to discourage mold formation on the egg, which must be exposed to an air supply. (From Halffter and Matthews, 1966.)

the adult feeding burrow. In both this and group II, there is no brood care. However, among dung beetles of group II the female parent protects the larval food supply from desiccation by modeling it into a sphere or pear with an outer shell of soil or clay. Sometimes both the egg and food mass are enclosed in separate clay-shelled compartments.

Members of the third group, which includes the well-known genera *Copris* and *Synapsis,* have a basic pattern which may be truly termed subsocial. A dung cake is accumulated, then cut up into several separate ovoids or spheres which are brooded until offspring pupate, by the female who cleans mold growth from the surface of the brood ovoids and thus increases brood survival significantly. Males of both *Copris* and *Synapsis* also play an unusually important role, not only helping dig and provision the nest but also aiding in compaction of the dung cake and, during this period of active participation, actively defending the nest (see also Klemperer and Boulton, 1976).

Group IV includes most of the tribe Scarabaeini, often called dung rollers. The ability to roll a ball of dung overland sharply distinguishes them from the previous groups and results in a quite homogeneous group behaviorally. However, because morphological adaptations to dung rolling often mean a partial or complete loss of digging ability, there are no elaborate nests or impressive excavations in this tribe, and some do not bury the brood ball at all. The brood pear is made by tearing apart the original ball, rebuilding it, then sometimes placing a layer of soil around the finished pear. No members of this group are known to pack food into a previously prepared burrow.

The Australian continent lacks native large mammalian herbivores. Consequently, native dung beetles have evolved as specialists on the small bits of excrement characteristic of marsupials such as wallabies and kangaroos. With the advent of sprawling cattle ranches in northern Australia, cattle dung became an unanticipated blight on the countryside as large amounts accumulated in the pastures. To restore the "balance of nature," Australian entomologists imported several species of dung beetles from Africa, a process that first necessitated detailed studies on dung beetle behavior (Waterhouse, 1974).

**Nesting Behavior in Solitary Bees and Wasps**

In early summer, sand dunes throughout the world are alive with large yellow and black wasps flying briskly about low over the dune surface. Later, these flying wasps seem to dwindle in number, but closer inspection will usually reveal wasps quietly at work, each digging, filling with prey, then closing nest burrows in the sand. A patient observer may be rewarded by the sight of a wasp suddenly appearing with a large, brown fly under her body held by her middle legs (see Fig. 1-8). The wasp alights in a nondistinctive spot, makes a few digging movements synchronously with her front legs, then quickly disappears through this previously invisible "door." If one then carefully digs along the burrow to its end several inches below the surface, one finds a chamber with a legless, helpless yellow-white grub feeding on the flies brought in by its mother. After eating 20 to 30 of them, the grub spins a cigar-shaped cocoon in which it will eventually metamorphose into an adult.

In their nest-building activities, the higher Hymenoptera (wasps, ants, and bees) exhibit some of the most complex behavior known in invertebrates. We have discussed some of these behaviors before: the "fixity of instinct" shown in the nesting activities of *Ammophila*

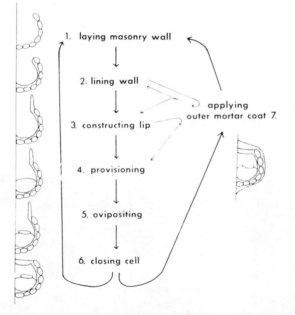

**Figure 10-6** Complex behavior in the mason bee, *Hoplitis anthocopoides*. The female builds a nest of pebbles and mud in protected locations and exclusively visits an immigrant roadside weed, *Echium vulgare* (Boraginaceae) for the nectar and pollen to provision the nest cells. The sequence of nest-building behaviors is shown: (1) laying a masonry wall of pebbles joined by mortar, (2) lining the walls with mortar, (3) constructing a mortar lip around the entrance, (4) provisioning with pollen and nectar, (5) oviposition, (6) closing the cell with mortar from the lip, (7) covering the cell with an outer mortar coat. The last phase may occur during or after any of the phases where mortar is used in the construction procedures. (Courtesy of G. C. Eickwort, from Eickwort, 1975b.)

*pubescens,* the remarkable ability of *Philanthus triangulum* to learn the locality of her nest, the interaction between predator and prey in the cricket-hunting *Liris nigra,* the sequence of prey catching by the bee wolf, *P. triangulum.* All of these are wasps in the family Sphecidae, a large and diverse group whose members exhibit some of the most elaborate and stereotyped behavior patterns to be found in the Insecta outside of the truly social insects. An example among the megachilid bees is *Hoplitis* (Fig. 10-6).

Being the focal point of activities, nests are in a sense the morphological expression of behavior, often the summation of the behavior of many individuals. Since unfortunately most nestbuilders choose concealed locations (e.g., soil) or else are social, the nests themselves can often be studied and compared more readily than the behavior patterns. They can also be measured and subjected to statistical analyses in the manner of a morphological character. Thus, an important part of the study of nesting behavior is the study of the nests themselves.

Among the solitary bees and wasps, nests belong to three broad types: those dug in a substrate such as soil, rotten wood, or plant pith (Fig. 10-7); those constructed in preexisting cavities such as hollow twigs (Fig. 10-8); and those constructed wholly for foreign materials such as plant pulp, mud, or resin (Evans and Eberhard, 1970). Interestingly, all truly social species construct nests of foreign materials. It should be noted that numerous reversals within groups and convergence between unre-

**Figure 10-7**  X-Ray of a portion of the nest of *Ceratina calcarata,* the little carpenter bee. Nests are excavated by the female burrowing into the pithy stems of plume grass (*Erianthus* sp.). One of the cells has been parasitized and contains several smaller larvae which will develop into tiny parasitic wasps. X-Ray photography is a useful technique for observing nest architecture and cell contents without physically harming the nest structure; similar techniques have also been applied to behavior studies of various wood-boring beetles. (Courtesy of C. J. Kislow.)

$\longrightarrow$ ENTRANCE

**Figure 10-8** *Trypargilum striatum* (Sphecidae) constructs nests in various performed cavities, separating off individual cells with mud partitions (see also Fig. 7-1). Shown here are four cells, two packed with spider prey and two with cocoons, whose shape happens to be species diagnostic. Species of wasps and bees that utilize preexisting cavities may be studied by using "trap nests," simulated hollow twigs made from blocks of wood into which a hole has been drilled. Such traps, with holes of differing diameters, are then set on fence posts, in tree crotches, on window ledges, etc., where a variety of Hymenoptera will utilize them. (Photograph by the authors.)

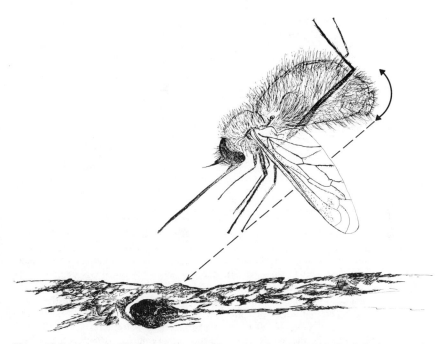

**Figure 10-9** A bombyliid fly hovers motionless over the entrance hole of the burrow of a solitary bee, into which she flings her eggs by repeated flicks (*arrow*) of her abdomen. The raised position of the hind legs is characteristic of these bee flies. (Drawing by J. W. Krispyn from a photograph.)

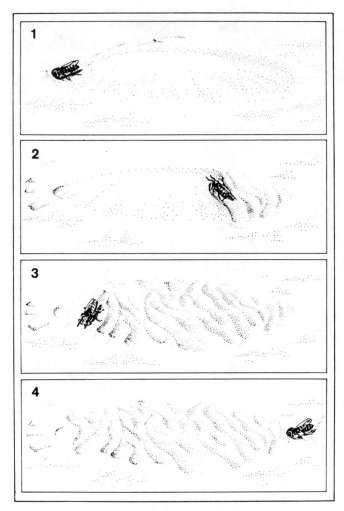

**Figure 10-10** Leveling behavior by females of the Australian sand wasp, *Bembix variabilis*. Throughout both digging and leveling, synchronous motion of the long-spined front tarsi serves to rake along the sand. (1) While digging her nest, the female gradually piles a mound of sand next to the nest entrance. (2) When the burrow is complete, the female closes the entrance and moves to the far end of the mound. Here she works forward, scraping sand with her front feet as she turns first to one side then to the other, thus producing a zigzag pattern in the sand. (3) As she repeats these movements over a slightly different course some 10–20 times, a characteristic pattern results in the now dispersed sand. (4) Finally the female digs a short accessory burrow, which is left open at the opposite end of the leveling pattern from the entrance. (Drawing by Sarah Landry; from Evans and Matthews, 1973; see also Evans and Matthews, 1975.)

lated groups with respect to nest type are known. One cannot assume that a series of nest types within a particular group necessarily represents an evolutionary progression. For example, nests made of mud have been independently evolved among certain Sphecidae, Eumenidae, Pompilidae, and Megachilidae. Likewise, among members of various genera (e.g., *Ammophila* and *Euodynerus*), both burrowers and species that utilize preexisting cavities appear.

Many of the details of nesting behavior appear to have been molded by biotic factors, notably various natural enemies and nest associates. For example, among ground-nesting sphecid wasps Evans (1966) has found several species which construct **accessory burrows,** blind-end false tunnels close beside the true nest burrow. In every case the accessory burrows are left open and the true nest entrance closed off. There is evidence that such accessory burrows divert the attention of parasites; bombyliid flies (Fig. 10-9) may lay their eggs in them, and velvet ants often spend time investigating and digging in the bottom of such holes.

Species nesting in the soil have another problem relative to pests, namely, soil removal. A mound of dirt at the nest entry can be a dead giveaway of the presence of that nest. So while some species simply let the displaced soil accumulate outside the nest entrance, others carry out the soil a bit at a time in their jaws, flying off a short distance before dropping it to the ground. Others push or rake the dirt backward out of the nest and then by a variety of behaviors remove or at least disperse the conspicuous mound. These **leveling** behaviors, species characteristic, typically consist of repeated passages over the mound while scraping with the front legs and turning slightly from side to side (Fig. 10-10). The resultant patterns, highly diagnostic of the species that made them, are often elaborate.

## SOCIAL ORGANIZATION

Most insects come together at least temporarily for mating; many also come together for other reasons, forming relatively permanent or only temporary groups. Some of these are based on relatedness, others are not. Sometimes such grouping results simply from a common response to some particular environmental factor, such as presumably occurs when swarms of butterflies form at a mud puddle (see Fig. 6-10). At other times, mutual attraction is clearly involved; some of the communicative devices involved have already been mentioned in Chapters 5, 6, and 7.

Might all of these be called "social responses"? The answer depends

**Table 10-1** A simple classification of the types of insect associations excluding those involving sexual activities[a]

| Association type | Basis of association | Degree of interaction among member individuals | Duration of association | Nature of association | Selected examples |
|---|---|---|---|---|---|
| Aggregations | Facultative "mutual attraction," mediated by either intrinsic or extrinsic (environmental) factors | Uncoordinated | Temporary (often of seasonal occurrence) | Open: members of other populations uncritically accepted | Feeding aggregations of fruit flies; communal roosts of butterflies; sleeping clusters of bees or wasps; hibernating associations of coccinellid beetles |
| Simple groups | As above | Coordinated movement occurs | Temporary and facultative | Open | Migrations of locusts and butterflies |
| Primitive societies | Often parent-offspring dependence | Reciprocal communication; often cooperative nest construction or defense | Often persistent through a particular life stage and typically centered around some type of nest | Open: members increasingly tend to share greater degrees of genetic relatedness and may often be siblings | Passalid beetles; tent caterpillars; cases of parental care of offspring; embiid webspinners |
| Advanced societies | Obligatory interdependence of all developmental stages | Highly integrated cooperative behavior, with efficient division of labor | Facultatively permanent | Closed: all members closely related genetically and members of other colonies excluded | All so-called "truly social" insects—ants eusocial bees, and wasps, termites |

[a] Overlap among the four major categories are not uncommon. Compiled from various sources, but primarily based on Wilson (1975a) and Lindauer (1974).

upon whom you ask. Biologists use the term *social* in a wide variety of ways. Some might speak of the social relations between sexes during courtship and mating; others speak of social interactions among animals in a herd, birds in a flock, or fish in a school. Ecologists sometimes refer to all of the organisms in a habitat or community as a society. However, a single thread is common to all these varied uses—the indication of adaptively significant, usually cooperative interactions between two or more individuals. Such organization by implication transcends the simple sum of the separate parts.

As Wilson (1975a) has aptly pointed out, such a broad usage of the terms social and society should be continued in order to keep from excluding from consideration many biologically interesting phenomena. At the same time, however, it should be recognized that to attempt to classify all the kinds and degrees of this broadly defined sociality into a single coherent system is nearly impossible. One attempt back in 1918 ended up with 40 categories, each of which had its own impossible terminology and often its own set of imprecise subdivisions!

Rather than getting mired in a bog of definitions, let us recognize that we are dealing with a continuum of relationships. The various labels and their meanings indicate not merely interactions among group members but interactions that produce effects that are, to a slight or great degree, qualitatively different from the mere summation of the independent activities of the individuals. Most attempts to classify the various kinds of animal associations have been made according to their form, basis for association, degree of interactions of members, and the nature and duration of association. One useful simple system is that represented in Table 10-1.

### Advantages of Group Behavior

Members of aggregations and simple groups share a number of distinct advantages. One that may operate in mating swarms is increased reproductive efficiency. Normally occurring only during a short time at a certain hour of the day or night, such swarming should theoretically be extremely advantageous to members of rare species and to those living in environments where the optimal time for mating is unpredictable. In addition, swarms may promote outcrossing and, by their very specificity in time and space, provide a premating isolating mechanism for different species.

Grouping also has many advantages in defense. First, any assemblage of individuals is likely to improve the chances of detection of potential predators. Additionally, simple gregariousness may enhance predator confusion. Sometimes, because of the aerodynamic and hydrodynamic advantages of being in close formation, grouping may increase speed and conserve energy during escape. Finally, many types of simple coordination of group behavior are involved in various defensive displays (see Fig. 5-9).

Group living also changes the nature of predator–prey relationships. One view of this is the so-called **gluttony principle** (Brown, 1975): to escape predation, it is better to live in an area where predators have full stomachs rather than where their stomachs are empty. Since predators of a given species tend to space themselves out (assuming little or no migration in response to the prey), highly clumped prey are probably at an advantage. Appearing intermittently in both space and time is even better. For example, periodical cicadas emerge as adults in synchrony every 13 or 17 years and, even though they are easily caught, the numbers of them present are beyond the capacity of predators to consume (Fig. 10-11).

Mollification of the environment is another function and advantage of group behavior. For example, during the dry season in Mexico some desert species of harvestmen gather in tight clumps in the lowest fork of the branches of candelabra cacti. Folding their legs over their backs, up to 70,000 of these invertebrates mass themselves into a single furlike pelt that retains the moisture given off by the cactus. This behavior, which prevents desiccation, is pheromonally mediated, and forcibly removed individuals will attempt to rejoin the aggregations from as far as 30 meters away (Wagner, 1954).

Food detection and/or utilization may also be more efficient in a group. When food supplies are patchy and transient but rich, a group may locate such food supplies more efficiently than isolated individuals, particularly if some sort of communication between the individuals has developed. (The social insects, of course, excel at recruitment of large numbers of workers to exploit food sources.) In addition, cooperative foraging and feeding may allow utilization of otherwise unavailable foods.

More subtle influences may also be at work in group life. Often, one can see in the behavior of group members well-developed patterns that appear to depend on mutual interaction. Many organisms consume more food when kept together than if housed singly, for example. Cockroaches, *Blattella germanica*, apparently grow larger and faster in groups than alone. Worker ants show more intense digging behavior when confined in groups than do isolated individuals given the same conditions. Human beings show analogous behavior. For example, it takes a great deal of resolution to avoid joining in when faced with a

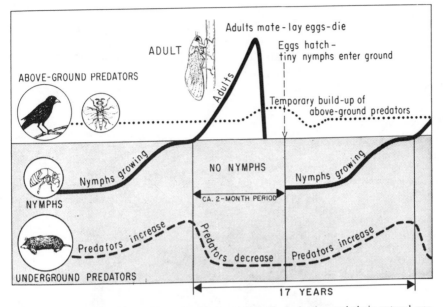

**Figure 10-11** Ecological relationships between periodical cicadas and their natural enemies, an unusual situation in which the life cycle of the prey is much longer than that of the predators. Through an aggregation in space and time, the periodical cicadas escape not predation itself, but the delayed density-dependent effects of predation on their population, for their emergence in great numbers always comes as a "surprise" to their enemies.

Every 17 years, significant numbers of adult cicadas appear above ground to mate and lay eggs. Birds and parasitic insects prey upon the abundant cicada supply, but the predators' resultant population increases cannot persist for 17 "lean" years until the next cicada bonanza. Underground, moles gradually increase their numbers as the biomass of periodical cicada nymphs increases. However, when the cicadas emerge from the ground, this food supply suddenly vanishes for 2 months, then is replaced by myriads of tiny nymphs too small to be practical as food for moles. Faced with this sudden crisis, "surplus" moles are driven into less suitable habitats. (From Lloyd and Dybas, 1966, in *Evolution* **20:** 133–149.)

yawning or laughing person. This tendency to follow suit is called **social facilitation.** More precisely, it may be defined as an increase in the pace or frequency of a given behavior due to the presence or activities of another conspecific individual. An example of its workings may be seen in the Australian sawfly, *Perga affinis*, where individual survival apparently depends on social facilitation. Females lay their eggs in pods within the tissue of eucalyptus leaves so that after hatching the young must first gnaw an exit hole before they can begin feeding. This is a

difficult task, but if one succeeds, the others follow. There is a strong
correlation between pod size and larval mortality. In one study (Carne,
1966), pods with fewer than 10 eggs suffered 66% mortality, but among
those containing more than 30 eggs mortality was only 43%. Social
facilitation is also involved when mature *Perga* larvae leave the tree to
pupate (Fig. 10-12).

**Figure 10-12** Australian sawfly larvae of the genus *Perga* crawling on the forest floor
prior to cocooning. In order to successfully construct cocoons and pupate they must first
penetrate the crusty soil. Although individuals are poorly adapted for such burrowing, in
larger aggregations at least one larva usually succeeds in breaking through and the others
are then able to follow. (Photograph by the authors.)

## Aggregations and Simple Groups

By flashlight, one can often find large clusters of bees or wasps assembled upon a branch to pass the night (Fig. 10-13). Up to 200 or 300 individuals may be involved, usually, but not always, of one species. Often they will return night after night to the same location, but they do not coinhabit a nest or rear young together, and the basis of their individual attraction to the sleeping roost is unknown (Linsley, 1962). Many soil-nesting bees and wasps group their nests in clusters; each female makes her own nest, but a given area may harbor an enormous number of burrows. Australian bush flies will gather on the downwind side of almost any animal (Fig. 10-14).

**Figure 10-13** Simple aggregation: a cluster of males and females of the solitary wasp, *Steniolia obliqua* (Sphecidae), on a branch of lodgepole pine in Wyoming. Clusters form each evening between 5:00 and 6:00 P.M. For more than six weeks over at least five consecutive summer seasons, the same branch was used every night; however, marking of individuals showed that the composition of such clusters changes nightly. (Courtesy H. E. Evans; from Evans and Gillaspy, 1964.)

**Figure 10-14**  An aggregation of Australian bush flies, *Musca vestustissima*, clustered on a man's back. The harmless flies, which thrive in the vast semidesert "outback," constitute a continual annoyance by their sheer numbers and tenacity. (Photograph by the authors.)

When an assemblage is composed of conspecific individuals including more than just a mated pair or family, all gathering temporarily in the same place but not internally organized or engaged in cooperative behavior, it is termed an **aggregation.** A diversity of such "uncoordinated" groups are common among various insects (see also Fig. 6-10).

The causal factors behind aggregations are often unknown. Do they result from limitations of suitable habitat? Some clearly do. In other cases, however, other factors are at work; one of these is social interaction. Thus, aggregations often grade into that slightly more sophisticated association where members exhibit at least some coordination in their movements. Such **simple groups,** again comprised of conspecific individuals remaining together for a period of time, are made up of group members that interact with one another to a distinctly greater degree than they do with other conspecific individuals. Swarms of midges, parasitic wasps (see Fig. 9-14), chorusing cicadas, migratory swarms of locusts or butterflies all provide good examples of simple coordinated groups which functionally may be divided into reproductive and nonreproductive groups.

## Primitive and Advanced Societies

*Odontotaenius (=Popilius) disjunctus* is a large, glossy black beetle living in well-decayed hardwood logs (Fig. 10-15). Unlike most other wood-consuming species, it apparently lacks digestive symbionts. Thus, an important part of its diet is fecal pellets and frass, which act as a substrate for bacterial and fungal development and are reingested. Adult bess beetles live two or more years; between raising their yearly generations of young, they are gregarious, many adults being found in

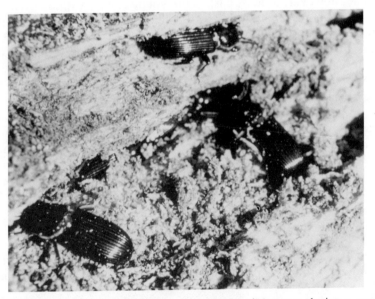

**Figure 10-15** Horned passalus beetles, *Odontotaenius disjunctus*, sharing a common tunnel system in a decaying oak log during the fall. During this nonreproductive season, both adult sexes are present but young are absent. Adults appear to have a rudimentary social organization including aggressive interactions which may often be accompanied by audible stridulatory noises. (Photograph by the authors; see also Mullen and Hunter, 1973.)

the same tunnel system. During the reproductive season, each pair of beetles maintains its own tunnel system in which 20 to 60 eggs are laid. The developing young feed on material prepared by the adults, and when ready to pupate they cooperate with the adults in construction of pupal chambers.

The diagnostic criterion of a **society** is reciprocal communication of a cooperative nature (Wilson, 1975a). The level of social organization recognized as a society includes most instances of parental care discussed in the first part of this chapter. It also embraces some groups of developing individuals of the same generation. Certain tent caterpillars, for example, spin a home web at which they gather gregariously to sleep and a feeding net which also serves as a protective community shelter; successful foragers deposit chemical trails that elicit preferential following by unfed individuals, a clear case of communication (Fitzgerald, 1976). Spider colonies may provide further examples (see Kullman, 1972, and Lubin, 1974; see also Fig. 4-18).

In most insect species parents die before progeny mature, or else offspring disperse and do not remain associated with the parental nest. In some, however, at least part of the young persist with their parent(s) for one or more generations as an extended family. A majority of the more highly developed insect societies are based upon such **kin groups**, that is, sets of individuals whose members are genetically more closely related than random because many have the same parents.

Traditionally, entomologists have reserved the word "social" for those species which live colonially in some type of nest and care for their progeny through to the adult stage. Yet even this seemingly narrow usage has been subject to confusion, compounded by the cumbersome terminology arising with new knowledge about the behavior and characteristics of "primitive" insect societies. Rather than ordering all the individual social examples found, the more useful systems have tended to catalogue social qualities such as group size, integration of behavior, and differentiation of roles. Advanced societies of "truly social," or **eusocial,** species are considered to include only those which meet three criteria: they live as groups of adults of different generations, with cooperative activity and with different individuals obligatorily performing different roles essential for the success of the group (Michener, 1969).

Insect societies may be viewed as mechanisms for massive reproduction. This high productivity has been made possible through the evolution of a variation in parental care—the extensive involvement of a major proportion of the colony inhabitants solely in the care and feeding of young. In all eusocial insects colony maintenance depends on a range of different jobs being carried out by nonreproductive individuals, or **workers**. In most cases they differ morphologically in little more than detail of proportions, but among the honey bees there is highly developed task-related specialization (see Table 10-2). Among the Hymenoptera, workers are female and usually sterile, although some may occasionally produce eggs. Termite workers, on the other hand, are of either sex. Members of insect societies typically represent one or more **castes** (see Fig. 10-20),which are sets of individuals in a given colony that are both morphologically distinct and specialized in behavior. Well-fed large ants of many species, for example, develop enormous heads and mandibles and come to constitute a "soldier" or "major" caste. Polymorphism reaches its most spectacular development in the Isoptera. Queens of some species develop huge abdomens so distended as to resemble a small sausage (Fig. 10-16), while soldiers exhibit either enlarged heads with powerful jaws or conical nozzlelike heads that eject a sticky substance (see Fig. 8-28). Weaver (1966) reviews the physiology of caste determination.

The range of duties surrounding the care of the young becomes correspondingly greater as colony complexity increases, and in some cases an age-related division of labor develops. For example, in the ant *Myrmica scabrinodis,* worker individuals which have emerged during the present season function as nurses; those which became adults during the previous season are builders, while even older individuals act as foragers. Such **age polyethism** reaches its greatest development among the honey bees (Lindauer, 1961). On emergence, a young bee works as a cleaner for about three days. After this, coincidental with labial gland development, she becomes a nurse, producing secretions with which she feeds larvae. About the tenth day of her life her abdominal glands begin to produce wax and her labial glands atrophy; she becomes a builder. At about the sixteenth day she begins to receive nectar and pollen loads from foragers and stores them in the comb. On about the twentieth day, she fills the post of guard. Finally, she becomes a forager, working at this for the rest of her life. Age polyethism, though well developed, is not inflexible. If a hive is divided so that one half contains only young bees and the other only older foraging bees, after a few days adjustment workers in each half carry out all normal hive tasks.

In addition to a well-developed system of castes, the major eusocial groups all have chemical communication systems for colony integration (Blum, 1974) and elaborate nest structures which provide temperature and humidity control within relatively precise tolerances. Once regarded

**Table 10-2** Comparison of some structural differences between worker and queen castes of the honey bee, *Apis mellifera*[a]

| Feature | Worker | | Queen |
|---|---|---|---|
| Relative area of antennal surface | 2* | | 1 |
| No. of chemoreceptive plates per antenna | 2400* | | 1600 |
| No. of facets of compound eyes[b] | 4000–6300* | | 3900–4920 |
| Antennal lobes of brain | Larger* | | Smaller |
| Mushroom bodies (coordination centers) of brain | Larger* | | Smaller |
| Hypopharyngeal glands | Large | * | Vestigial |
| Mandibular glands | Large* | | Very large |
| Wax glands | Present | * | Absent |
| Nasanov's gland | Present | * | Absent |
| Ovaries | Small | * | Enormous |
| Number of ovarioles | 2–12 | 4* | 150–180 |
| Spermatheca | Small or rudimentary | * | Large |
| Sting barbs | Strong | | Minute* |
| Sting axis | Straight | | Curved |
| Mandible with subapical angle | Absent | | Present* |
| Mandibular form | Slender | | Robust |
| Mandibular groove from near opening of mandibular gland | Present | | Absent |
| Proboscis | Longer* | | Shorter |
| Corbicula | Present* | | Absent |
| Pollen press (auricle, pecten, rostellum, etc.) | Present* | | Absent |

[a] At least 53 distinct morphological differences between queens and workers have been discovered in addition to physiological and behavioral differences. In the table, the probable ancestral condition is followed by an asterisk; an asterisk between the columns indicates an intermediate ancestral condition. (Reprinted by permission of the publishers from *The Social Behavior of the Bees: A Comparative Study* by Charles D. Michener, Cambridge, Mass.: The Belknap Press of Harvard University Press, Copyright © 1974 by the President and Fellows of Harvard College.)

[b] Estimates differ, but all agree that the queen has fewer facets than the workers.

**Figure 10-16** Social insect castes. Above, two castes of the African termite *Macrotermes*. The soldier's huge head serves in colony defense, but soldiers must depend upon the smaller workers to feed them. (Courtesy of R. E. Silberglied.) Below, the sausagelike abdomen of this army ant queen (*Eciton*) dominates the picture, greatly overshadowing her diminuitive head and thorax. Her abdominal tergites have become pulled widely apart by the stretching of the intersegmental membranes and appear as crescent-shaped islands on a sea of whitish membrane. Attending the queen are numerous workers. (Courtesy of C. W. Rettenmeyer.)

as the "social medium," trophallaxis (see Chapter 4) occurs in nearly all; in addition to communicative and nutritional functions, it may facilitate the maintenance of a colony-specific odor. Finally, altruistic behaviors—from simple sharing of labor to the suicidal defensive acts of individual honey bees—are commonplace in all eusocial groups. These then are some of the important ways in which eusocial insects differ fundamentally from their solitary relatives.


## THE INSECTAN SOCIAL REGISTER

While it seems that all levels of intraspecific interaction, from completely solitary behavior to highly complex coordinated societies, are found in insects, comparatively few species are truly social. They number a few thousand species and are included in only two orders, Isoptera (termites) and Hymenoptera (ants, bees, and wasps).

In addition to the major sources listed in the summary, the following references will help to provide access to the voluminous literature. Michener (1974) gives detailed accounts of the eusocial bees, including the infrasocial groups. Primitively social bees are treated by Plateaux-Quénu (1972). Chauvin (1968) and Morse (1975) focus on honey bee biology. Bumble bees are reviewed by Alford (1974) and Free and Butler (1959). For wasps, Evans and Eberhard (1970) offer a very readable general account; see also Iwata (1971). Spradbery (1973) and Kemper and Döhring (1967) provide good accounts of eusocial vespids, especially the European fauna.

The diversity of ant biology has tended to encourage specialization by various authorities. Sudd (1967), however, offers an informative general account, and the classic work of Wheeler (1910, reprinted in 1960) still contains a wealth of valuable information. Books by Weber (1972) on fungus-growing ants and by Schneirla (1971) on army ants provide good summaries. The biology of termites is treated by Krishna and Weesner (1969, 1970); Howse (1970) also provides a valuable survey.

Social life is unknown among the lower Hymenoptera; however, among the so-called "stinging" (aculeate) Hymenoptera, eusociality occurs in the wasps (Vespoidea, Sphecoidea), bees, and ants.

### The Wasps

Vespoid wasps, which can be distinguished from most others by the notched inner margins of the compound eyes and a characteristic lengthwise folding of the front wings, include both solitary and social species. Two subgroups are eusocial: the polistines and the vespines. All eusocial vespoids construct nests of paper made from pulp fiber chewed off weathered wood and/or dead plants.

The polistine most often encountered in temperate regions is *Polistes*, the paper wasp which commonly builds small open nests beneath house eaves and in outbuildings such as barns and garages (see Fig. 6-11). Colonies are annual; in early spring, inseminated females emerge from hibernation and initiate nests either individually or jointly. Dominance hierarchies exist among the foundresses, and eventually a single queen dominates. Throughout the first half of the summer, only workers are produced. In early fall, sexual forms start to appear; concurrently, the queen ceases to lay eggs, and the colony slowly declines, finally abandoning the nest entirely. West Eberhard (1969) treats polistine biology in detail.

Tropical *Polistes* are also found and live a life quite similar to this, except that no queen hibernation occurs. However, the most conspicuous social wasps in the Tropics are the polybiine wasps, relatives of *Polistes*. Unlike most other polistines, polybiine wasps begin new nests through swarming as large groups of morphologically similar individuals split off periodically from the perennial parent nest. Queens are difficult to distinguish from workers, and in many species there is evidence that several queens coexist in a single nest. Polybiine wasp colonies are frequently raided by foraging army ants; when this occurs, the adult population typically abandons the nest to start anew elsewhere (Naumann, 1975).

In contrast to the open cells of *Polistes*, most polybiine species (Fig. 10-17) surround their brood combs with a papery carton, providing protection from the elements and a degree of internal nest environment regulation. By restricting access to the brood to an easily guarded entrance tunnel, the envelope enhances protection from predators and parasites while concurrently providing camouflage, since enveloped nests are difficult to discern against the lichen-covered trees and rocks where they are commonly found.

Such enveloped nests are universal to the second group of eusocial wasps, the vespines, a predominantly temperate group which includes the hornets and yellow jackets. Their annual nest cycle (Fig. 10-18) is basically similar to that of *Polistes*, except that the solitary queens differ morphologically from the workers who later populate their nests. Most species of yellow jackets construct subterranean nests; hornet nests are typically aerial or built within hollow trees. Most wasps feed upon a variety of arthropods caught alive and carried back to the nest piecemeal in their jaws, but certain yellow jacket species have become predomi-

**Figure 10-17**   Representative polybiine wasp nests. Left, *Brachygaster* sp. from Brazil. In response to disturbance, workers exhibiting typical alarm defense postures with abdomens pointing outwards cover the outer envelope. Right, *Synoeca surinama* from Mexico. Built on the side of a small tree, this nest was 1 meter long. Nests of both *Brachygaster* and *Synoeca* are perennial and may persist in the same location for many years. (Photographs courtesy of R. L. Jeanne; see also Jeanne, 1975.)

nantly scavengers, making them nuisances around picnic and recreation areas, especially in late summer.

Outside of the Vespidae, primitive eusociality has been achieved in other wasps only rarely. One example is a Central American sphecid wasp, *Microstigmus comes* (Fig. 10-19), that preys exclusively upon Collembola. Inside the pendent baglike nests (unique among the Sphecidae), as many as 11 females cooperate in nest defense and cell provisioning; available evidence indicates that one serves as egglayer, the others as workers (Matthews, 1968). In another tropical wasp, *Zethus* (Eumenidae), nests established by a single female last for more than three years staffed by her female descendants.

Evans (1958), Richards (1971), and Jeanne (1975) review various aspects of social wasp biology.

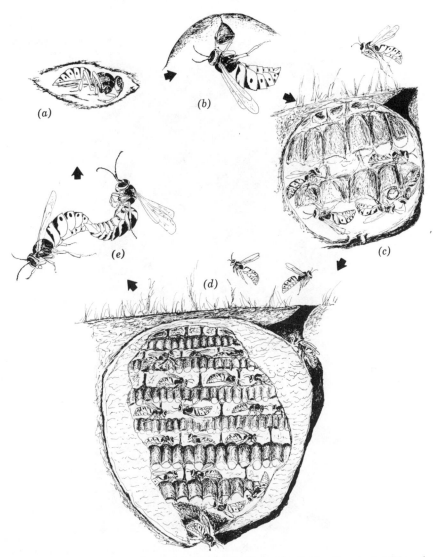

**Figure 10-18** Stages in the annual colony cycle of a generalized yellow jacket (*Vespula* sp.). (*a*) A fertile female wasp hibernates in leaf litter with her wings tucked under her abdomen. (*b*) In early spring she awakens and begins a nest by constructing paper cells hung from the top of a preexisting soil cavity such as an abandoned rodent burrow. (*c*) By early summer the nest, now the size of a tennis ball, is covered by a paper envelope and contains over a hundred workers. From this point it grows rapidly. In some species it may ultimately reach the size of a beach ball and contain over 5000 workers. (*d*) In a mature late summer nest, cells of the bottom one to four (here two) combs are larger than the upper, worker-producing cells. These bottom cells produce males and virgin queens, both of which leave the nest. (*e*) Reproductives mate in late autumn (here, female *left,* male *right*). Males, like workers, do not survive the winter, but most mated females hibernate successfully to become the next season's queens. (Drawings by J. W. Krispyn; not to same scale.)

**Figure 10-19** Nest of the primitively eusocial sphecid wasp, *Microstigmus comes,* suspended from a frond of a palm tree. The thimble-sized nest is constructed of plant fibers bound together by silk thread produced by glands in the adult's abdomens and has a single entrance at the top. (Photograph courtesy of C. W. Rettenmeyer; see Matthews, 1968.)

## The Ants

Of all eusocial groups ants are the most widespread and numerically abundant and contain more known genera and species than all other eusocial groups combined. In variety of ecological and social adaptations they are unparallelled. Primitive ants, beginning as predators upon other arthropods, were not bound like termites by a cellulose diet dependent on intestinal symbionts. Nesting in the soil and leaf mold (as most of their descendants still do), the ants were in a position to exploit an extremely rich microhabitat and to build and maintain long-term nests in a protected location. The development of a wingless worker caste, an innovation already present as far back as the mid-Cretaceous period, increased ease of access into soil and plant crevices. Acids secreted from the metapleural glands, an anatomical development possessed by all ants but by no other Hymenoptera (see Fig. 5-1), inhibited growth of microorganisms in the moist nest chambers.

Most ants live in soil still, either excavating chambers or constructing elaborate mound structures. Many others, however, have adopted an arboreal existence with tough carton-covered nests or homes in hollow twigs or specialized plant parts such as hollow thorns (see Figs. 4-7 and 4-9). Others cultivate gardens of fungi. Weaver ants of the Old World Tropics are unique in their use of larval silk to fasten leaves together to form large baglike nests. The largest genus of ants the carpenter ants (*Camponotus*), with about 1500 species, excavate nests in wood as their name implies. In a number of wood nesters, larger workers possess distinctive disc-shaped heads used to form living plugs at nest entrances (Fig. 10-20).

Some ants make no permanent nest. The term "army ant" popularly includes any of the 200 or more carnivorous ant species which, living as nomads, send out sorties of great numbers of workers. In a stricter sense, it applies just to members of the ant subfamily Dorylinae, awesome predators which Wheeler called the "Huns and Tartars of the insect world." Nearly blind, these predominantly tropical ants feed

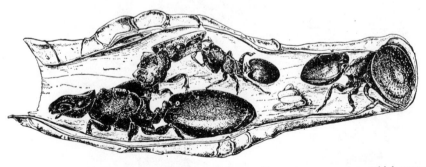

**Figure 10-20** A colony fragment of the myrmecine ant *Zacryptocerus varians* which nests in a hollowed out stem of red mangrove. Three female castes are shown: the queen rests on the floor of the nest to the left, while on the right a large major worker blocks the nest entrance with its saucer-shaped head. Behind the queen another major worker receives regurgitated liquid from a minor worker. (Drawing by Sarah Landry; from Wilson, 1976.)

almost entirely upon live arthropods captured during mass expeditions or raids. Moving like great armies across the land, overcoming every obstacle in their path, even bridging small streams with living suspensions of workers, they seem to have military precision. Columns of ants maintain infantry formation while, like a real army, accompanied by both scouts and fighting soldiers. The first investigator to study army ant behavior objectively in detail was Schneirla (1971), who spent over 35 years studying *Eciton* in both laboratory and field.

CASE STUDY: THE COLONY CYCLE OF *Eciton* ARMY ANTS

Each day within the tropical forest, *Eciton* workers stream out at dawn to begin their raids anew, their branching columns quickly overrunning areas up to 100 meters from their nest. At the front line, biting, stinging ants attack insects and other arthropods, tearing their prey apart and carrying the softer pieces back to the nest so that the forest floor soon has a series of two-lane highways traveled by steady streams of advancing raiders and returning victors (Fig. 10-21). So efficiently does the massive raiding progress that one colony may haul in more than 100,000 other arthropods in a single day. Then, as night begins to fall, the whole colony begins to emigrate along one of the day's principal raiding trails. Moving in solemn procession, sometimes through most of

**Figure 10-21** A raiding column of army ants, *Eciton hamatum*, moves across a log along the Neotropical forest floor. Both soldiers, distinguishable by their light-colored enlarged heads and ice-tonglike mandibles, and workers carrying booty are visible. (Courtesy of C. W. Rettenmeyer.)

the night, the colony finally settles in a new **bivouac**, or temporary camp, often under a low-hanging branch or vine. After about two weeks, however, the colony's behavior abruptly changes. It appears to quiet down and enter a *statary* phase. Few workers go out on raids; when they do, the forays are much smaller. No longer do nightly migrations occur. Remaining at the same site for about three weeks, the colony acts as though it had gone partially dormant. Then, just as abruptly, a new *nomadic* phase of intense foraging activity begins.

What factors underly such spectacular phasic behavior? Most investigators, Schneirla found, accepted a straightforward and seemingly logical explanation: depletion of food supply. The army ants, they argued, simply stay in one place until the food supply is exhausted and then move on to new hunting grounds. However, an alternative explanation also had some appeal. Perhaps the cycles were cued to some environmental phenomenon, such as phases of the moon, or perhaps changes in temperature, humidity, or air pressure.

Some detailed observational data on colonies in the field were sorely needed. Schneirla began following a single colony, then another and another, through one or more complete cycles. Painstakingly, he logged a dozen armies through whole cycles and more than a 100 more through partial cycles, taking down data in the field and repeatedly sampling the internal colony composition at various stages of the behavioral cycle. One thing quickly became clear. Within a single environment, several colonies were often present; of this number, some were generally in the nomadic phase, others in the statary phase. This seemed to rule out major environmental factors as the determinant of the nomadic cycle. The second hypothesis was more difficult to substantiate or reject, but persistent observation eventually provided a clue. Schneirla was able to confirm that a nesting site vacated by one colony would sometimes be moved into by a second colony the following night; in some cases the newcomers would remain for the whole three-week statary period. These observations seemed clear evidence that the food supply around the bivouac had not been depleted.

This, of course, left Schneirla without an established hypothesis for the colony cycle. As he began to analyze brood samples from different stages in the behavioral cycle, however, a new picture began to take shape, implicating the breeding cycle within a particular colony as the determinant of its cyclical behavior. The nomadic phase, it appeared, always coincided with the period when a larval brood was developing in the colony. The statary phase, on the other hand, always seemed to begin at the point when mature larvae started to spin their cocoons. In fact, the statary phase coincided with a point in which only pupae or newly laid eggs were present, neither of which required daily feeding by the workers.

A sentimental explanation was this: while helpless young exist, the workers protect them with a fixed bivouac, but once eggs hatch and pupae eclose, the workers must suddenly feed this massive number of new mouths, so they become nomadic. But however this might explain the adaptive significance of such behavioral cycles, it does little to elucidate the actual mechanisms involved. Impressed by the role of trophallactic interactions among all social insects, Schneirla turned to that phenomenon for his answer. Although by the end of his first trip to the tropics he could only phrase it loosely, Schneirla was able to theorize that the regular fluctuations in *Eciton* colony activity were regulated by some sort of stimulative interactions between the colony's large developing broods and the adult workers. What might these "stimulative interactions" be?

In the field and in the lab, Schneirla watched as workers cared for pupae of different ages. During their maturation in their cocoons, pupae appeared to become progressively more attractive to the workers, who handled them more frequently, even partially tearing the cocoons open. Upon emergence, the new adults (callows) became the objects of much licking, stroking, and handling by the older adults; in turn, the callows themselves became quite hyperactive. This reciprocal stimulation resulted in a rising crescendo of excitement which soon seized the whole community and led invariably to colony emigration initiating the nomadic phase. Although he had no evidence of pheromone involvement at this time, Schneirla had no trouble calling this a trophallactic interaction, for he interpreted trophallaxis much more broadly than did Wheeler, to include not only the actual exchange of food but the mutual exchange of all "equivalent" stimuli, including tactile. Whatever the mode, some sort of bonds among colony members were obviously being mediated. Colonies from which he removed all callow brood appeared lethargic in contrast and invariably failed to initiate nomadic behavior at the expected time.

But while trophallaxis between workers and callows was appealing enough in theory, Schneirla's observations at this point did not rule out a second possibility, namely, that the necessity to feed developing larvae and new adults was causative. In fact, during later stages of the nomadic phase when the larvae had grown and developed voracious appetites, they seemed to become the source of the colony "drive." To test the role of larval brood in stimulation of the workers' activity, he split a colony into two parts of comparable size. In one group he left the larvae intact; in the other he removed them. The workers in contact with larvae continued to show considerable activity, but those in the broodless portion were much less active. As another experiment, Schneirla removed the entire larval brood from a colony which was in the nomadic phase. True to expectations, the colony stopped emigrating, and the intensity of its daily raids diminished.

Thus, it appears that the interrelationships among all colony members provide the driving force for the cyclical behavior of army ants (Fig. 10-22). Nomadic phases are triggered by worker–callow interactions and maintained by worker–larval interactions. As cocooning of mature larvae begins, intensity of mutual stimulation between adults and brood declines temporarily and the colony lapses back into the statary condition until the pupal brood emerges.

In recent years, it has become increasingly clear that pheromonal mediation takes place in the army ants and that the adaptive significance of the alternation of quiescent and nomadic phases is to protect not only the helpless young but the queen as well through a time when both are most vulnerable. While little mention of the queen's role in the two phases has been made, Schneirla observed early that queens captured during the nomadic phase had considerably less swollen abdomens than those captured during the statary phase (see Fig. 10-16). The queen lays her clutch of thousands of eggs all during just a few days in the midstatary phase. As the colony enters the nomadic phase, her ovaries cease activity.

What controls the queen's reproductive cycle? This question remains unanswered, for so far work on *Eciton* physiology is virtually nonexistent. Schneirla postulated that the queen was stimulated to feed in excess because of the intense worker activity which occurred when mature larvae began to reduce feeding demands prior to spinning cocoons, thus freeing an abundance of food. However, physiological events could equally well be timing queen reproductive behavior, and increased feeding might be a side effect of this. Rettenmeyer (1963), Schneirla (1971), and Tophoff (1972) provide additional information on army ant behavior.

### The Bees

Few insects are more familiar than the ubiquitous honey bee. Elsewhere, we have discussed its unique dance language, its recently discovered ability to sense the earth's magnetic field and other sensory capabilities, and the multiple roles of the queen substance pheromone in colony life. At this point it is appropriate to put the honey bee back into

456

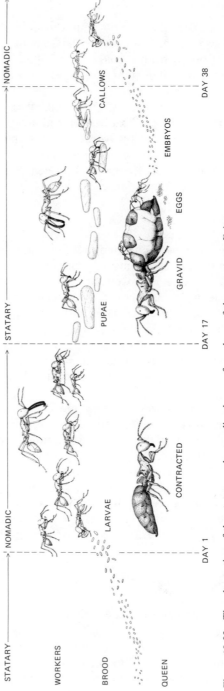

**Figure 10-22** The alternation of the statary and nomadic phases of a colony of the army ant, *Eciton burchelli*. The cycle is closely synchronized to the reproductive status of the colony. (From "The Social Behavior of Army Ants" by H. Tophoff. Copyright © 1972 by Scientific American, Inc. All rights reserved. See also Schneirla and Piel, 1948.)

perspective as but one of several hundred species of eusocial bees, surrounded by many times that number of other close relatives showing all manner of presocial behaviors, for the bees, more than any other insect group, display the full spectrum of social evolution.

Bees, the superfamily Apoidea, have arisen from a different trunk of the hymenopteran phylogenetic tree than ants and vespoid wasps, but are so like sphecoid wasps that they are sometimes characterized as being simply sphecoids that have specialized on collecting pollen instead of insect prey as larval food. For over 50 million years they have evolved in close contact with the angiosperms upon whose flowers they depend. Among the nine families of bees recognized today, social behaviors of various levels crop up repeatedly, but eusociality appears independently within three.

The Halictidae are a behaviorally diverse family of bees that are often called "sweat bees" because of their sometimes annoying habit of lapping up perspiration from people's skin during hot weather. Most of them nest in soil; a few nest in rotting wood. On at least five different occasions, a primitive type of eusociality has been achieved in this family; in addition, a wide variety of presocial behaviors have been reported (Michener, 1974; Plateaux-Quénu, 1972).

Primitively eusocial halictids characteristically have an annual colony cycle which is begun by one or more fertilized females of the same generation which behave more or less like solitary individuals except for sometimes sharing a communal nest space. However, as their offspring begin to come of age, these daughters do not disperse, but remain with the nest as functional workers. By summer, the colony typically has a single egg laying "queen" who remains in the nest while her daughters do all the foraging. As each cell is filled with a pollen–nectar mixture, a single egg is laid inside, and the cell is sealed. Occasionally, the cell may be opened and inspected by the workers, but no additional food is provided. Sometimes the queen continues to function throughout the life of the colony. In other cases she eventually dies and is replaced by one of her daughters. In some species there is a slight worker–queen size difference (Batra, 1966).

Primitively eusocial species are also found in the so-called allodapine bees (Michener, 1971), which are related to carpenter bees but confined to the south temperate regions of the Old World Tropics. Unlike the nests of all other known bees, those of allodapines have no cells. Larvae are kept together on the nest floor in the manner of ant larvae, and like the young of ants they are moved about the nest by the mother and arranged in groups according to age (Fig. 10-23). Importantly, allodapine larvae are fed progressively, being given more frequent pollen meals as

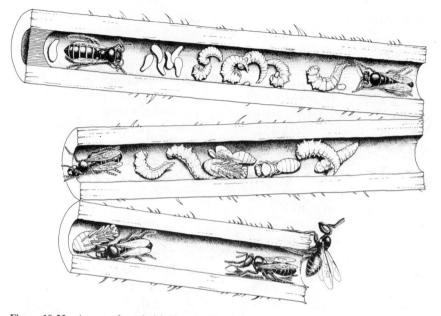

**Figure 10-23** A nest of a primitively eusocial allodapine bee, *Braunsapis sauteriella*, in a pithy stem of *Lantana camara* in Formosa. The brood is arranged by age, with eggs clustered in the bottom while the mother queen rests nearby; progressively older larvae and pupae are situated in order toward the entrance. Eversible projections and tubercles on the larval body help in maintaining position and in feeding progressively on the small pollen balls brought by the several adult bees. (Drawing by Sarah Landry; based on K. Iwata. In S. F. Sakagami, Ethological peculiarities of the primitive social bees, *Allodape* Lepeltier and allied genera. *Insectes Sociaux* 7(3): 231–249, 1960. Reprinted by permission of the publishers from *The Insect Societies* by Edward O. Wilson, Cambridge, Mass.: The Belknap Press of Harvard University Press, Copyright © by the President and Fellows of Harvard College.)

they grow, a behavior which promotes mother–offspring contact. In most species, the mother dies before her young emerge. In some, however, the mother's life slightly overlaps the adulthood of her offspring, who often remain in the parental nest and help rear a second generation.

A third group entirely composed of primitively eusocial bees is familiar to almost everyone—the robust hairy bees of the genus *Bombus*, commonly called bumble bees. Bumble bees are characteristic of cool north temperate climates around the world. With few exceptions, they have not succeeded in invading the tropics, except at high elevations where they may become very abundant locally. A behaviorally cohesive group, all of the more than 200 species have relatively small annual colonies (Fig. 10-24) founded by single overwintered fertile queens. Abandoned rodent burrows are preferred nest sites; inside, a queen fashions her nest from wax secreted by abdominal glands. Rather than directly feeding her brood as do other primitively eusocial groups, the bumble bee characteristically stocks pollen and honey separately from the brood cells. At first, the queen fashions a separate storage pot (see Fig. 3-15); later in the season, cast-off pupal cocoons are used. The queen lays several eggs together in a single distensible wax cell and, like the allodapines, rears her larvae in groups. However, since the bumble bee's larvae are contained in a capped cell with their entire food supply, there is no trophallactic contact between parent and offspring.

The true honey bees, four species of *Apis*, are the best-known

**Figure 10-24** A colony of the European bumble bee, *Bombus lapidarius,* in an abandoned mouse nest. The larger individual at the top right is the queen, resting on a cluster of pupae-containing cocoons. At the upper and lower left are three communal larval cells; the waxen envelopes of the lower two have been opened to reveal the cluster of developing larvae inside. At the right and lower right are clusters of empty cocoons now used to store pollen. Larger vessels (center and left) are waxen honey pots. (Drawing by Sarah Landry, based on photographs by Sladen and by Free and Butler. Reprinted by permission of the publishers from *The Insect Societies* by Edward O. Wilson, Cambridge, Mass.: The Belknap Press of Harvard University Press, Copyright © 1971 by the President and Fellows of Harvard College.)

**Figure 10-25** One of the species of stingless honey bees of the genus *Trigona* from Costa Rica builds large aerial nests (*right*) of secreted wax mixed with large amounts of resin to form a tough material called cerumen. Inside the nest, pollen and honey pots (*left, below and above respectively*) fill the brood chamber. While *Trigona* cannot sting, they actively defend their nests by vigorously biting intruders. (Photographs courtesy of R. L. Jeanne.)

members of the Apinae; however, the subfamily also includes a great tribe of stingless honey bees, the Meliponini. Exclusively tropical, these latter are among the most conspicuous and numerous bees in these regions. All the highly eusocial bees in the Apinae differ from the primitively eusocial species above in having perennial colonies, with new colonies produced by swarms; individual queens or workers cannot survive alone for long. Furthermore, the female castes of the highly eusocial bees are strikingly different from each other—behaviorally, physiologically, and morphologically (see Table 10-2). All highly eusocial bees have colonies with large populations, up to 180,000 workers in some stingless bee species. Colony integration and communication are necessarily complex; while none of the primitively eusocial bees is known to be able to communicate food source location, both the honey bees and meliponine bees have evolved elaborate means of doing so (see Chapter 7). Also, workers of highly eusocial species typically pass through a series of age-correlated behavior patterns, or temporal division of labor.

Both honey bees and stingless bees build nests with wax, with those of the latter possessing the greater architectural complexity (Fig. 10-25). Larvae of both are reared in individual cells arranged in combs within the nest, while separate cells are used to store honey and pollen. Meliponine larvae are mass provisioned; those of *Apis* are fed progressively. In addition, the existence of distinctive morphological differences between the stingless bees and honey bees has led Winston and Michener (1977) to suggest an independent origin of the two groups. Kerr (1969) and Michener (1969) review the evolution of eusocial bees.

## The Termites

The more than 2000 species of termites are all highly social. A cosmopolitan but predominantly tropical group, termites are classified in six families; the Termitidae is by far the largest and contains the most advanced species. Along with the closely related cryptocercid cockroaches, termites are the only wood-eating insects that depend on symbiotic intestinal protozoans to break down cellulose into usable sugars (see Chapter 4), and it appears that termite societies began as feeding communities and evolved social brood care later, in a sequence that is the reverse of social evolution in the Hymenoptera.

Because termites have a soft cuticle and are easily desiccated if exposed directly to the outside environment, they are largely restricted to the closed environment of their nests in which conditions are regulated and relatively stable (Fig. 10-26). Foraging is typically done

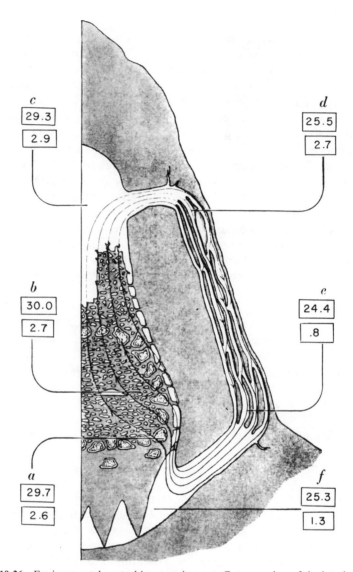

**Figure 10-26** Environmental control in a termite nest. Cutaway view of the interior of the nest of *Macrotermes bellicosus,* an African fungus-growing termite, showing the temperature and percentage carbon dioxide concentration at different positions. The metabolic heat from the huge biomass of colony members concentrated in the central core of the nest rises by convection to the large upper hollow cavity, then diffuses toward the sides of the nest where it flows into a network of narrow channels close to the surface where the air is cooled and gaseous exchange occurs by diffusion through the thin dry walls. Refreshed air sinks to the lower passages of the nest and eventually recycles. (From "Air-conditioned Termite Nests" by M. Lüscher. Copyright © 1961 by Scientific American, Inc. All rights reserved.)

along radiating covered tunnels built out from the nest. Constructed of soil, excrement, and saliva, nests often show a striking parallel to those of ants, some being completely subterranean, others in wood, and still others entirely arboreal. The most conspicuous nests are begun underground and as they grow begin to protrude progressively, eventually assuming impressive species-characteristic shapes (Fig. 10-27; see also Fig. 3-7).

In a very literal sense, termites are "social cockroaches," so similar in a wide variety of characteristics as unrelated as wing venation,

**Figure 10-27** Two unusual giant termite mounds, distinctive and conspicuous features of tropical landscapes all over the world. Mounds may persist for over 50 years and reach heights exceeding 14 feet. Termites are among the major decomposers of dead wood and other cellulose detritus in tropical ecosystems. Left, *Nasutitermes triodiae* from northern Australia. (Courtesy of H. E. Evans). Right, *Amitermes sp.* from Taperinha, Brazil. (Courtesy of R. L. Jeanne.)

**Table 10-3** Comparison of the major similarities and differences between the termites and highly eusocial Hymenoptera[a]

| Similarities | Differences | |
|---|---|---|
| | Termites | Eusocial Hymenoptera |
| 1. The castes are similar in number and kind, especially between termites and ants | 1. Caste determination in the lower termites is based primarily on pheromones; in some of the higher termites it involves sex, but the other factors remain unidentified | 1. Caste determination is based primarily on nutrition, although pheromones play a role in some cases |
| 2. Trophallaxis occurs and is an important mechanism in social regulation | | |
| 3. Chemical trails are used in recruitment as in the ants, and the behavior of trail laying and following is closely similar | 2. The worker castes consist of both females and males | 2. The worker castes consist of females only |
| 4. Inhibitory caste pheromones exist, similar in action to those found in honey bees and ants | 3. Larvae and nymphs contribute to colony labor, at least in later instars | 3. The immature stages (larvae and pupae) are helpless and almost never contribute to colony labor |
| 5. Grooming between individuals occurs frequently and functions at least partially in the transmission of pheromones | 4. There are no dominance hierarchies among individuals in the same | 4. Dominance hierarchies are commonplace but not |

| | |
|---|---|
| 6. Nest odor and territoriality are of general occurrence | colonies |
| 7. Nest structure is of comparable complexity and, in a few members of the Termitidae, of considerably greater complexity. Regulation of temperature and humidity within the nest operates at about the same level of precision | 5. Social parasitism between species is almost wholly absent<br>6. Exchange of liquid anal food occurs universally in the lower termites, and trophic eggs are unknown | universal<br>5. Social parasitism between species is common and widespread<br>6. Anal trophallaxis is rare, but trophic eggs are exchanged in many species of bees and ants |
| 8. Cannibalism is widespread in both groups (but not universal, at least not in the Hymenoptera) | 7. The primary reproductive male (the "king") stays with the queen after the nuptial flight, helps her construct the first nest, and fertilizes her intermittently as the colony develops; fertilization does not occur during the nuptial flight<br>8. Normal diploid method of sex determination | 7. The male fertilizes the queen during the nuptial flight and dies soon afterward without helping the queen in nest construction<br>8. Haplodiploid method of sex determination |

---

[a] Modified from Wilson (1971). Reprinted by permission of the publishers from *The Insect Societies* by Edward O. Wilson, Cambridge Mass.: The Belknap Press of Harvard University Press. Copyright © 1971 by the President and Fellows of Harvard College.

endocrine system, and intestinal fauna that some specialists have placed termites, cockroaches, and mantids in the same order (Dictyoptera). Others retain the more traditional classification which places termites in an order of their own, the Isoptera. Either way, it is apparent that termites have attained their eusociality from a base extremely remote in evolution from the Hymenoptera.

How fundamentally does their social organization differ from that of the Hymenoptera? Surprisingly, the differences are not as great as one might expect. A great many convergent similarities exist not only in social mechanisms but in level of complexity attained (Table 10-3). The most striking differences in the two groups concern the nature and care of their young. Because the larvae of the eusocial Hymenoptera are helpless grubs which must be nursed continuously, their care requires a great deal of adult labor. The young stages of termites, on the other hand, are active creatures quite similar to the adults and quite able to fend for themselves, particularly among the so-called "lower" termites. Many termite workers are nymphs that will eventually develop into winged adults. Others, **pseudergates,** are workers that for chemical reasons will probably never grow up, although they retain the potential for developing into any of the castes. Thus, up to a point it may be said that termites rely upon "child labor" for the maintenance of their societies. Another difference is that the male sex in termites participates in colony labor, which is never true for male hymenopterans. Wilson (1971) provides an account of termite natural history.

## ORIGINS OF INSECT SOCIALITY

Since all living members of both the termites and the ants are totally eusocial, they can tell us little about the *origins* of social behavior. Attention has therefore centered about the bees and wasps, for both of these major groups exhibit a continuum of apparent stages leading to social life.

Among halictid bees the following generalized evolutionary sequence seems consistent with much of the available data, although there are many variations on the pattern. First, groups of females of the same generation frequently tend to use a common nest, each, however, behaving more or less like a solitary female. Among such **communal** nesting groups there seems to be a recurring tendency for females to cooperate in nest and cell construction and even to occasionally open brood cells and inspect them. Such apparent cooperation among females in care of the brood is the criterion for the next level of social evolution, the **quasisocial** stage of behavior. Since it would not be unusual among a communally nesting group of bees to have variation in ovarian development, a reproductive division of labor could gradually result, with some poorly developed individuals tending to assume the role of workers, others becoming mainly egg layers (queens). When this reproductive division of labor exists colonies are said to be **semisocial.** When the life of the adult bees lengthens to the point where two adult generations overlap, the threshold of eusociality is crossed. Correlated with this evolutionary progression is an increasing tendency toward polymorphism in size, with sharply differentiated worker and queen castes the ultimate endpoint and a tendency to delay male production until the end of the nesting season. Michener (1974) has thoroughly reviewed these developments in the various halictid bees. The progression of social stages from communal to quasisocial to semisocial to eusocial is termed the **parasocial** route to eusociality. Parasocial insect societies are simply any colonies in which the adults consist of a single generation rather than two generations as are ordinarily present in eusocial forms. Typically small in numbers, it appears that their most important advantage to their inhabitants is improved defense.

The situation among most temperate wasps and bumble bees, as well as among the allodapines, is different: nests are founded by a single adult female. Even when nesting may be gregarious, each female does all her own work. Where cooperation exists, it is between a mother and her offspring, not between sisters. This phenomenon forms the basis of the **subsocial** or familial route to sociality (Fig. 10-28) first conceived by

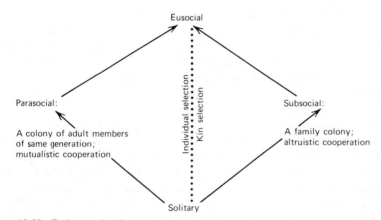

**Figure 10-28** Pathways leading to complex social behavior in Hymenoptera. (Modified after Lin and Michener, 1972.)

Wheeler (1923), in which eusocial life is the culmination of a series of steps of increasing intimacy between mother and offspring coupled with her increasing longevity until ultimately she lives long enough to gain her offsprings' cooperation when they reach adulthood.

Regardless of which evolutionary route has been taken, the resultant society seems to present an evolutionary paradox in the existence of **altruistic,** or self sacrificial, behaviors. One aspect of these is apparent in the awesome willingness of most wasps and hornets to throw themselves into battle. Taking the chance of mortality to the ultimate, honey bees have evolved a barbed sting which upon delivery remains in the victim and disembowels the bee, thereby ensuring the bee's death. In *The Origin of Species,* Darwin wrestled with another aspect of the paradox: if natural selection favors the individual able to produce the greatest number of viable offspring that live to reproduce, how could the worker caste of insect societies ever have begun? How can one evolve to become sterile and leave no offspring? Social insects must represent a special case of what is now called **group selection,** he reasoned, an example where natural selection was operating on the level of the colony, or family group, rather than on the single organism.

Theoretically, group selection is plausible, and there appears to be no reason why selection above the individual level might not sometimes affect gene frequencies, thus contributing to the persistence or extinction of a social trait (Fig. 10-29). By the time one reaches the level at which an entire breeding population is the unit (interdemic selection), such selection must be far less common. In addition, before group selection can occur, an allele still must first become established by selection at the individual level. Therefore, attention in recent years has turned toward developing better explanations of social behavior and altruism at the individual level. One particularly persuasive hypothesis has been the **kin-selection theory** advanced by Hamilton (1964, 1972) and expanded on by others (Alexander, 1974; Alexander and Sherman, 1977; Michener and Brothers, 1974; Trivers and Hare, 1976; West Eberhard, 1974). Wilson (1975a) provides a good synthesis of kin selection.

Ethics aside for a moment, suppose that you had two choices: you could die without heroism or, by the process of your death, save an otherwise endangered group that included some of your relatives. In an evolutionary sense, it would clearly be preferable to save at least some relatives, for in doing so some genes like your own would still be passed along after your death. The more closely related the survivors were to you, the more of "your own" genes, that is, the ones identical for you and the relatives, would be maintained in the gene pool. In fact, if you could rescue your identical twin, it would be the most ideal situation of

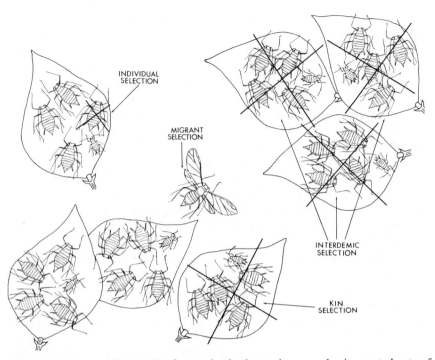

**Figure 10-29**  Ascending levels of natural selection acting on enlarging nested sets of related individuals. In this simplified example, plants under aphid attack are viewed as a group of islands, each containing a whole interbreeding aphid population. Whenever selection affects two or more members of a lineage group as a unit, it is group selection. On a given plant, each leaf is assumed to hold a family group, the offspring of a single female. When the aphid group on one leaf is selected for or against relative to the group on another, kin selection is operating. When the entire plant with all its aphid colonies (the entire breeding unit) is selected relative to other such units (the entire population of one or more other plants), it is interdemic (interpopulation) selection. The differential tendency of individuals of different genetic constitution to disperse is referred to as migrant selection. (Drawing by J. W. Krispyn, inspired by Wilson, 1973).

all, for saving that person would be *genetically* indistinguishable from saving yourself.

The kin-selection theory adds significantly to classical evolutionary theory in recognizing that the fitness of an individual has not just one component, but two. The classical component was an individual's *personal fitness,* that is, his total lifetime effect on the gene pool of succeeding generations through production of his own offspring. To this, Hamilton added a *kinship component*—his effect on the gene pool of succeeding generations through effects on the reproduction of other individuals possessing genes to varying degrees like his own; this could be expressed as the individual's effects on the fitness of his neighbors multiplied by his respective fractional relatedness to them. Together, these components comprise the individual's **inclusive fitness.**

Kin selection can operate at any level of relatedness; although the large proportion of unlike genes carried by individuals outside the immediate family group slows the progress of selection, it does not alter its direction. Under certain situations altruism is advantageous even toward quite distant relatives. The probability of its occurrence is increased, for example, if the beneficiary stands to gain a great deal (as in emergency) and/or if the cost is low (as when the altruist is excluded from reproduction on his own or is in control of an abundant resource). The same is true if the donor is particularly efficient at giving aid and/or if the beneficiary is particularly efficient at using it.

It is often easiest to understand a phenomenon by looking for instances where it is most extreme. In the Hymenoptera, many different degrees of sociality occur, and eusocial behaviors appear to have evolved independently several times. Why has eusociality evolved so often in this one insect order and only once (termites) in the other two dozen or so orders of insects? Why also does altruisim appear so often among the wasps, bees, and ants?

In most insects, males and females both arise from fertilized eggs; each sex is *diploid*, that is, possesses two sets of genes, one from each parent. Within the immediate family, sons and daughters are equally related, sharing an average of one half of their genes with each other and with either parent. However, in the Hymenoptera all fertilized eggs develop into females, while unfertilized eggs develop parthenogenetically into males. As a result, while females have two sets of genes like "normal" species, males are *haploid*, that is have only one gene set. This reproductive system, called **haplodiploidy**, means that the degree of genetic relatedness between family members is dependent upon the sex of those members (Table 10-4).

Since the hymenopteran father is haploid, the half of their genes which female offspring receive from him are all identical. Within the gene set they receive from their diploid mother, only an average of one half of the genes are identical. Thus, hymenopteran sisters share through common descent an average of three fourths of their genes. As a result, under haplodiploidy a daughter is more closely related to her full sisters than she is to her own mother, a biologically unusual situation. Just as unusual is the male hymenopteran. He has a grandfather but no father. Because he receives all his genes—in one of two alleles—from his diploid mother, he shares an average of one half of his genes with his

**Table 10-4**   Maximal average degrees of relationship ($r$) among close kin in hymenopteran groups[a]

|        | Mother | Father | Sister | Brother | Son | Daughter |
|--------|--------|--------|--------|---------|-----|----------|
| Female | ½      | ½      | ¾[b]   | ¼       | ½   | ½        |
| Male   | 1      | 0      | ½      | ½       | 0   | 1        |

[a] From Trivers and Hare (1976).
[b] In reality, a female's relatedness to sisters (and/or nieces) will be lower than shown if their mothers have mated more than once, which is often true in eusocial Hyenoptera.

brothers but possesses only an average of one fourth of his genes in common with his sisters. This equivalent of the average fraction of genes shared by common descent is called $r$, the **coefficient of relationship** (Table 10-4).

In theory, an individual's own altruistic sacrifice in fitness could be counterbalanced by an increase in the fitness of some group of relatives. Hamilton proposed that this increase must be by a factor greater than the reciprocal of the coefficient of relationship to that group (i.e., greater than $1/r$) if an altruistic trait is to evolve. In the majority of animals, when an altruistic female sacrifices her life or reproductive success for a sister or daughter, there is one half chance that the latter shares the gene for that altruistic trait. For it to be statistically probable that the altruistic gene will be fixed, the reproductive success of the sister or daughter must be at least doubled as a result of the sacrifice. However, among haplodiploid sisters where there is a three quarter chance that the gene is shared, altruistic behavior is statistically probable when the recipient's gain in fitness is equal to only four thirds, or 1.33, times the donor's loss.

As a result, when the hymenopteran mother lives to reproduce beyond the adulthood of her first female offspring, these offspring may increase their inclusive fitness more by care of their younger sisters than by an equal amount of care given to their own offspring. Likely evolutionary results include sterile female castes and a tight colonial organization centered around a single fertile female. Furthermore, since sisters are more closely related than are brothers, altruism will be more favored among daughters than among sons. This may also help explain the fact that hymenopteran males never perform colony labor in the manner of diploid termite males. Haploid males have more to gain by fathering daughters, with whom they share an average of one half of their genes, than by assisting in the production of more sisters, with whom they share an average of only one fourth of their genes.

All other things being equal, Hymenoptera should all tend to become social! Yet the truth is that although apparently all of the thousands of solitary Hymenoptera species have this same haplodiploid reproductive system, most show no trace of sociality. Why aren't all of them social?

The answer involves the concept of **preadaptation,** that is, for a trait to be a selective advantage, the animal must already be equipped to use it in advance of its occurrence. Previously existing behavior patterns, physiological processes, and morphological structures that are already functional in some other context become available as stepping stones to

**Table 10-5** Some important behavioral, physiological, and morphological preadaptations for development of insectan eusociality

| Preadaptation | Leading to |
|---|---|
| 1. Haplodiploid method of sex determination | Development of altruistic behaviors depending on coefficient of relationship |
| 2. Construction or possession of a nest in which young are reared and to which female returns repeatedly | Potential family gathering place and behavioral equipment for gathering there |
| 3. Plasticity in stereotyped nesting patterns | |
|     Altering sequences in nest and cell construction | Provisioning several cells simultaneously, resulting in more or less synchronous brood development |
|     Omission of certain behaviors | Construction and provisioning of cells without oviposition |
|     Placement of numerous cells at one site | Clumping of brood and increased chances of neighbors being relatives |
| 4. Increased longevity of females | Overlapping lifespan of mother with young |
| 5. Control over sex ratio | Strong female bias (i.e., workers); postponement of male production until late in nest cycle |
| 6. Effective channels for intraspecific communication (especially chemical) | Means for integrating and regulating multiple interactions among large numbers of individuals |
| 7. Chewing mouthparts | Manipulation of nest materials and other objects such as brood |
| 8. Mutual tolerance | Possibility of cooperation |

new adaptations. In addition to haplodiploidy, several other preadaptations are generally considered important for the evolution of eusocial behavior among the Hymenoptera (Table 10-5).

Nor can the total biology be ignored. For example, while Hamilton's theory very nicely fits the subsocial path to eusociality, there are some difficulties if one applies it to explain semisocial behavior in halictid bees (Lin and Michener, 1972). Here, individuals join in groups of females which at best may be sisters but most often are more distantly related. Suppose they are sisters; if a joiner becomes a worker, then she is enhancing the production of nieces to whom her coefficient of relationship is only $\frac{3}{8}$. Evolutionarily, the better strategy would seem to be to live alone and produce offspring with a relationship of $\frac{1}{2}$. However, the balance may be tipped by enhanced nest defense and protection from parasites resulting from mutualistic adult cooperation and by the manner in which early-season adult cooperation effectively produces an "instant" colony, with the head start it makes possible. Such mutualistic behaviors simultaneously increase the fitness and productivity of all individuals involved. Because they are activities that directly promote the perpetuation of each individual's own genes, such behaviors are "selfish." They fit well into classical evolutionary theory and do not require any close relationship of the partners, although sometimes such does occur.

## INTERSPECIFIC SOCIAL ALLIANCES

Certain ants practice slavery, capturing workers of another species and forcing them to do the work of their captors. Some stingless bees make a living by plundering the nests of other species, stealing the honey and provisions for their own nest. Queens of some wasps intrude into established nests of related wasps and either usurp the queen's position or lay their own queen-producing eggs alongside those of the host, to be raised by the host's worker force. We have already discussed (Chapter 4) a number of symbiotic interspecific alliances in the context of feeding strategy. Now we focus on symbioses between fully social species.

Nest building and the care and feeding of an individual's young are usually thought to be very conservative, physiologically deep-seated behavior patterns. However, even these may be selected against if increased fitness results. For example, among solitary Hymenoptera various species (see Fig. 4-11) have lost the behaviors of nest building and provisioning and rely entirely upon the nests and labors of others. Various preadaptations might be expected to favor the evolution of such brood parasitism. One is the use of open nests and/or reuse of old nests or preformed burrows; another is low survival rates of young fed by their own parents. Others include having an egg incubation period slightly shorter than the host's, tolerance of the host's diet, and a wide range of behaviors facilitating location of and approach to the host's nest.

Among the eusocial Hymenoptera, **social parasitism,** which occurs when the young of one species are raised by another, has evolved independently many times and in many ways. The term "social parasite" is unfortunately somewhat misleading, for the parasite *need* not be social nor does the label refer to *any* parasite of a social species. Rather, a social parasite is a species that uses its host as a work force (something social insect species are particularly suited for) rather than as a direct source of food. However, because many social parasites are closely related to their hosts, they are often themselves social.

Ants show a greater diversity of forms of social parasitism than any other group of animals, with over 160 species of social parasites no two of which are exactly alike in their adaptations for this mode of life. The name *Teleutomyrmex* means "final ant," a name most appropriate for what is probably the ultimate social symbiont. In the Swiss and French Alps, one may find this ant in small isolated populations which include no workers. While it is surrounded by quite an assemblage of ant species, *Teleutomyrmex* is parasitic upon only one—its closest phylogenetic neighbor, *Tetramorium*. It has never been found outside the nests of its hosts. *Teleutomyrmex* queens spend much of their time riding upon the backs of the queens of their host colony. Very delicate, the symbionts seldom move and apparently feed only upon regurgitates passed from workers to the host queen. Placed in an artificial nest they cannot survive, even if host workers are present. Their brains, mandibles, nervous system, and skin, in fact nearly all of their morphology, shows extensive degeneration, except in their reproductive system, for each of the tiny physogastric (see p. 154) parasites lays an average of one egg every 30 seconds (Wilson, 1971). Having up to six or eight ectoparasitic ants riding upon her back may slow down a host queen; infested colonies tend to be somewhat smaller than uninfested ones. Significantly, infested colonies produce no sexual forms of the host. *Teleutomyrmex* adults, especially older females, apparently produce a very potent attractant, for host workers lick them continuously. Perhaps this substance as it is circulated throughout the host colony functions to impose "reproductive castration" upon it. Such a castration phenomenon has been demonstrated in a number of similar parasite–host colony relationships, but its physiological mechanism has yet to be determined.

This most extreme form of social parasitism in insects is a condition of permanent and complete dependence, in which the parasite's entire life cycle is carried out within the host's nest. Commonly, when the parasite is a social species, its own workers are nonexistent or conspicuously degenerate. Such extreme social parasites, or **inquilines,** appear to have arisen convergently along various evolutionary routes (Fig. 10-30). One of these is slavery, or **dulosis.**

For example, Amazon ants with their saberlike mandibles are fierce

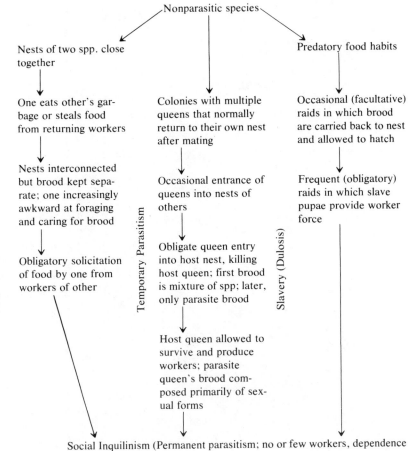

Nonparasitic species

Nests of two spp. close together

One eats other's garbage or steals food from returning workers

Nests interconnected but brood kept separate; one increasingly awkward at foraging and caring for brood

Obligatory solicitation of food by one from workers of other

*Trophic Parasitism*

Colonies with multiple queens that normally return to their own nest after mating

Occasional entrance of queens into nests of others

Obligate queen entry into host nest, killing host queen; first brood is mixture of spp; later, only parasite brood

Host queen allowed to survive and produce workers; parasite queen's brood composed primarily of sexual forms

*Temporary Parasitism*

Predatory food habits

Occasional (facultative) raids in which brood are carried back to nest and allowed to hatch

Frequent (obligatory) raids in which slave pupae provide worker force

*Slavery (Dulosis)*

Social Inquilinism (Permanent parasitism; no or few workers, dependence on host species for food)

**Figure 10-30**   Hypothesized major pathways in the convergent evolution of social inquilinism in Hymenoptera (Based on Wilson, 1971, 1975a).

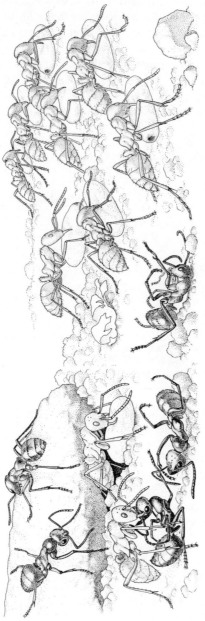

**Figure 10-31** Amazon ants of the species *Polyergus rufescens* (light) conducting a raid upon a colony of the slave species *Formica fusca* (dark) nesting in dry soil beneath a stone. Killing resistors by piercing them with their saberlike mandibles, the Amazons rush off with captured brood. Such behavior occurs quite often among cold temperate species, where slave labor is found in at least 35 ant species from six independently evolved groups. While the vast majority of ants live in the Tropics and warm temperate zones, behavior even remotely approaching slavery is unknown in any of these species. (Drawing by Sarah Landry; from "Slavery in Ants" by E. O. Wilson. Copyright © 1975 by Scientific American, Inc. All rights reserved.)

fighters but totally inept housekeepers (Wilson, 1975c). In their home nest, their only activities are grooming themselves and begging for food. Living as a pure colony, they would surely perish, for they neither excavate nests nor care for their own young. What keeps them alive is a unique slave-making habit. Periodically, the Amazon ants swarm out of their nest, marching swiftly to the nests of *Formica fusca* to launch a raid (Fig. 10-31). With feverish haste, they pour over the colony in a body; any defenders who resist the attack are punctured and killed. Then, like pillaging soldiers, the Amazons proceed to cart off *fusca* pupal cocoons.

Back in the Amazon nest, the pupae soon begin to hatch; genetically programmed to perform various housekeeping tasks, they begin work. Some bring food into the nest, while others tend the eggs, larvae, and pupae of their captors. Still others actually feed the adult Amazons who have made them slaves, responding to their begging by regurgitating liquid droplets. Throughout all this activity, the *fusca* slaves make no distinction between their genetic siblings and the Amazons, fully accepting their captors as sisters. Eventually, their numbers dwindle, because as members of the worker caste they cannot reproduce. In response, the slave-making Amazons set out to pillage alien colonies once again.

The nest-raiding techniques of slave makers are among the most sophisticated behavior patterns found in the insect world. Not all physically overpower their victims, for example. Some depend on various chemical ruses. Two species of *Formica* spray certain acetates at resisting nest defenders; these chemicals act as "propaganda substances," imitating the alarm pheromones of the slave species so powerfully that they throw the resistors into utter helpless panic (Regnier and Wilson, 1971). In addition to this long-lasting disruptive effect, the acetates serve as an attractant to workers of the slave maker, quickly assembling them where fighting has broken out.

How has ant slavery arisen? Many ants are quite aggressive predators; perhaps the first step toward slavery was simple predation. Suppose that some ants, while raiding other species for food, were to carry away the immature forms to eat at home. When not eaten soon enough, a few pupae might emerge as workers; if accepted as nestmates, these accidental captives would join the work force. In some cases, they might prove more valuable in this capacity than as food. When this occurred, selection would favor bringing home more pupae after raids and evolving adaptations to raid even more successfully. Wilson (1975b) further discusses the evolution of slavery in ants.

In southern United States, workers of *Conomyrma* ants collect the dead insects and colony corpses thrown out by *Pogonomyrmex*. In

India, certain *Crematogaster* ants lie in wait like highwaymen for grain-laden *Holycomyrmex*. Several ant genera (*Solenopsis* and relatives) are called "thief ants" because of their habit of stealthily entering the nests of larger species of ants or termites to prey upon the inhabitants and/or their food. Nor is such robbery, or **trophic parasitism,** limited to ants. Some stingless bees pilfer stored supplies from other free-living species, filling their own storage pots with honey–pollen mixtures they can no longer independently gather, having lost their own basitarsal pollen baskets. Of 150 species of termites studied in Africa, 70% at least occasionally had other species encroach upon their nests. Nest robbery provides another possible starting point for social inquilinism. In a subtle behavioral shift, trophic parasites may become tolerated guests. Wilson (1975a) gives references to these and other examples of trophic parasitism.

Sexual forms of most social Hymenoptera leave their home colony for their mating flights. Some mated queens fail to find their own colony again following such flights, or after hibernation fail to successfully initiate their own nest the following season. Should they locate a colony of another species and somehow eliminate the host queen, they gain an established nest complete with a work force. For a brief period, when the usurper queen has begun to reproduce but not all the original queen's brood have reached old age and died, the colony consists of workers of both species. Eventually, however, the nest will come to contain only the usurper and her offspring. This situation, termed **temporary parasitism** (Fig. 10-32), may evolve from a facultative affair to an obligatory one where usurper queens depend entirely upon this method to found new colonies. Temporary parasites have increased their chances of successfully entering the host nest in a variety of ways, from stealth to aggression to deceptions involving chemistry, morphology, or behaviors such as "playing dead." If the host workers are sufficiently capable to rear the intruder's new sexual forms adequately, selection should favor usurper queens who put all their energies into the production of sexual forms, relying exclusively upon the host for workers. By allowing the host queen to live and produce workers, true inquilinism may be achieved.

The apparent ease by which symbionts have repeatedly intruded themselves into their host's colony is somewhat surprising, particularly in view of the exceptional ability of social insects to defend themselves from larger predators. The clue to understanding this paradox lies in the organization and integration of insect societies. As stressed elsewhere, insect societies are integrated primarily by nonpersonal colony odors, an easier code to break than the visually based individual recognition or learned role dominance-based interactions characteristic of vertebrate

**Figure 10-32** Temporary parasitism in yellow jackets. A queen of *Vespula squamosa* (*arrow*) has been introduced into a young laboratory colony of *V. maculifrons*. *V. maculifrons* workers respond in an aggressive manner, biting and attempting to sting the intruder. Eventually the *V. squamosa* queen dismembers and kills her attackers and their queen and usurps the colony and brood within. *V. squamosa* represents an intermediate stage in the evolution of social parasitism because the normal worker caste is always produced and usurpation of nests of other species seems to be facultative rather than obligatory. (Photograph by the authors; see also MacDonald and Matthews, 1975).

societies. Furthermore, organization into castes results in role specialization in which the individuals of one caste largely lack a broad awareness of the roles of other colony members. These combine to lend an impersonal nature to social insect organization, which has apparently made it relatively easy for social symbionts to insert themselves into the colony regime. Wilson (1971, 1975a, b) and Eickwort (1975a) provide more background on the evolution of social parasitism.

## SUMMARY

Insects live in various intraspecific groups whose existence depends on rather different selection pressures. Simple aggregations are usually temporary and result from facultative mutual attractions. Parent-offspring groups are those concerned with any form of parental care, from egg guarding to the large societies consisting of kin groups. The behavior of nest building crops up repeatedly in diverse insect groups from earwigs and crickets to beetles and wasps. It is most complex in various wasps and bees.

The most advanced groups behaviorally are the eusocial insects, in which there is obligatory interdependence of all developmental stages. Eusocial insect societies are characterized by (1) being essentially permanent (with some exceptions); (2) mutual cooperation, in which individual activities are subservient to colony needs; (3) offspring being the product of a single female (queen) or a very few, but tended by many; (4) a queen which survives over generations of her offspring; and (5) individuals possessing a definite mutual attraction and depending entirely upon the colony for their existence.

Eusocial insects are restricted to the Isoptera and Hymenoptera. For the latter, the fact that males are haploid and females diploid seems to be a major preadaptation because of the resultant greater degree of genetic relatedness among sisters as compared to that between mothers and daughters. Mutualistic cooperation among females of the same generation is also a plausible evolutionary route to eusociality.

Various groups of social Hymenoptera have changed from a free-living to a socially parasitic existence which in its extreme form, inquilinism, results in total dependence of the parasite on the host's labors.

A vast literature of insect sociobiology exists. Wilson (1971, 1975a) provides the most comprehensive overview of eusocial and presocial insects. Prior to his work, Wheeler's books (1923, 1928) are classics. Of the several popular books on social insects, those of Michener and Michener (1951), Richards (1953), and Larson and Larson (1968) are especially useful. Malyshev (1968) provides another view of hymenopteran evolution.

### SELECTED REFERENCES

Alexander, R. D. 1974. The evolution of social behavior. *Annu. Rev. Ecol. Syst.* **5:** 325–383.

Alexander, R. D. and P. W. Sherman. 1977. Local mate competition and parental investment in social insects. *Science* **196:** 494–500.

Alford, D. V. 1974. *Bumblebees.* Davis-Poynter, London, 352 pp.

Batra, S. W. T. 1966. The life cycle and behavior of the primitively social bee, *Lasioglossum zephyrum* (Halictidae). *Kans. Univ. Sci. Bull.* **46:** 359–422.

Blum, M. S. 1974. Pheromonal bases of social manifestations in insects. In *Pheromones*, M. C. Birch (ed.). North-Holland, Amsterdam, pp. 190–199.

Brown, J. L. 1975. *The Evolution of Behavior*. Norton, New York, 761 pp.

Carne, P. B. 1966. Primitive forms of social behaviour and their significance in the ecology of gregarious insects. *Proc. Ecol. Soc. Aust.* **1:** 75–78.

Chauvin, R. (ed.). 1968. *Traité de Biologie de l'Abeille*, 5 Vols. Masson, Paris, 2099 pp.

Dias, B. F. 1975. Comportamento presocial de Sinfitas do Brazil Central. I. *Themos olfersii* (Klug) (Hymenoptera: Argidae). *Studia Entomol.* **18:** 401–432.

Eberhard, W. G. 1975. The ecology and behavior of a subsocial pentatomid bug and two scelionid wasps: strategy and counterstrategy in a host and its parasites. Smithsonian Contrib. Zool. No. 205, 39 pp.

Eickwort, G. C. 1975a. Gregarious nesting of the mason bee *Hoplitis anthocopoides* and the evolution of parasitism and sociality among megachilid bees. *Evolution* **29:** 142–150.

Eickwort, G. C. 1975b. Nest building behavior of the mason bee *Hoplitis anthocopoides* (Hymenoptera: Megachilidae). *Z. Tierpsychol.* **37:** 237–254.

Evans, H. E. 1956 (1958). The evolution of social life in wasps. *Proc. 10th Int. Congr. Entomol.* **2:** 449–457.

Evans, H. E. 1966. The accessory burrows of digger wasps. *Science* **152:** 465–471.

Evans, H. E. and J. E. Gillaspy. 1964. Observations on the ethology of digger wasps of the genus *Steniolia* (Hymenoptera: Sphecidae: Bembicini). *Amer. Midl. Natur.* **72:** 257–280.

Evans, H. E. and M. J. W. Eberhard. 1970. *The Wasps*. Univ. of Michigan Press, Ann Arbor, 265 pp.

Evans, H. E. and R. W. Matthews. 1973. Systematics and nesting behavior of Australian *Bembix* sand wasps (Hymenoptera: Sphecidae). Mem. Amer. Entomol. Inst. No. 20, 386 pp.

Evans, H. E. and R. W. Matthews. 1975. The sand wasps of Australia. *Sci. Amer.* **233:** 108–115 (December).

Fitzgerald, T. D. 1976. Trail marking by larvae of the eastern tent caterpillar. *Science* **194:** 961–963.

Free, J. B. and C. G. Butler. 1959. *Bumblebees*. Collins, London, 208 pp.

Halffter, G. and E. G. Matthews. 1966. The natural history of dung beetles of the subfamily Scarabaeinae (Coleoptera: Scarabaeidae). *Folia Entomol. Mexicana* **12–14:** 1–312.

Halffter, G., V. Halffter, and I. Lopez G. 1974. *Phanaeus* behavior: food transportation and bisexual cooperation. *Environ. Entomol.* **3:** 341–345.

Hamilton, W. D. 1964. The genetical theory of social behaviour, I, II. *J. Theoret. Biol.* **7:** 1–52.

Hamilton, W. D. 1972. Altruism and related phenomena, mainly in social insects. *Annu. Rev. Ecol. Syst.* **3:** 193–232.

Hinton, H. E. 1977. Subsocial behaviour and biology of some Mexican membracid bugs. *Ecol. Entomol.* **2:** 61–79.

Howse, P. E. 1970. *Termites: A Study in Social Behaviour*. Hutchinson Univ. Library, London. 150 pp.

Iwata, K. 1971. *Evolution of Instinct. Comparative Ethology of Hymenoptera.* Amerind Publ. Co., New Delhi (Translated from Japanese, 1976), 535 pp.

Jeanne, R. L. 1975. The adaptiveness of social wasp architecture. *Quart. Rev. Biol.* **50:** 267–287.

Kemper, H. and E. Döhring. 1967. *Die Sozialen Faltenwespen Mitteleuropas.* Paul Parey, Berlin, 180 pp.

Kerr, W. E. 1969. Some aspects of the evolution of social bees. *Evol. Biol.* **3:** 119–175.

Klemperer, H. G. and R. Boulton. 1976. Brood burrow construction and brood care by *Heliocopris japetus* (Klug) and *H. hamadryas* (Fabricius) (Coleoptera, Scarabaeidae). *Ecol. Entomol.* **1:** 19–29.

Krishna, K. and F. M. Weesner (eds.). 1969. *Biology of Termites,* Vol. 1. Academic Press, New York, 598 pp.

Krishna, K. and F. M. Weesner (eds.). 1970. *Biology of Termites,* Vol. 2. Academic Press, New York, 643 pp.

Kullman, E. J. 1972. Evolution of social behavior in spiders (Araneae; Eresidae and Theridiidae). *Amer. Zool.* **12:** 419–426.

Larson, P. P. and M. W. Larson. 1968. *Lives of Social Insects.* World, Cleveland, 226 pp.

Lin, N. and C. D. Michener. 1972. Evolution of sociality in insects. *Quart. Rev. Biol.* **47:** 131–159.

Lindauer, M. 1961. *Communication Among Social Bees,* revised ed. Harvard Univ. Press, Cambridge, Mass., 173 pp.

Lindauer, M. 1974. Social behavior and mutual communication. In *The Physiology of Insecta,* 2nd ed., Vol. 3, M. Rockstein (ed.). Academic Press, New York, pp. 149–228.

Linsley, E. G. 1962. Sleeping aggregations of aculeate Hymenoptera. *Ann. Entomol. Soc. Amer.* **55:** 148–164.

Lloyd, M. and H. S. Dybas. 1966. The periodical cicada problem: I, population ecology. *Evolution* **20:** 133–149.

Lubin, Y. D. 1974. Adaptive advantages and the evolution of colony formation in *Cyrtophora* (Araneae: Araneidae). *Zool. J. Linn. Soc.* **54:** 321–339.

Lüscher, M. 1961. Air-conditioned termite nests. *Sci. Amer.* **205:** 138–145 (July).

MacDonald, J. F. and R. W. Matthews. 1975. *Vespula squamosa,* a yellowjacket wasp evolving toward parasitism. *Science* **190:** 1003–1004.

Malyshev, S. I. 1968. *Genesis of the Hymenoptera and the Phases of Their Evolution.* (Translated from the Russian by B. Haigh; O. W. Richards and B. Uvarov, eds.). Methuen, London, 319 pp.

Matthews, R. W. 1968. *Microstigmus comes:* Sociality in a sphecid wasp. *Science* **160:** 787–788.

Michener, C. D. and M. H. Michener. 1951. *American Social Insects.* Van Nostrand, New York, 261 pp.

Michener, C. D. 1969. Comparative social behavior of bees. *Annu. Rev. Entomol.* **14:** 299–334.

Michener, C. D. 1971. Biologies of African allodapine bees (Hymenoptera: Xylocopinae). *Bull. Amer. Mus. Nat. Hist.* **145:** 219–302.

Michener, C. D. 1974. *The Social Behavior of the Bees: A Comparative Study.* Belknap Press of Harvard Univ. Press, Cambridge, Mass., 404 pp.

Michener, C. D. and D. J. Brothers. 1974. Were workers of eusocial Hymenoptera initially altruistic or oppressed? *Proc. Nat. Acad. Sci., USA* **71:** 671-674.

Milne, L. J. and M. Milne. 1976. The social behavior of burying beetles. *Sci. Amer.* **235:** 84-89 (August).

Morse, R. A. 1975. *Bees and Beekeeping.* Cornell Univ. Press, Ithaca, N.Y., 295 pp.

Mullen, V. T. and P. E. Hunter. 1973. Social behavior in confined populations of the horned passalus beetle (Coleoptera; Passalidae). *J. Ga. Entomol. Soc.* **8:** 115-123.

Naumann, M. G. 1975. Swarming behavior: evidence for communication in social wasps. *Science* **189:** 642-644.

Plateaux-Quénu, C. 1972. *La Biologie des Abeilles Primitives.* Masson, Paris, 200 pp.

Regnier, F. E. and E. O. Wilson. 1971. Chemical communication and "propaganda" in slave-maker ants. *Science* **172:** 267-269.

Rettenmeyer, C. W. 1963. Behavioral studies of army ants. *Univ. Kans. Sci. Bull.* **44:** 281-465.

Richards, O. W. 1953. *The Social Insects.* Macdonald, London, 219 pp.

Richards, O. W. 1971. Biology of the social wasps. *Biol. Rev.* **46:** 483-528.

Schneirla, T. C. 1971. *Army Ants: A Study in Social Organization.* H. R. Tophoff. (ed.) W. H. Freeman, San Francisco, 349 pp.

Schneirla, T. C. and G. Piel. 1948. The army ant. *Sci. Amer.* **178:** 16-23 (June).

Spradbery, J. P. 1973. *Wasps: An Account of the Biology and Natural History of Solitary and Social Wasps.* Sidwick and Jackson, London, 408 pp.

Sudd, J. H. 1967. *An Introduction to the Behaviour of Ants.* St. Martin's Press, New York, 200 pp.

Tophoff, H. 1972. The social behavior of army ants. *Sci. Amer.* **227:** 70-79 (November).

Trivers, R. L. 1971. The evolution of reciprocal altruism. *Quart. Rev. Biol.* **46:** 35-57.

Trivers, R. L. and H. Hare. 1976. Haplodiploidy and the evolution of social insects. *Science* **191:** 249-263.

Wagner, H. O. 1954. Massenansammlungen von Weberknechten. *Z. Tierpsychol.* **11:** 348-352.

Waterhouse, D. F. 1974. The biological control of dung. *Sci. Amer.* **230:** 100-109 (April).

Weaver, N. 1966. Physiology of caste determination. *Annu. Rev. Entomol.* **11:** 79-102.

Weber, N. A. 1972. *Gardening Ants: The Attines.* Mem. Amer. Phil. Soc. no. 92, Philadelphia, 146 pp.

West, M. J. and R. D. Alexander. 1963. Sub-social behavior in a burrowing cricket, *Anurogryllus muticus* (De Geer). Orthoptera: Gryllidae. *Ohio J. Sci.* **63:** 19-24.

West Eberhard, M. J. 1969. The social biology of polistine wasps. *Misc. Publ. Mus. Zool. Univ. Mich.* **140:** 1-101.

West Eberhard, M. J. 1974. The evolution of social behavior by kin selection. *Quart. Rev. Biol.* **50:** 1-33.

Wheeler, W. M. 1923. *Social Life Among the Insects.* Harcourt, Brace, New York, 375 pp.

Wheeler W. M. 1928. *The Social Insects, Their Origins and Evolution.* Harcourt, Brace, New York, 378 pp.

Wheeler, W. M. 1910 (1960). *Ants.* Columbia Univ. Press, New York, 663 pp.

Wilson, E. O. 1971. *The Insect Societies.* Belknap Press of Harvard Univ. Press, Cambridge, Mass., 548 pp.

Wilson, E. O. 1973. Group selection and its significance for ecology. *BioScience* **23:** 631–638.

Wilson, E. O. 1975a. *Sociobiology: The New Synthesis.* Belknap Press of Harvard Univ. Press, Cambridge, Mass., 697 pp.

Wilson, E. O. 1975b. *Leptothorax duloticus* and the beginnings of slavery in ants. *Evolution* **29:** 108–119.

Wilson, E. O. 1975c. Slavery in ants. *Sci. Amer.* **232:** 32–36 (June).

Wilson, E. O. 1976. A social ethogram of the Neotropical arboreal ant *Zacryptocerus varians* (Fr. Smith). *Anim. Behav.* **24:** 354–363.

Winston, M. L., and C. D. Michener. 1977. Dual origin of highly social behavior among bees. *Proc. Natl. Acad. Sci. USA* **74:** 1135–1137.

Wood, T. K. 1976. Alarm behavior of brooding female *Umbonia crassicornis* (Homoptera: Membracidae). *Ann. Entomol. Soc. Am.* **69:** 340–344.

# Author Index

Pages on which the author's work is fully cited are shown in brackets; pages in *italics* refer to citations in figure captions.

# Subject Index

Page numbers in **bold face** refer to definitions of terms; pages in *italics* refer to citations in figure captions.